*Modeling and Simulation in Science, Engineering and Technology*

# Modeling of Biological Materials

Francesco Mollica
Luigi Preziosi
K.R. Rajagopal
*Editors*

Birkhäuser
Boston • Basel • Berlin

Francesco Mollica
Dipartimento di Ingegneria
Università di Ferrara
Via Saragat 1
44100 Ferrara
Italy

Luigi Preziosi
Dipartimento di Matematica
Politecnico di Torino
Corso Duca degli Abruzzi 24
10129 Torino
Italy

K.R. Rajagopal
Department of Mechanical Engineering
Texas A&M University
College Station, TX 77843
USA

Mathematics Subject Classification: 92B05, 92C10, 92C30, 92C35, 92C50

**Library of Congress Control Number:** 2006936550

ISBN-10: 0-8176-4410-5          e-ISBN-10: 0-8176-4411-3
ISBN-13: 978-0-8176-4410-9      e-ISBN-13: 978-0-8176-4411-6

Printed on acid-free paper.

9  8  7  6  5  4  3  2  1

*www.birkhauser.com*                                    (Lap/SB)

# *Table of Contents*

# Preface

One of the primary purposes and obligations of science, in addition to understanding nature in general and life in particular, is to assist in enhancing the quality and longevity of life, indeed a most daunting challenge. To be able to meaningfully meet the last of the above expectations, it is necessary to provide the practitioner of medicine with diagnostic and predictive capabilities that science will accord when its seemingly disparate parts are melded together and brought to bear on the problems that they face. The development of interdisciplinary activities involving the various basic sciences—biology, physics, chemistry, and mathematics, and their applied counterparts, engineering and technology—is a necessary key to unlocking the mysteries of medicine, which at the moment is a curious admixture of art, craft, and science.

Significant strides have been taken during the past decades for putting into place a methodology that takes into account the interplay of the various basic sciences. Considerable progress has been made in understanding the role that mechanics has to play in the development of medical procedures. This collection of survey articles addresses the role of mechanics with regard to advances in the medical sciences. In particular, these survey articles bring to one's attention the central role played by mathematical modeling in general and the modeling of mechanical issues in particular that have a bearing on the biology, chemistry, and physics of living matter.

This book is written with the intent to highlight the fact that it is necessary to expend considerable effort in bringing together scientists, engineers, and doctors to address important problems in the medical sciences. The aim is to foster the notion that it is necessary to train professionals who are able to understand the medical needs, as well as the basic sciences, in sufficient depth to be able to apply them to the problem at hand; to develop a reasonable model; to have the capability to carry out numerical simulations; and finally to develop tools that can be used by those in the practice of medicine. Even more challenging is the ability to suggest new technological innovations on the basis of such modeling efforts. Completing the crucial circle of observations in the medical field to the development of phenomenological models, the use of computer simulations, and the development of diagnostic and predictive tools, as well as new medical technologies that can be used for further medical observations, is the ultimate aim of the medical research scientist, and the chapters in this book address an important part of this circle.

The articles that appear in this volume are not merely conjectural in nature. There is a genuine effort to test the theories under realistic situations. Explicit theoretical results are extracted either in analytical or in numerical form, and a comparison with experimental findings is carried out. The success of the models that have been developed can be judged by the agreement between the predictions of the theories and the experiments.

It ought to be borne in mind that the modeling and subsequent mathematical analysis of many biomechanical problems relevant to the different parts of the human body share some basic characteristics; however, as the applications are very different, spanning a variety of complex problems, the modes of approach, the ways of thinking, and the semantics, are widely different. As a result the pertinent literature is spread across journals with widely varying styles and often involves mutually nonintersecting communities of readership, for example, engineers, biologists, and medical doctors. Given such a situation, another ambitious aim of this book is to alleviate this issue to a certain degree, by publishing several survey papers of those actively working in a variety of areas and gathering them in one place so that the diverse audience will have the opportunity to recognize the similarities as well as the differences in the approaches that are used. Articles by specialists working in some of the most important areas of biomedicine—tumor growth, blood flow, mechanics of the circulatory system, cell rheology, mechanical properties of hard tissues, modeling of biological membranes, and the modeling of natural tissue substitutes (i.e., biomaterials)—are featured in this book. There is also a conscious effort to address both theoretical and experimental issues.

Each chapter focuses on a specific biomedical application and explains as simply as possible both the biological and mechanical background, together with the basic ideas of the appropriate models. One can easily discern features that are peculiar to each model as well as those that are in common. An attempt is made to explain the rationale behind the model, and in most of the contributions theoretical predictions are compared with experimental data. The authors have avoided cumbersome mathematical techniques to illustrate their ideas; however, at the same time they provide references to literature where more complicated methods have been used to attack specific problems.

The first chapter focuses on the description of the mechanical properties of biological materials both from a macroscopic and a microscopic point of view and of the different techniques for measuring cell and tissue properties. The second chapter concerns biological and mechanical aspects involved in blood flow. The next three chapters deal with other important problems involving the cardiovascular system. In particular, the third chapter deals with the formation and growth of aneurysms, the fourth chapter with the mechanics of biodegradable stents, and the fifth chapter with the process of hemostasis, including the effects of hemodynamic conditions and the mechanical properties of the vasculature. The last three chapters deal with the mechanical properties of different biological materials–bones, tumors, and cell membranes, respectively.

It is our hope that the book will be of interest to graduate students in applied mathematics, engineering, biomedicine, and to young researchers with an aptitude for multidisciplinary research in biomedicine.

January 2007
Francesco Mollica<br>Luigi Preziosi<br>K.R. Rajagopal

# 1

## Rheology of Living Materials

R. Chotard-Ghodsnia and C. Verdier

*Laboratoire de Spectrométrie Physique, UJF-CNRS, UMR 5588*
*BP87, 140 avenue de la Physique*
*F-38402 Saint-Martin d'Hères, France*

Abstract. In this chapter, the properties of biological materials are described both from a microscopic and a macroscopic point of view. Different techniques for measuring cell and tissue properties are described. Models are presented in the framework of continuum theories of viscoelasticity. Such models are used for characterizing experimental data. Finally, applications of such modeling are discussed in a few situations of interest.

## 1.1 Introduction

### 1.1.1 What Is Rheology ?

Rheology (in Greek, *rheos*: to flow, *-logy*: the study of) is a pluridisciplinary science describing the flow properties of various materials, i.e. the study of the stresses that are needed to produce certain strains or rate of strains within a given material. It consists of different approaches that are all needed to understand the complexity of fluids or materials possessing viscoelastic or viscoplastic properties. The main approaches are:

- Measurements of the rheological properties in simple flows (shear and elongation)

1

- Simultaneous description of the underlying microstructure as a function of deformation

- Constitutive modeling using generalized continuum or molecular models based on the previous observations

- Applications to real situations (complex flows and geometries)

- Numerical simulations when analytical works are not possible

Typical fluids or materials under investigation in the framework of rheology can be polymers (including polymer solutions, rubbers, polymer emulsions), suspensions of particules or deformable objects (e.g. blood cells), and other complex systems (foams, gels, tissues, etc.) as described by Larson [LAa].

### 1.1.2 Importance of Rheology in the Study of Biological Materials

Biological materials (see the textbook by Fung [FUa] or the review paper by Verdier [VEa]) start at the cell level (order of a few microns) and can reach sizes up to the size of an assembly of cells or tissue (a few millimeters). Rheology is important for describing such materials because they are usually made of the above-mentioned systems. The cell (Figure 1.1), to start with, possesses an elastic nucleus, a viscous or viscoplastic cytoplasm, and is surrounded by an elastic membrane made of a lipid bilayer, where adhesion proteins coexist [ALa].

An example of a biological fluid is blood, which is a suspension of proteins (i.e. polymers) and cells (erythrocytes, as well as various white blood cells) inside a fluid (plasma). Another example is the one of a biological

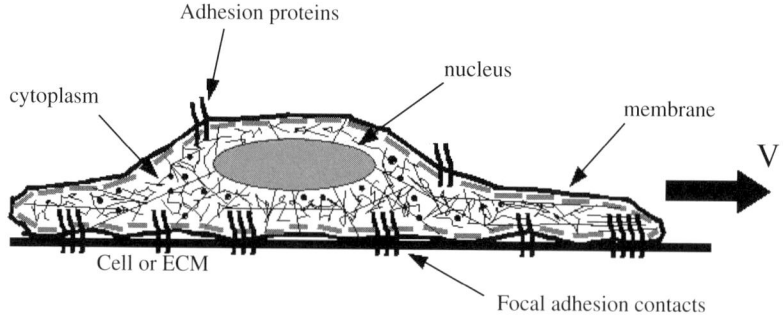

Figure 1.1. Sketch of a cell modifying its contacts and microrheology to undergo migration, at velocity $V$.

tissue. It contains an assembly of cells connected to each other by the extracellular matrix (ECM) and adhesion proteins. These complex systems lead to viscoelastic properties due to the presence of fluids (viscous effects), and the presence of elastic components (particles, inclusions with interfacial tensions, elastic membranes). Several rheological models can already predict such behaviors. The complexity of the biological tissue or fluid relies on the fact that such media are active, and can rearrange their microstructure to produce different local (or nonlocal) properties or stresses in order to resist, or to achieve a precise function; in other words, they are intelligent materials.

In this chapter, we describe a few rheological models useful for modeling complex media. Then a brief description of biological media is presented. Measuring techniques to investigate the properties of cells and tissues is then explained. Finally, we present a few examples of predictions achieved through rheological modeling.

## 1.2 Rheological Models

In this section, a simple 1-D viscoelastic model, the Maxwell model, and a viscoplastic 1-D one, the Bingham model, are presented. These models are quite interesting and can provide valuable information for the interpretation of simple experiments. A generalization of such models is made in three dimensions.

### 1.2.1 One-Dimensional Models

#### 1.2.1.1 *The Maxwell Fluid*

We consider a simple viscoelastic 1-D model, which is depicted in Figure 1.2 where a spring (spring constant $G$) and dashpot (viscosity $\eta$) are assembled in series. The resulting stress $\tau$ is the same in the two elements whereas the deformation $\gamma$ is the sum of the deformations in each element. This leads to the following constitutive equation,

$$\lambda\dot{\tau} + \tau = \eta\dot{\gamma}, \tag{2.1}$$

where $\lambda = \eta/G$ is the relaxation time, and $\dot{\gamma}$ is the rate of deformation. This form is the differential form of the Maxwell model. It leads to a decrease in stress $\tau(t) = \tau_0 \exp(-t/\lambda)$, when motion is stopped, where $\tau_0$ is the initial stress.

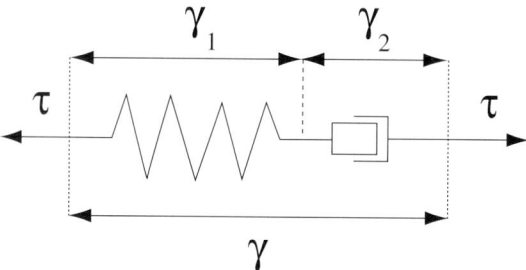

Figure 1.2. Schematic representation of the Maxwell element.

Limiting cases of elastic and viscous materials can be recovered:

- When $t \ll \lambda$ (short times), $\tau = G\gamma \Rightarrow$ the material is elastic $(G = \eta/\lambda)$.

- When $\lambda \ll t$ (long times), $\tau = \eta\dot{\gamma} \Rightarrow$ the material is Newtonian.

Equation (2.1) has an exact solution which is

$$\tau(t) = \int_{-\infty}^{t} G\, e^{-((t-t')/\lambda)} \dot{\gamma}(t')\, dt'. \tag{2.2}$$

Relation (2.2) is the integral version of the Maxwell model. It has the advantage of showing that the stress is a function of the history of deformation, and also shows that stress and rate of strain are related via this integral form with a kernel $G(t) = G \exp(-t/\lambda)$. This function is called the relaxation function. Similarly, a relation between the strain $\gamma(t)$ and the stress $\tau(t)$ can be obtained in integral form through another kernel $J(t) = 1/G + t/\eta$ which is called the compliance.

**Remark 2.1**     *The other simple model considered in the literature is the Kelvin–Voigt model consisting of a spring and dashpot in parallel. This model exhibits a constitutive equation of the type $\tau = G\gamma + \eta\dot{\gamma}$. In this case $G(t) = G$ and $J(t) = 1/G(1 - \exp(-t/\lambda))$, where $\lambda$ is defined similarly but is a retardation time.*

A generalization to multiple-mode Maxwell models can easily be made through the use of the following distribution $(G_i, \lambda_i)$ appearing in the relaxation function, corresponding to the use of several Maxwell elements in series,

$$G(t) = \sum_{i=1}^{n} G_i\, e^{-(t/\lambda_i)} \tag{2.3}$$

This is sometimes enough to obtain a good idea of the dynamic moduli, often used in oscillatory rheometry [BIc]. Otherwise, one may use a continuous distribution of relaxation times:

$$G(t) = \int_0^\infty \frac{H(\lambda)}{\lambda} \exp\left(-\frac{t}{\lambda}\right) d\lambda. \tag{2.4}$$

Such formulae have been used successfully for predicting the dynamic rheology of molten polymers, in particular Baumgaertel and Winter [BAc] who used two specific empirical formulae for $H(\lambda)$ for small times and long times.

### 1.2.1.2  *The Bingham Fluid*

Bingham fluids are an example of the so-called viscoplastic materials. A Newtonian fluid can flow under the action of any stress but it is unlikely that a Bingham fluid will do the same. Indeed, it usually requires that a certain stress, i.e. the so-called "yield stress," is applied. For example, under the action of its weight alone, a fluid element might or might not flow. This relies on the fact that the typical stress (shear, elongation) dominates the effect of the yield stress. Yield stresses are due to complex interactions taking place at the microscopic level, which link the material particles. A solution of F-actin (essential component of the cytoplasm), as an example, can flow only if the actin concentration is not too large. If it is large, then interactions between the actin proteins are such that temporary links exist throughout the fluid, showing the existence of a yield effect.

The simplest 1-D model (in the case of shear) is the Bingham fluid [MAa], given by

$$\tau \leq \tau_s \qquad \dot{\gamma} = 0 \quad \text{or} \quad \tau = G\gamma$$

$$\tau \geq \tau_s \qquad \tau = \tau_s + \eta\dot{\gamma}. \tag{2.5}$$

This relation explains that a certain stress $\tau_s$ needs to be overwhelmed by the shear stress ($\tau$) to achieve flow. Under this threshold, no flow is obtained, but a simple elastic relation can be verified generally. Above this threshold, the material exhibits a Newtonian behavior with viscosity $\eta$. Other viscoplastic relations exist, such as the Casson model (useful for describing the behavior of blood, at different hematocrit contents), or the Herschel–Bulkley model [MAa]. Finally, note that the cell cytoplasm may be modeled using the Bingham model or Herschel–Bulkley model, based on the constituents in presence. Another model introduced by He and Dembo [HEa], is the sol–gel model: it has been found to be successful to predict cell division. Sol–gel models can help in predicting transitions from a liquid state ("sol") to a quasi-solid state ("gel") when a certain constituent's concentration is reached.

### 1.2.2    Three-Dimensional Models

To start with, we recall the principal relations used in classical fluid mechanics and elasticity, before presenting the 3-D viscoelastic models. Let us define the total stress tensor $\underline{\underline{\Sigma}}$ with an isotropic part, and an extra stress $\underline{\tau}$:

$$\underline{\underline{\Sigma}} = -p\underline{\underline{I}} + \underline{\tau}, \tag{2.6}$$

where $p$ is the pressure, and $\underline{\underline{I}}$ is the identity tensor. To define constitutive equations and generalize the 1-D ones in the previous section, we need to define laws valid for all observers (principle of material frame-indifference). Indeed, two observers in given reference frames should be able to measure the same material properties or laws. Consider a motion $\vec{x} = \vec{x}(\vec{X}, \mathbf{t})$ and stress tensor $\underline{\underline{\Sigma}} = \underline{\underline{\Sigma}}(\vec{X}, \mathbf{t})$. Another observer with clock $t^* = t - a$, in a frame defined by $\vec{x}^* = \vec{x}^*(\vec{X}, \mathbf{t}^*) = \vec{c}(\mathbf{t}) + \underline{\underline{Q}}\vec{x}(\vec{X}, \mathbf{t})$, should observe a stress $\underline{\underline{\Sigma}}^* = \underline{\underline{Q}}(t)\underline{\underline{\Sigma}}\,\underline{\underline{Q}}^{\mathbf{T}}(t)$, following Malvern [MAd], for example. In the previous relations, $a$, $\vec{c}$, and $\underline{\underline{Q}}$ are, respectively, any constant, vector, and orthogonal tensor. $\underline{\underline{\Sigma}}$ is said to be frame-indifferent.

### *1.2.2.1    General Fluids*

For the classical Newtonian fluid, the extra stress is expressed in terms of $\underline{\underline{D}}$, the symmetric part of the velocity gradient tensor, which is frame-indifferent. This defines the isotropic Newtonian fluid:

$$\underline{\underline{\Sigma}} = -p\underline{\underline{I}} + \lambda tr(\underline{\underline{D}})\underline{\underline{I}} + 2\eta\underline{\underline{D}}, \tag{2.7}$$

where $\eta$ and $\lambda$ are the viscosity and second viscosity coefficients, respectively. For incompressible fluids, the equation simply reduces to $\underline{\underline{\Sigma}} = -p\underline{\underline{I}} + 2\eta\underline{\underline{D}}$.

More general fluids can be defined, such as the Reiner–Rivlin fluid, where one makes use of the fact that an expansion of the stress in terms of the powers of $\underline{\underline{D}}$ is also a good model and can be reduced to powers of $\underline{\underline{D}}$ and $\underline{\underline{D}}^2$ only.

$$\underline{\underline{\Sigma}} = -p\underline{\underline{I}} + 2\eta\underline{\underline{D}} + 4\eta_2\underline{\underline{D}}^2, \tag{2.8}$$

where $\eta_2$ is considered to be a second viscosity, but has different units $(Pa.s^2)$. In Eq. (2.8), the constants $\eta$ and $\eta_2$ depend on the invariants of $\underline{\underline{D}}$, in particular $I\!I_D$. Note that the first invariant $I_D$ is 0 for incompressible materials. Equation (2.8), associated with the assumption that $\eta_2 = 0$, has been used extensively to predict the shear thinning (and thickening) behavior of polymer solutions and suspensions [MAa].

### 1.2.2.2  Elastic Materials

Similarly to fluids, the constitutive equation of an isotropic elastic material gives the stress $\underline{\underline{\Sigma}}$ in terms of the linear part of deformation $\underline{\underline{\epsilon}}$ (symmetric part of $\vec{grad\vec{u}}$, where $\vec{u} = \vec{x} - \vec{X}$ is the displacement between the initial and present position, in Lagrangian coordinates):

$$\underline{\underline{\Sigma}} = \lambda tr(\underline{\underline{\epsilon}})\underline{\underline{I}} + 2\mu\underline{\underline{\epsilon}}, \tag{2.9}$$

where $\lambda$ and $\mu$ are the Lamé coefficients (in Pa). The generalization of this relation to large deformations is made possible through the use of the frame-indifferent strain tensor $\underline{\underline{B}} = \underline{\underline{F}}\underline{\underline{F}}^T$, where $\underline{\underline{F}}$ is the deformation gradient. Again, by assuming the stress to be a polynomial function of $\underline{\underline{B}}$, and making use of the Cayley–Hamilton theorem, we obtain a relation for large deformations, the so-called Mooney–Rivlin model (MR):

$$\underline{\underline{\Sigma}} = \alpha\underline{\underline{I}} + 2C_1\underline{\underline{B}} - 2C_2\underline{\underline{B}}^{-1}, \tag{2.10}$$

where $C_1$ and $C_2$ are constants in the MR model but may also be functions of the invariants of $\underline{\underline{B}}$. $\alpha$ is a constant that needs, like a pressure, to be determined, for example, through boundary conditions.

More generally, any function of the invariants of $\underline{\underline{B}}$ can be used, via a strain energy function $W(I_B, I\!I_B)$. Here, for an incompressible material $I\!I\!I_B = 1$. This defines a more general hyperelastic material:

$$\underline{\underline{\Sigma}} = \alpha\underline{\underline{I}} + 2\frac{\partial W}{\partial I_B}\underline{\underline{B}} - 2\frac{\partial W}{\partial I\!I_B}\underline{\underline{B}}^{-1}. \tag{2.11}$$

### 1.2.2.3  Viscoelastic Materials

If one wants to generalize the case of the Maxwell element in Eq. (2.1), it is natural to replace the 1-D stress $\tau$ by the tensor $\underline{\underline{\tau}}$, and then to replace $\dot{\gamma}$ by its tensor form $2\underline{\underline{D}}$. Still one needs to be careful about the generalization of $\dot{\tau}$, because $\underline{\underline{\dot{\tau}}}$ is not frame-indifferent. In fact, there are a limited number of frame-indifferent derivatives [MAd], such as the following upper and lower convected derivatives given by, respectively,

$$\overset{\triangledown}{\underline{\underline{\tau}}} = \underline{\underline{\dot{\tau}}} - \vec{grad\vec{v}}\,\underline{\underline{\tau}} - \underline{\underline{\tau}}(\vec{grad\vec{v}})^T, \tag{2.12}$$

$$\overset{\triangle}{\underline{\underline{\tau}}} = \underline{\underline{\dot{\tau}}} + \underline{\underline{\tau}}\vec{grad\vec{v}} + (\vec{grad\vec{v}})^T\underline{\underline{\tau}}. \tag{2.13}$$

The co-rotational derivative is also another one, $\overset{\circ}{\underline{\underline{\tau}}} = \frac{1}{2}(\overset{\triangledown}{\underline{\underline{\tau}}} + \overset{\triangle}{\underline{\underline{\tau}}})$, as well as other linear combinations of the above two. Note that we are now using the extra stress $\underline{\underline{\tau}}$.

Following this, we obtain the upper convected Maxwell model, as the generalized differential form of (2.1):

$$\lambda \overset{\triangledown}{\underline{\underline{\tau}}} + \underline{\underline{\tau}} = 2\eta \underline{\underline{\mathbf{D}}}. \tag{2.14}$$

There is also an integral version of this equation:

$$\underline{\underline{\tau}}(t) = \int_{-\infty}^{t} M(t - t')(\underline{\underline{\mathbf{B}}}(t, t') - \underline{\underline{\mathbf{I}}})\, dt', \tag{2.15}$$

where $M(t) = (\eta/\lambda^2)\exp(-(t/\lambda)) = -(dG/dt)$, and $\underline{\underline{\mathbf{B}}}(t, t')$ is the relative deformation tensor. Formula (2.15) is called Lodge's formula and can include general functions $G(t)$, for example, such as the ones in (2.3), or more generally decreasing convex functions that have a finite value for $G(0)$.

Finally, a generalization of Lodge's model, as well as ideas developed in the previous part on elasticity, in particular, Eq. (2.11), lead to another more general form, known as the K–BKZ equation (factorized form):

$$\underline{\underline{\tau}}(t) = \int_{-\infty}^{t} M(t - t') \left[ 2\frac{\partial W}{\partial I_B}(I_B, I\!I_B)\underline{\underline{\mathbf{B}}}(t, t') - 2\frac{\partial W}{\partial I\!I_B}(I_B, I\!I_B)\underline{\underline{\mathbf{B}}}^{-1}(t, t') \right] dt'. \tag{2.16}$$

This model has been used quite a lot for predicting the elongational properties of polymers or polymer solutions, and gives excellent agreement [WAa]. On the other hand, it is not very good for predicting shear data [BIc].

### 1.2.2.4  *Viscoplastic Models*

The generalization of Eq. (2.5) to a tensor form gives:

$$\begin{aligned} I\!I_\tau \le \tau_s \qquad & \underline{\underline{\mathbf{D}}} = \underline{\underline{\mathbf{0}}} \quad \text{or} \quad \underline{\underline{\tau}} = G\underline{\underline{\mathbf{B}}}, \\ I\!I_\tau \ge \tau_s \qquad & \underline{\underline{\tau}} = 2\left(\eta + \frac{\tau_s}{\sqrt{I\!I_{2D}}}\right)\underline{\underline{\mathbf{D}}}, \end{aligned} \tag{2.17}$$

where $I\!I_\tau = \frac{1}{2}(tr(\underline{\underline{\tau}})^2 - tr(\underline{\underline{\tau}}^2))$ is the second invariant of the extra stress tensor $\underline{\underline{\tau}}$. This way of defining the inequality implies that all components of the stress may contribute to overwhelm the yield effect. Other possible models (Herschel–Bulkley, Papanastasiou, Casson [MAa]) can be used, in particular, using the fact that the viscosity $\eta$ in (2.17) may be taken to be a function of the invariants of $\underline{\underline{\mathbf{D}}}$, in particular, $I\!I_D$. As an example, models that characterize blood are described in the part concerning rheological modeling of tissues and biofluids.

## 1.3   Biological Materials

The response of a material to applied loads depends upon its internal consti-
tution and interconnections of its microstructural components. Biological
tissues are composed of the same basic constituents: cells and extracellular
matrix.

### 1.3.1   Cells

Cells are the fundamental structural and functional unit of tissues and organs. It has been known over the last two decades that many cell types
change their structure and function in response to changes in their mechanical environment. A typical cell consists of a cell membrane, a cytoplasm,
and a nucleus [ALa].

The *cell membrane* consists of a phospholipid bilayer with many embedded proteins that function as channels, receptors for target molecules, and
anchoring sites.

The *cytoplasm* is a fluid containing the cytoskeleton and dispersed organelles.

The *cytoskeleton* is a network of protein filaments extending throughout
the cytoplasm. The cytoskeleton provides the structural framework of
the cell, determining the cell shape and the general organization of the
cytoplasm. The cytoskeleton is responsible for the movements of entire cells
and for intracellular transport. It is formed by three structural proteins:
actin filaments, intermediate filaments, and microtubules. Actin filaments
are extensible and flexible (5–9 nm in diameter). Intermediate filaments
are ropelike structures (10 nm in diameter). Microtubules are long cylinders (25 nm in diameter) with a higher bending stiffness than the other
filaments. These three primary structural proteins perform their functions
through interactions with accessory cytoskeletal proteins, such as actinin,
myosin, and talin.

The *organelles* play various roles. For example, the Golgi apparatus plays
a role in the synthesis of polysaccharides and in the transport of various
macromolecules. The endoplasmic reticulum is the site of the synthesis of
proteins and lipids.

The *extracellular matrix* consists of proteins (collagens, elastin, fibronectin,
vitronectin, etc.), glycosaminoglycans, and water. Collagen, the most
abundant protein in the body and a basic structural element for soft and
hard tissues, and elastin, the most linearly elastic and chemically stable

protein, are primary structural constituents of the extracellular matrix, from a mechanical point of view. The extracellular matrix has many functions. It serves as an active scaffold on which cells can adhere and migrate. It serves as an anchor for many substances (growth factors, inhibitors) and provides an aqueous environment for the diffusion of nutrients between the cell and the capillary network. It maintains the shape of a tissue, giving it strength and mechanical integrity. To summarize, the extracellular matrix controls cell shape, orientation, motion, and function.

### 1.3.2   Tissues

A tissue is a group of cells that perform a similar function. There are four basic types of tissues in the human body: epithelium, connective tissue, muscle tissue, and nervous tissue [FAb].

*Epithelium* is a tissue composed of a layer of cells. It can line internal (e.g. endothelium which lines the inside of blood vessels) or external (e.g. skin) free surfaces of the body. Functions of epithelial cells include secretion, absorption, and protection.

*Connective tissue* is any type of biological tissue with extensive extracellular matrix that holds everything together. There are several basic types.

- Bone: contains specialized cells, osteocytes, embedded in a mineralized extracellular matrix and functions for general support, protection of organs, and movements. It is a relatively hard and lightweight composite material. It is formed mostly of calcium phosphate which has relatively high compressive strength though poor tensile strength. Although bone is essentially brittle, it has a degree of elasticity contributed by its organic components (mainly collagen).

- Loose connective tissue: holds organs in place and attaches epithelial tissue to other underlying tissues. It also surrounds blood vessels and nerves. Fibroblasts are widely dispersed in this tissue and secrete strong fibrous proteins (e.g. collagen and elastin) and proteoglycans as an extracellular matrix.

- Fibrous connective tissue: has a relatively high tensile strength due to a relatively high concentration of collagenous fibers. This tissue is primarily composed of polysaccharides, proteins, and water and does not contain many living cells. Such tissue forms ligaments and tendons.

- Cartilage: is a dense connective tissue primarily found in joints. It is composed of chondrocytes that are dispersed in a gellike extracellular matrix mainly composed of chondroitin sulfate.

- Blood: is a circulating tissue composed of cells (hematins, leukocytes, and platelets) and fluid plasma (its extracellular matrix). The main function of blood is to supply nutrients (e.g. glucose, oxygen) and to remove waste products (e.g. carbon dioxyde). It also transports cells and different substances (hormones, lipids, amino acids) between tissues and organs.

- Adipose tissue: is an anatomical term for loose connective tissue composed of adipocytes. It provides cushioning, insulation, and energy storage.

*Muscle* is a contractile form of tissue. Muscle contraction is used to move parts of the body and substances within the body. There are three types of muscles:

- Cardiac muscle

- Skeletal muscle (or "voluntary"): attached to the skeleton and used for movement

- Smooth muscle (or "involuntary"): found within intestines, throat, and blood vessels.

The *nervous tissue* has the function of communication between parts of the body. It is composed of neurons, which transmit impulses, and the neuroglia, which assists the propagation of the nerve impulse and provides nutrients to the neurons.

## 1.4 Measurements of Rheological Properties of Cells and Tissues

### 1.4.1 Microrheology

Conventional rheological measurements are not always adapted to biological materials because they require large quantities of rare materials and provide an average measurement and do not allow for local measurements in inhomogeneous systems. Microrheological methods address this issue by probing the material on a micrometer-length scale using microliter sample volumes. Two microrheological approaches can be found in the literature: active tests and passive tests.

#### 1.4.1.1 Active Tests

In this class of microrheological measurements, the stress is locally applied to the material by active manipulation of the probe through the use of electric or magnetic fields or micromechanical forces. Then, the resultant strain is measured to obtain the local shear moduli. We describe three active manipulation techniques in the following paragraphs.

*Optical tweezers.* Optical tweezers consist of a focused laser directed at a micrometer-scale dielectric object, such as beads or organelles, and used to control their position [ASa]. For typical object sizes (0.5–10 $\mu$m in diameter) used, the force (limited to the pN range) is generated by the refraction of the laser within the bead coupled with the differences in photon density from the center to the edge of the beam. Very small objects do not trap well because the trapping force decreases with decreasing object volume. By moving the focused laser beam, the trapped particle is forced to move and applies a local stress to the sample. The resultant particle displacement reports strain from which rheological properties can be obtained. Elasticity measurements are possible by applying a constant force with the optical tweezers and measuring the resultant displacement of the particle. The membrane elastic modulus of red blood cells was measured using this approach [HEc]. Frequency-dependent rheological properties can be measured by oscillating the laser position with an external steerable mirror and measuring the amplitude of the bead motion and the phase shift with respect to the driving force. Microrheology of soft materials was studied using this approach [HOa].

*Magnetic tweezers and magnetic twisting.* Magnets are used to apply either a linear force (magnetic tweezers) or a twisting torque (magnetic twisting) to embedded magnetic particles. The resultant particle displacements measure the rheological response of the surrounding material. Particle selection is critical because its magnetic contents influence the applied force. Ferromagnetic beads can generally exert large forces but they retain a part of their magnetization each time they are exposed to a magnetic field. Paramagnetic beads are less susceptible to magnetization but they exert smaller forces (typically 10 pN to 10 nN). Videomicroscopy is used to detect the displacements of the particles under applied forces. The spatial resolution is typically in the range of 10–20 nm and the temporal range is 0.01–100 Hz. Three modes of operation are available: (1) a viscometry measurement by applying a constant force; (2) a creep response measurement by applying a pulselike excitation: using this method Bausch et al. found linear three-phasic creep responses consisting of an elastic domain, a relaxation regime, and a viscous flow behavior for different cell types [BAd, BAe]; and (3) a measurement of the viscoelastic moduli in response to an oscillatory stress: using this method Fabry et al. [FAa] suggested that the cytoskeleton may behave as a soft glassy material [BOa, SOa], existing close to a glass transition, rather than as a gel.

*Atomic force microscopy.* The atomic force microscope [BIb] has been widely used to study the structure of soft biological materials, in addition to imaging information about the surface topology. A soft cantilever is pushed into the sample surface. The cantilever deflection is measured by a laser detection system (reflecting laser beams off the cantilever and position-sensitive photodetector). Knowing the cantilever's spring constant, the cantilever deflection gives the force required to indent a surface and this has been used to measure the local elasticity and viscoelasticity of thin samples and living cells. For elasticity measurements [RAa, ROa], a modified Hertz model is used to describe the elastic deformation of the sample by relating the indentation and the loading force. Elastic mapping of cells was possible using this technique [AHa]. A common source of error for very thin or soft materials is the contribution of the elastic response of the underlying substrate on which the sample is placed. This contribution is higher for higher forces and indentations and can be reduced by using a spherical tip. This increases the contact area and allows application of the same stress to the sample with a smaller force [MAb].

For viscoelastic measurements, an oscillating cantilever tip is used [AHa, MAb, ALb, MAc]. A small amplitude (5–20 nm) sinusoidal signal is applied normal to the surface around the initial indentation position at frequencies ranging from 20 to 400 Hz. Using this test, Mahaffy et al. [MAc] characterized the lamellipodium of fibroblast cells and obtained a rubber plateaulike behavior. Smith et al. [SMa] characterized the microscale viscoelasticity of smooth muscle cells with AFM indentation modulation in their fibrous perinuclear region. They found that the complex shear modulus, measured in response to nanoscale oscillatory perturbations, exhibited soft glassy rheology. The elastic (storage) modulus of these cells scaled as a weak power law and their loss modulus scaled with the same power law dependence on frequency.

### 1.4.1.2 Passive Tests

In this class of microrheological measurements, the passive motion of the probes, due to thermal or Brownian fluctuations, is measured [MAe]. In this case, no external force is applied to the material. Phenomenologically, embedded probes exhibit larger motions when their local environments are less rigid or less viscous. Both the amplitude and the time scale are important for calculating the mechanical moduli. The mean squared displacement (MSD) of the probe's trajectory is measured over various lag-times to quantify the probe's amplitude of motions over different time scales. In purely viscous materials, MSDs of probes vary linearly with lag-times. In purely elastic materials, MSDs are constant regardless of lag-times. A viscoelastic material can be modeled as an elastic network that is viscously coupled to and embedded in an incompressible Newtonian fluid. A natural way to incorporate the elastic response is to generalize the standard Stokes–Einstein equation for a simple, purely viscous fluid with a complex shear modulus to materials that also have a real component of the shear modulus [MAf]:

$$\tilde{G}(s) = \frac{k_B T}{\pi a s < \Delta \tilde{r}^2(s) >}, \tag{4.1}$$

where $s$ is the complex Laplace frequency, $k_B$ is the Boltzmann constant, $T$ is the absolute temperature, $a$ is the radius of the probe, and $< \Delta \tilde{r}^2(s) >$ is the unilateral Laplace transform of the two-dimensional MSD $< \Delta r^2(t) >$. To compare with bulk rheology measurements, $\tilde{G}(s)$ can be transformed into the Fourier domain to obtain the complex shear modulus $G^*(\omega)$.

For most soft materials, the temperature cannot be changed significantly. Thus, the upper limit of the measurable elastic modulus depends on both the size of the embedded probe and on our ability to resolve small particle displacements. The resolution of detecting particle centers depends on the particle tracking method and ranges from 1 to 10 nm. This allows measurements with micron-sized particles of materials with an elastic modulus of 10 to 500 Pa. This is smaller than the range accessible by active tests but sufficient to study many soft materials. Moreover, passive measurements have the advantage of dealing with the linear viscoelastic regime because no external stress is applied.

To use the generalized Stokes–Einstein relation (4.1) to obtain macroscopic viscoelastic shear moduli of a material, the medium should be treated as a continuum material around the embedded particle. Thus, the size of the embedded particle must be larger than any structural length scales of the material. Moreover, this relation is valid for a large frequency range (10 Hz to 100 kHz) which is much higher than traditional methods where inertial effects become significant around 50 Hz. The MSD of embedded particles should be measured with a very good temporal and spatial resolution to take full advantage of the range of frequencies and complex moduli measurable in a passive microrheology test. The MSD can be obtained from methods that directly track the particle position as a function of time or from light-scattering experiments.

The rheological properties of living cells have been measured using particle tracking microrheology. Yamada et al. [YAa] showed that the cytoplasmic viscoelasticity of kidney epithelial cells varied within subcellular regions and was dynamic. At low frequencies, lamellar regions ($820 \pm 520$dyne/cm$^2$) were more rigid than viscoelastic perinuclear regions ($330 \pm 250$dyne/cm$^2$) of the cytoplasm, but the spectra converged at high frequencies ($>1000$rad/s).

Finally, other groups have used improved particle-tracking methods: in particular, Crocker et al. [CRa] have developed a two-point microrheology technique, based on cross-correlation of the motion of two particles. Tseng et al. [TSa], on the other hand, used multiple-particle-tracking microrheology to spatially map mechanical heterogeneities of living cells.

To summarize, microrheological techniques allow us to characterize complex materials on length scales much shorter than those measured with macroscopic techniques. In an incompressible homogeneous material, the response of an individual probe due to an external force (active tests) or to thermal fluctuations (passive tests) is an image of the bulk viscoelastic properties of the surrounding medium. In heterogeneous materials, the motion of individual probes gives the local properties of the material, and cross-correlated motion of the probes gives its bulk properties. Moreover, viscoelastic properties can be measured in a larger frequency range (0.01 Hz to 100 kHz) than in macroscopic measurements.

### 1.4.2  Macroscopic Tests

There are different macroscopic tests that can be carried out with biological tissues. They are classified here into two categories: the ones giving access to shear properties (transient assays, steady shear, oscillations) such as pure shear and compression, or the ones giving access to tensile or elongational properties.

#### 1.4.2.1  *Shear*

Classical definitions of shear experiments need to be given first. Usually, it is common to carry out experiments in a rotational rheometer when dealing with biological fluids or materials. This instrument allows us to have access to stresses (or torques) and strains while measuring strains and stresses, respectively. The basic idea is that operating in a circular geometry allows us to keep the fluid (material) in the same device, whereas capillary rheometry requires systems that push the fluid, therefore they require a larger amount of material. Usual rotational rheometers include different geometries such as the plate–plate, cone–plate, and Couette geometry (usually used for less viscous fluids). There is a different working formula for each one.

*Transient motions.* In classical rheometry, one applies a steady shear rate $\dot{\gamma}$ (in fact, startup from 0 to $\dot{\gamma}$) and the stress $\tau = \sigma_{12}$ is measured. Then the transient viscosity $\eta^+(t, \dot{\gamma})$, the first and second normal stress coefficients $\psi_1^+(t, \dot{\gamma})$, and $\psi_2^+(t, \dot{\gamma})$, can be determined:

$$\eta^+(t, \dot{\gamma}) = \frac{\sigma_{12}(t,\dot{\gamma})}{\dot{\gamma}}, \tag{4.2}$$

$$\psi_1^+(t, \dot{\gamma}) = \frac{(\sigma_{11}-\sigma_{22})(t,\dot{\gamma})}{\dot{\gamma}}, \tag{4.3}$$

$$\psi_2^+(t, \dot{\gamma}) = \frac{(\sigma_{22} - \sigma_{33})(t, \dot{\gamma})}{\dot{\gamma}}. \tag{4.4}$$

They are named the viscosimetric functions. If these behaviors correspond to typical viscoelastic materials, one expects a rise of $\eta^+(t, \dot{\gamma})$, $\psi_1^+(t, \dot{\gamma})$, and $\psi_2^+(t, \dot{\gamma})$, until a plateau is reached. The limiting values will then be $\eta(\dot{\gamma})$, $\psi_1(\dot{\gamma})$, $\psi_2(\dot{\gamma})$. When elastic effects are important at high shear rates, for example, there might be an overshoot in the time evolution of the previous functions [BIc] until a steady state is found.

*Steady-state functions.* As previously discussed, the steady-state functions $\eta(\dot{\gamma})$, $\psi_1(\dot{\gamma})$, $\psi_2(\dot{\gamma})$ are the limits of the time-dependent functions as time becomes large. These limits are usually reached in times inversely proportional to the velocity of deformation (shear rate $\dot{\gamma}$ here). Materials or fluids with a decreasing $\eta(\dot{\gamma})$ are said to be shear-thinning whereas the opposite is the shear-thickening behavior, as observed for certain suspensions of particles. When the stress goes to a limit at small shear rates, the material exhibits a yield stress as explained previously for Bingham fluids.

The first and second normal stress differences are quite important, because they are related to elastic effects, not usually encountered with Newtonian fluids. They correspond to the fact that the application of a shear in the 1–2 plane can give rise to normal stresses $\sigma_{22}$ and $\sigma_{33}$ in the other directions (rod-climbing effect, etc.).

*Dynamic rheometry.* This is the most common test used for characterizing biological materials, when large quantities of materials are available (collagen, actin solutions, etc.) and when such properties can be considered to be homogeneous, and not local. In shear, one applies a sinusoidal deformation $\gamma(t) = \gamma_0 \sin(\omega t)$. The stress $\tau$ is assumed to vary as does $\tau(t) = \tau_0 \sin(\omega t + \varphi)$. One of its components is in phase with $\gamma$ (elastic response), and the other one varies as does $\dot{\gamma}$ (viscous part). One defines, respectively, the elastic and viscous moduli $G'$ and $G''$. They are defined by $G'\gamma_0 = \tau_0 \cos\varphi$ and $G''\gamma_0 = \tau_0 \sin\varphi$. The loss angle $\varphi$ is given by $\tan\varphi = G''/G'$. In complex variables, the complex modulus $G^*$ is: $G^* = G' + iG''$ and the dynamic complex viscosity is $\eta^* = G^*/i\omega = \eta' - i\eta'' = G''/\omega - iG'/\omega$. Moduli $G'$ and $G''$ are determined for small deformations (linear domain), i.e. the domain where they remain constant for small enough $\gamma_0$. An example of the dependence of $G'$ and $G''$ versus frequency $\omega$ is given in Figure 1.3.

As can be seen, two different behaviors can be obtained:

- Newtonian behaviors at low rates: Case of usual fluids with respective slopes 2 and 1 for $G'$ and $G''$.

- Yield stress effects: $G'$ and $G''$ have a limiting value (fluid with a yield stress).

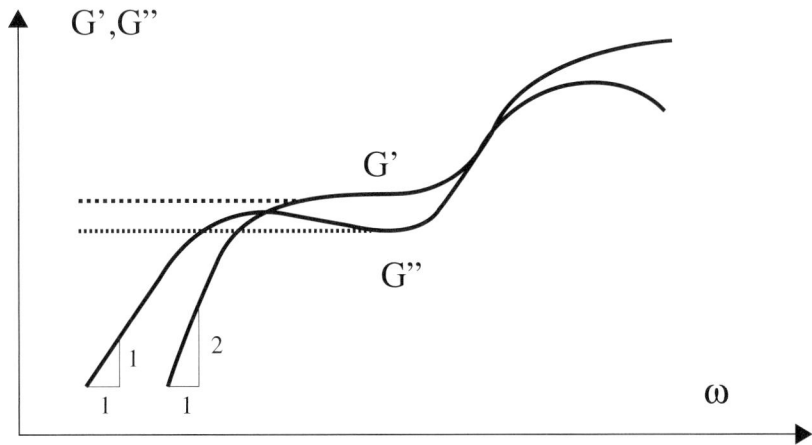

Figure 1.3. Dependence of G'(Pa) and G''(Pa) on frequency $\omega$(rad/s). At low frequencies, the fluid exhibits yield effects (dotted lines), such as in the case of concentrated actin solutions [SCa] or decreases with slopes 2 and 1 respectively, as in the limiting case of the Newtonian fluid (solid lines).

At moderate frequency values, $G'$ usually exhibits a plateau value (case of polymer solutions) and at high frequencies, $G'$ and $G''$ increase similarly as $\omega^n$, where $n$ is about 0.5–0.7, until the solid-high frequency regime is obtained.

**Remark 4.1** *The Maxwell model (Fig. 1.2) has a complex modulus $G^* = G\left(i\omega\lambda/(1 + i\omega\lambda)\right)$ therefore $G'(\omega) = G(\omega^2\lambda^2/(1 + \omega^2\lambda^2))$ and $G''(\omega) = G(\omega\lambda/(1 + \omega^2\lambda^2))$. This allows us to recover the typical behavior at low frequencies corresponding to the slopes of 2 and 1 for $G'$ and $G''$.*

### 1.4.2.2 Extension

The extensional properties of viscoelastic biological materials have been usually characterized using traction machines [FUa], or more subtle systems when biofluids are used. Usually constant rates of extension should be used, although this is rarely the case. In a constant stretching experiment at rate $\dot{\epsilon}$, the fluid element length increases exponentially, which is difficult to achieve. The elongational stresses of interest are combined to eliminate pressure effects in the form $\sigma_{11} - \sigma_{22}$, and the transient elongational viscosity is defined by:

$$\eta_E^+(t, \dot{\epsilon}) = \frac{\sigma_{11} - \sigma_{22}}{\dot{\epsilon}}(t, \dot{\epsilon}). \qquad (4.5)$$

If formula (4.5) has a limit for infinite times, then the elongational viscosity $\eta_E(\dot{\epsilon})$ can be defined. This limit exists usually at small enough elongational rates $\dot{\epsilon} < 1/2\lambda$ ($\lambda$ is the relaxation time defined previously), but it happens (as in Maxwell's fluid) that there is no limit above this value.

Typical instruments for obtaining constant stretch experiments are the traction experiment, the spinning fiber method, or the opposed jet method and four-roll mill apparatus for less viscous fluids [MAa].

Usually, the cases of biological materials under investigation has led to results that give rise to hyperelasticity, as depicted by sharply increasing stress–strain curves, but there is always a small effect due to the rate of stretch [FUa] which is usually neglected, unlike with biofluids.

## 1.5   Applications of Rheological Models

### 1.5.1   Cells

In this section, a few examples of successful predictions of cell modeling corresponding to real situations are presented, in particular in the case of cell motion under flow, and in the case of cell migration.

#### 1.5.1.1   *Cell Behavior Under Flow*

*Basic ideas.* Red blood cells have a very precise size (diameter 8 to 10 µm) whereas leukocytes are usually bigger (around 15 µm). Cells are able to travel through arteries or veins and also through small capillaries that are of the order of a few µm. They must therefore possess very special rheological properties to achieve these features; sometimes they can also migrate through the endothelial barrier (in the case of leukocytes or cancer cells). Let us consider the case of a single cell traveling in a vessel, and assume that the plasma is Newtonian. When subjected to an applied pressure, it takes an equilibrium position depending on the viscosity ratio and on the capillary number (ratio of viscous forces over surface effects). For leukocytes in capillaries, for example, assuming the Reynolds number is small and that the viscosity ratio is usually larger than one, the cell will take an equilibrium position between the wall and the centerline, and its deformation will basically depend on the capillary number [FUa].

To model a cell, several possibilities exist. The first authors to model cells used a membrane with a cortical tension surrounding a Newtonian fluid [YEa]. This model is already a good one for red blood cells. Membranes can also be considered to be linear elastic or nonlinear elastic sheets [SKa] (deriving from a strain energy function). They usually have a large

2-D elastic modulus so that their surface is almost inextensible. They only exhibit a bending energy [HEb]. Other possibilities (dropletlike models) such as a Newtonian fluid surrounded by a cortical layer [YEa] are also possible, and have proved to be efficient for describing micropipette experiments, for example. Finally, viscoelastic cells with a cortical tension have been proposed recently [KHa,VEb], and seem to be good candidates for describing cell behavior under flow.

Usually, due to the constraints, or simply to the fact that cells do not really travel in a linear fashion, they will eventually get close to the vascular wall and interact with it. This is studied next.

*Modeling cell interactions.* Cells are known to exhibit proteins (LFA–1, MAc–1, ICAM–1, etc.) on their surface, also named ligands (ICAM–1, VCAM–1, etc.). These ligands might interact with receptors present at the wall, on the endothelial lining. A simple reaction between a ligand ($L$) and a receptor ($R$) can give rise to a bond ($L$–$R$) according to

$$L + R \rightleftharpoons L - R. \tag{5.1}$$

Assume that $N_L^0$ and $N_R^0$ are the initial concentrations of ligands and receptors, respectively, and that $N$ is the number of bonds formed; then the rate-equation for $N$ is:

$$\frac{dN}{dt} = k_f(N_L^0 - N)(N_R^0 - N) - k_r N. \tag{5.2}$$

This is called a kinetics equation for cell-mediated adhesion. The solution of this equation starts at an initial prescribed value and then decreases until a plateau is reached. This model (microscopic) can be coupled with the usual macroscopic equations describing the cell behavior [DEb], therefore it is a way to couple the microscopic and macroscopic descriptions. Indeed, the forward and backward constants $k_f$ and $k_r$ are respectively known through

$$k_f = k_f^0 \exp\left[-\frac{\sigma_{ts}(x_m - \lambda)^2}{2k_B T}\right], \tag{5.3}$$

$$k_r = k_r^0 \exp\left[\frac{(\sigma - \sigma_{ts})(x_m - \lambda)^2}{2k_B T}\right], \tag{5.4}$$

where $k_f^0$ and $k_r^0$ are constants, $x_m$ is the bond length, $\lambda$ the equilibrium length, $k_B$ is the Boltzmann constant, and $T$ is temperature. $\sigma$ and $\sigma_{ts}$ (transition state) are the spring constants.

Then the force within a bond is simply given by $f_B = \sigma(x_m - \lambda)$, and finally the macroscopic force $F_B$ is equal to the single force times the bond density $N$: $F_B = N f_B$ [DEb]. Other simple models may use a force $F_B$ which is attractive and derive from a simple attractive potential [VEb].

*Models combining cell viscoelasticity and interactions.* There are limited number of studies devoted to the motion of cells close to a wall. The most interesting ones are studies by Dembo et al. [DEb], N'Dry et al. [NDa], Liu et al. [LIa], Jadhav et al. [JAa], Khismatullin et al. [KHa], or Verdier et al. [VEb]. The first studies are 2-D analyses of the motion of cells close to walls using kinetic models described previously. Let us discuss the cases of the works of Khismatullin et al. [KHa] and Jadhav et al. [JAa] dealing with 3-D problems.

Khismatullin et al. [KHa] used a nonlinear viscoelastic model for the cell description. This model is a Giesekus model which is the same as the one in Eq. (2.14), except that the nonlinear term $\kappa \underline{\underline{\tau}}^2$ on the left-hand side has been added. Note that this type of nonlinear equation is useful for predicting shear transient motions during startup. The originality of the work is also that a composite cell is considered. Indeed, the cell consists of a viscoelastic nucleus and a viscoelastic cytoplasm. A kinetic law of attachment–detachment is used as previously described. Finally, cortical tensions are imposed at the boundaries. The problem of the motion of a cell close to a wall (with receptors) is considered in the presence of microvilli. Deformability of the cell is calculated under physiological conditions (shear stresses of 0.8 to 4 Pa), as well as inclination angle, flow field, contact times, and microvilli number of attachments. Typically, cells are deformed quite a lot and exhibit a very small contact area.

Let us now compare this analysis with a quite different model [JAa], where viscoelasticity is introduced through the combined effects of a non-linear elastic membrane with a Newtonian cytoplasm. The same kinetics of bond formation is used but a stochastic process is used to model receptor–ligand interactions. For example, the probability of bond formation $P = 1 - \exp(-k_{on}\Delta t)$, where $k_{on}$ is a formation constant, is introduced and compared with a random number between 0 and 1. If it is larger, the bond will form, otherwise not. Differences with the previous model are obtained. Indeed, the cells are less deformed and exhibit a round shape, but adhere very strongly and form a much larger contact area, increasing with decreasing membrane elasticity. This can be understood because the stronger the membrane, the more spherical the cell, therefore the smaller number of bonds is formed.

To compare the models, one needs to compute a capillary number $C_a = \eta V/\sigma$ in the first case [KHa] ($\eta$ is the viscosity of the carrying fluid and $V$ a typical stream velocity), and $Ca = \eta V/Eh$ in the second one [JAa], because the elastic component acting against the flow to stabilize the cell shape is either the cortical tension $\sigma$ or the elastic 2-D modulus $Eh$ ($E$ is the elastic modulus and $h$ the membrane thickness). We find that the case considered by Khismatullin et al. [KHa] leads to very large capillary numbers, and thus to large deformations whereas the model of Jadhav et al. [JAa] has capillary numbers of order 1, and thus smaller deformations. Still, the flow field plays a role, as well as the model used and the other parameters. More studies are still needed to understand such problems better; they might be very important for understanding cancer cell extravasation, especially because cancer cells are considered to be less rigid as compared to other cells.

### 1.5.1.2 Cell Migration

*Principle of migration.* Cell migration is a complex mechanism, which involves both the adhesive properties but also the rheological properties of the cell, as depicted in Figure 1.1. Under chemotaxis or haptotaxis, a cell can polarize and develop a lamellipodium which extends far to the front [COb] in the case of a fibroblast on a rigid surface. Inside the cell, changes in the local actin concentration can generate changes in the microrheological properties, allowing the cell to deform and move. Actin filaments can form cross-links at the front (as in a gel), whereas they become less densely packed (sol) at the uropod (tail), in order to preserve the total actin concentration. In order for the cell to move, it requires the generation of traction forces to pull itself forward. These forces are generated by focal adhesion plaques, such as integrin clusters. Some cells can migrate very quickly as do the neutrophils of the immune system (mm/hour) whereas other cells, such as cancer cells, reach velocities of only a few tenths of μm/hour. There is a complex machinery involving actin binding proteins (ABP) together with myosin to form actin units. Other disassembly proteins are also needed to break actin units. Integrins bind to the cytoskeleton, which is made of parallel bundles of actin filaments thus creating a reinforced structure that allows the cell to generate traction forces. Such traction forces can be measured on deformable substrates in the case of fibroblasts, for example [DEa]. Other methods also exist based on wrinkle patterns on deformable substrates as well [BUa] and provide interesting information in the case of keratocytes.

*Models of cell migration.* In order to migrate efficiently, a cell must develop strong traction forces, but they should not be too large, otherwise they will be difficult to break at the rear. Indeed there is an optimal velocity of migration [PAa] depending on the typical bond forces or cell–substrate affinity. One way to model adhesion is through a distribution of bonds, as seen previously. This idea comes from observations (RICM) of adhering cells showing unattached cell parts. The model of Dickinson and Tranquillo [DIa] assumes such a distribution of receptor–ligand bonds. Adhesion gradients can also be considered that influence cell motility. A stochastic model is assumed to show how migration is affected by the forces and the distribution of ligands on the cell. Adhesion receptors undergo rapid binding, and this results in a time-dependent motion. Mean speed, persistence time, and random motility coefficients can then be obtained. A bell-shaped curve is finally obtained showing a maximum in velocity as a function of the adhesion concentration factor, as shown experimentally [PAa].

Another approach by DiMilla et al. [DIb] includes cell polarization, cytoskeleton force generation, and dynamic adhesion to create cell movement. A model for cell viscoelastic properties (1-D) is also included. Similar effects for the velocity of migration as a function of force are obtained, but further effects such as force and cell rheology as well as receptor–ligand dynamics can be added. The maximum in the speed of migration is related to the balance between cell contractile force and adhesiveness.

*Cancer cell migration.* Cancer cell migration is different from the previous cells studied (fibroblasts, leukocytes). Friedl and coworkers [FRa] have shown that tumor cells develop migrating cell clusters. They also seem to develop stronger interactions (and pulling forces) and are more polarized (direction). Most cancer cells are usually bigger and slower than migrating leukocytes. They are also capable of reorganizing the extracellular matrix (ECM) easily. Therefore, cancer cell migration is still hard to model and requires more experimental data.

## 1.5.2   Tissues

Biological tissues are complex structures subjected to a number of external stimuli (e.g. mechanical forces, electrical signals, and heat). The structure of these tissues determines their response to the stimuli. In addition, cells within the tissues can sense the stimuli and adapt or change the tissue matrix structure. Biological tissues differ in many ways from typical engineering materials. They are extremely heterogeneous within a single body and between individuals. They always have hierarchical structures with many different scales. And they are able to change their structure in response to external stimuli.

In this section, a few examples of connective tissue modeling such as blood and soft tissues under physiological loads are presented.

### 1.5.2.1   Blood

Blood is a circulating tissue. It is a complex fluid composed of red blood cells (RBC or erythrocytes), white blood cells (WBC or leukocytes), and platelets suspended in plasma (an aqueous solution of electrolytes and proteins such as fibrinogen and albumin). Plasma is the extracellular matrix of blood cells). Blood cells' volumic concentration (hematocrit) is about 38–45% corresponding to $5.10^6/\text{mm}^3$ of RBC, $5.10^3/\text{mm}^3$ of WBC and $3.10^5/\text{mm}^3$ of platelets. Plasma behaves as a Newtonian fluid of 1.2 mPa.s viscosity at 37°C. The whole blood behaves as a non-Newtonian fluid. Its viscosity varies with the hematocrit, with the temperature, and with the disease state [CHa].

When looking at blood flow in large vessels, it can be considered as a homogeneous fluid. This can be analyzed using a Couette flow viscometer where the width of the flow channel is much larger than the diameter of blood cells. Using Couette flow viscometry, Cokelet et al. [COa] found that blood has a finite yield stress in shear. For a small shear rate ($\dot{\gamma} < 10 s^{-1}$) and for hematocrit less than 40%, their data are approximately described by Casson's law [CAa]:

$$\sqrt{\tau} = \sqrt{\tau_s} + \sqrt{\eta \dot{\gamma}}, \tag{5.5}$$

where $\tau$ is the shear stress, $\dot{\gamma}$ the shear rate, $\tau_s$ is the yield stress in shear ($\approx 5.10^{-3}$ Pa), and $\eta$ is a constant. At high shear rates (about $100 s^{-1}$), the whole blood behaves as a Newtonian fluid with a constant viscosity ($4-5$ mPa.s). The whole blood flow in a cylindrical tube follows a plug flow profile [FUa]. This behavior can be explained by the fact that human RBCs form aggregates (known as rouleaux) which are more important under low shear rates. When the shear rate tends to zero the whole blood becomes like a big aggregate with a solidlike behavior (a viscoplastic behavior as described in Section 2.1). When the shear rate increases, the aggregates tend to break and the viscosity of blood decreases. For further increase in shear rate, RBCs become elongated and align with flow streamlines [GOa] inducing a very low viscosity (3–4 mPa.s) for such a concentrated suspension.

When looking at blood flow in capillaries, it can be considered as a nonhomogeneous fluid of at least two phases: blood cells and plasma. Indeed in capillaries, whose diameter is in the range of blood cell diameter (4–10 μm), blood cells have to squeeze and arrange themselves in single file [FUa]. Thus, mechanical properties of RBCs play a predominant role in the microcirculation. These cells are nucleus-free deformable liquid capsules enclosed by a nearly incompressible membrane that exibits elastic response to shearing and bending deformation. As an application of rheological measurements to determine RBC mechanical properties, we can refer to the work of Drochon et al. [DRa]. They measured the rheological properties of a dilute suspension of RBCs and interpreted their experimental data based on a microrheological model, proposed by Barthès-Biesel et al. [BAa]. This model illustrated the effect of interfacial elasticity on capsule deformation and on the rheology of dilute suspensions for small deformations. Thus, Drochon et al. determined the average deformability of a RBC population in terms of the mean value of the membrane shear elastic modulus.

## 1.5.2.2 Soft Tissues

Most biological tissues exhibit a time- and history-dependent stress–strain behavior that is a characteristic property of viscoelastic materials.

Viscoelastic models for soft tissues can be divided into two groups: microstructural and rheological models.

*Microstructural models* are based on mechanical behavior of the constituents of the tissue. The mechanical response of the components is generalized to produce a description of the tissue's gross mechanical behavior. For example, Lanir introduced a microstructural model of lung tissue [LAa]. He considered lung tissue as a cluster of a very large number of closely packed airsacks (alveoli) of irregular polyhedron shape, bounded by the alveolar wall membrane. Lanir employed a stochastic approach to tissue structure in which the predominant structural parameter is the density distribution function of the membrane's orientation in space. Based on this model, the behavior of the alveolar membrane and its liquid interface was related to general constitutive properties of lung tissue.

Another microstructural 2-D model of lung tissue consisted of a sheet of randomly aligned fibers of various orientations embedded in a viscous liquid ground substance [BAb]. The fiber orientations constantly change due to thermal motion. When the sheet is stretched, the fibers align in the direction of strain and there is a net transfer of momentum between the fibers and the ground substance, due to the constant thermal motion of the fibers. This model also predicts that any stress generated within the tissue will decay asymptotically to zero as the fibers reorient.

Microstructural models were applied to other tissues. Guilak and Mow [GUa] modeled the articular cartilage based on a biphasic theory in which the tissue is treated as a hydrated soft material consisting of two mechanically interacting phases: a porous, permeable, hyperelastic, composite solid phase composed of collagen, proteoglycans, and chondrocytes; and a viscous fluid phase, which is predominantly water and electrolytes. Both phases are intrinsically incompressible and diffusive drag forces between the two phases give rise to the viscoelastic behavior of the tissue. Such models are well suited to study the connection between the structure and the mechanical properties (stress, strain, fluid flow, and pressurization).

Tensegrity models have been developed [FUb] based on the ideas of deformable structures (i.e. civil engineering) made of sticks and strings in tension or compression. They can be applied to cells [INa, INb] because the cell cytoskeleton can be depicted as an assembly of rods and springs (various cytoskeleton filaments). Similar ideas have been developed at a higher scale, by considering homogenization methods in the case of cardiomyocytes, assumed to form discrete lattices [CAb] of bars linked together. When the components are elastic, one can recover an elastic constitutive model; also hyperelasticity can be obtained.

*Rheological models* describe the gross mechanical behavior of the tissue in the simplest possible terms. Sanjeevi et al. [SAa] proposed a 1-D rheological model of viscoelastic behavior for collagen fibers. The quasi-linear viscoelastic models (see, for example, Eq. (2.2)) have been useful to describe various tissues such as the heart muscle [PIa, HUa] and the cervical spine [MYa]. Bilston et al. [BIa] developed a constitutive model that accurately reproduces the strain-rate dependence of brain tissue and its linear stress–strain response in shear.

On the other hand, Dehoff [DEa] described the nonlinear behavior of soft tissues by adapting a continuum-based formulation previously used to characterize polymers. Phan-Thien et al. [PHa] also used a nonlinear Maxwell model to predict the behavior of kidney under large-amplitude oscillatory squeezing flow. They added a nonlinear stress-dependent viscosity in front of $\underline{\underline{\mathbf{D}}}$ in Eq. (2.14). Nasseri et al. [NAa] developed a multimode upper-convected Maxwell (see Section 1.2) model with variable viscosities and time constants for viscoelastic response, coupled to a hyperelastic response (see Eq. (2.11)). In their model, the sum of the elastic and the viscoelastic contributions were modified by a nonlinear damping function to control the nonlinearity of stress–strain profiles: (1) in the limit of small strain (0.2%), the damping function reduced to unity and their model reduced to a multimode Maxwell model with shear-rate-dependent viscosities; (2) in high strain rate loading, this model gave a rubberlike response. Their model predicted well the rheological properties of the kidney cortex under strain sweep, small amplitude oscillatory motion (dynamic testing), stress relaxation, and constant shear rate (viscometry) tests. They showed that this tissue was highly shear thinning, and at higher strain amplitudes this phenomenon was more significant. The damping function was strain-dependent and could be determined to match well various nonlinear features of the shear tests.

## 1.6 Conclusions

In this chapter, we have made an attempt to investigate the rheological properties of biological systems in a nonexhaustive manner. There are at the moment good methods for characterizing tissues and fluids, and also cellular elements. Although these techniques are very promising, there is still a lack of characterizations of tissues or cells, coupled with microscopic observations, in particular, the ones based on the use of fluorescence.

Some of these experimental procedures have been correlated successfully with existing models, which have already been developed in the field of classical rheology. It has also been shown that cell–cell interactions are very important when modeling cell or tissue behavior.

What are still missing today are actual models including the active response of the tissues or cells, which can in return induce changes to the cell cytoskeleton or membrane. Some progress has been done recently but there is a need for taking into account signalization and its effects on the rheological properties, in other words, understanding mechanotransduction.

**Acknowledgments**
This work has been made possible thanks to two European networks (HPRCT–2000–00105 and MRTN–CT–2004–503661) devoted to cancer modeling.

## 1.7   References

[AHa]   A-Hassan, E., Heinz, W.F., Antonik, M.D., D'Costa, N.P., Nageswaran, S., Schoenenberger, C., and Hoh, J.H., Relative microelastic mapping of living cells by atomic force microscopy, *Biophys. J.*, **74** (1998), 1564–1578.

[ALa]   Alberts, B., Bray, D., Lewis, J., Raff, M., Roberts, K., and Watson, J.D. **Molecular Biology of the Cell, 3rd edition**, Garland, New York (1994).

[ALb]   Alcaraz, J., Buscemi, L., Grabulosa, M., Trepat, X., Fabry, B., Farre, R. and Navajas, D., Microrheology of human lung epithelial cells measured by atomic force microscopy, *Biophys. J.*, **84** (2003), 2071–2079.

[ASa]   Ashkin, A., Forces of a single-beam gradient laser trap on a dielectric sphere in the ray optics regime, *Biophys. J.*, **67** (1992), 569–589.

[BAa]   Barthès-Biesel, D. and Rallison, J.M., The time dependent deformation of a capsule freely suspended in a linear shear flow, *J. Fluid Mech.*, **113** (1981), 251.

[BAb]   Bates, J.H.T., A micromechanical model of lung tissue rheology, *Ann. Biomed. Eng.*, **26** (1998), 679–687.

[BAc]   Baumgaertel, M. and Winter H.H., Determination of discrete relaxation and retardation time spectra from dynamic mechanical data, *Rheol. Acta*, **28** (1989), 511–519.

[BAd]   Bausch, A.R., Ziemann, F., Boulbitch, A.A., Jacobson, K., and Sackmann, E., Local measurements of viscoelastic parameters of adherent cell surfaces by magnetic bead microrheometry, *Biophys. J.*, **75** (1998), 2038–2049.

[BAe] Bausch, A.R., Moller, W., and Sackmann, E., Measurement of local viscoelasticity and forces in living cells by magnetic tweezers, *Biophys. J.*, **76** (1999), 573–579.

[BIa] Bilston, L.E., Liu, Z., and Phan-Thien, N., Linear viscoelastic properties of bovine brain tissue in shear. *Biorheology*, **34** (1997), 377–385.

[BIb] Bining, G., Quate, C.F., and Gerber, C., Atomic force microscope, *Phys. Rev. Lett.*, **56** (1986), 930–933.

[BIc] Bird, R.B., Armstrong, R.C., and Hassager, 0., **Dynamics of Polymeric Liquids. Fluid Mechanics, Vol. I**, Wiley Interscience, New York (1987).

[BOa] Bouchaud, J.–P., Weak ergodicity–breaking and aging in disordered systems, *J. Phys. I. France*, **2** (1992), 1705–1713.

[BUa] Burton, K., Park, J.H., and Taylor, D.L., Keratocytes generate traction forces in two phases, *Molecular Biol. Cell*, **10** (1999), 3745–3769.

[CAa] Casson, M., A flow equation for pigment-oil suspensions of the printing ink type, in: Mills, C.C., ed., **Rheology of Disperse Systems**, Pergamon, Oxford (1959).

[CAb] Caillerie, D., Mourad, A., and Raoult, A., Cell-to-muscle homogenization. Application to a constitutive law for the myocardium, *Math. Model. Numer. Anal.*, **37** (2003), 681–698.

[CHa] Chien, S., Shear dependence of effective cell volume as a determinant of blood viscosity, *Science*, **168** (1970), 977–979.

[COa] Cokelet, G.R., Merrill, E.W., Gilliand, E.R., Shin, H., Britten, A., and Wells, R.E., The rheology of human blood measurement near and at zero shear rate, *Trans. Soc. Rheol.*, **7** (1963), 303–317.

[COb] Condeelis, J., Life at the leading edge: the formation of cell protrusions, *Ann. Rev. Cell Biol.*, **9** (1993), 411–444.

[CRa] Crocker, J.C., Valentine, M.T., Weeks, E.R., Gisler, T., Kaplan, P.D., Yodh, A.G., and Weitz, D.A., Two-point microrheology of inhomogeneous materials *Phys. Rev. Lett.*, **85** (2000), 888–891.

[DEa] Dehoff, P.H., On the nonlinear viscoelastic behavior of soft biological tissues. *J. Biomech.*, **11** (1978), 35–40.

[DEb] Dembo, M. and Wang, Y.-L., Stresses at the cell-substrate interface during locomotion of fibroblasts, *Biophys. J.*, **76** (1999), 2307–2316.

[DEc] Dembo, M., Torney, D.C., Saxman, K., and Hammer, D., The reaction–limited kinetics of membrane to surface adhesion and detachment, *Proc. R. Soc. Lond. B*, **234** (1988), 55–83.

[DIa] Dickinson, R.B. and Tranquillo, R.T., A stochastic model for adhesion-mediated cell random motility and haptotaxis, *J. Math. Biol.*, **31** (1993), 563–600.

[DIb] DiMilla, P.A., Barbee, K., and Lauffenburger, D.A., Mathematical model for the effects of adhesion and mechanics on cell migration speed. *Biophys. J.*, **60** (1991), 15–37.

[DRa] Drochon, A., Barthès-Biesel, D., Lacombe, C., and Lelièvre, J.C., Determination of the red blood cell apparent membrane elastic modulus from viscometric measurements, *J. Biomech. Eng.*, **112** (1990), 241–249.

[FAa] Fabry, B., Maksym, G.N., Butler, J.P., Glogauer, M., Navajas, D., and Fredberg, J.J., Scaling the microrheology of living cells, *Phys. Rev. Lett.*, **87** (2001), 148102:1–4.

[FAb] Fawcett, D. W., **Bloom and Fawcett: A Textbook of Histology**, 11th edn. W. B. Saunders, Philadelphia (1986).

[FRa] Friedl, P., Brocker, E.B., and Zanker, K.S., Integrins, cell matrix interactions and cell migration strategies: fundamental differences in leukocytes and tumor cells. *Cell Adhes. Commun.*, **6** (1998), 225–236.

[FUa] Fung, Y.C., **Biomechanics. Mechanical Properties of Living Tissues**, Springer–Verlag, New York (1993).

[FUb] Fuller, R.B., Tensegrity, *Portfolio Artnews Annual*, **4** (1961), 112–127.

[GOa] Goldsmith, H.L., The microrheology of human erythrocyte suspensions, in: Becker, E. and Mikhailov, G.K., eds., **Theoretical and Applied Mechanics, Proc. 13th IUTAM Congress**, Springer, New York (1972).

[GUa] Guilak, F. and Mow, V.C., The mechanical environment of the chondrocyte: a biphasic finite element model of cell-matrix interactions in articular cartilage. *J. Biomech.*, **33** (2000), 1663–1673.

[HEa] He, X. and Dembo, M., A dynamic model of cell division, in: Alt, W., Deutsch, A., and Dunn, G. eds, **Mechanics of Cell and Tissue Motion**, (1997), 55–66.

[HEb] Helfrich, W., Elastic properties of lipid bilayer: Theory and experiments, *Z. Naturforsch*, **C28** (1973), 693–703.

[HEc] Hénon, S., Lenormand, G., Richert, A., and Gallet, F., A new determination of the shear modulus of the human erythrocyte membrane using optical tweezers, *Biophys. J.*, **76** (1999), 1145–1151.

[HOa] Hough, L.A. and Ou-Yang, H.D., A new probe for mechanical testing of nanostructures in soft materials, *J. Nanoparticle Res.*, **1** (1999), 495–499.

[HUa] Huyghe, J.M., Van Campen, D.H., Arts, T., and Heethar, R.M., The constitutive behaviour of passive heart muscle tissue: A quasi–linear viscoelastic formulation, *J. Biomech.*, **24** (1991), 841–849.

[INa] Ingber, D.E., Cellular tensegrity: Defining the rules of biological design that govern the cytoskeleton. *J. Cell Sci.*, **104** (1993), 613–627.

[INb] Ingber, D.E., Heidemann, S.R., Lamoureux, P., and Buxbaum, R.E., Opposing views on tensegrity as a structural framework for understanding cell mechanics, *J. Appl. Physiol.*, **89** (2000), 1663–1678.

[JAa] Jadhav, S., Eggleton, C.D., and Konstantopoulos, K., A 3–D computational model predicts that cell deformation affects selectin–mediated leukocyte rolling, *Biophys. J.*, **88** (2005), 96–104.

[JOa] Johnson, G.A., Livesay, G.A., Woo, S.L.Y., and Rajagopal, K.R., A single integral finite strain viscoelastic model of ligaments tendons, *J. Biomech. Eng.*, **118** (1996), 221–226.

[KHa] Khismatullin, D.B. and Truskey, G.A., A 3D numerical study of the effect of channel height on leukocyte deformation and adhesion in parallel-plate flow chambers, *Microvasc. Res.*, **68** (2004), 188–202.

[LAa] Lanir, Y., Constitutive equations for the lung tissue, *J. Biomech. Eng.*, **105** (1983), 374–380.

[LAb] Larson, R.G., **The Structure and Rheology of Complex Fluids**, Oxford University Press, New York (1999).

[LIa] Liu, X.H. and Wang, X., The deformation of an adherent leukocyte under steady flow: A numerical study, *J. Biomechanics*, **37** (2004), 1079–1085.

[MAa] Macosko, C.W., **Rheology, Principles, Measurements and Applications**, Wiley-VCH, New York (1994).

[MAb] Mahaffy, R.E., Shih, C.K., MacKintosh, F.C., and Käs, J., Scanning probe-based frequency-dependent microrheology of polymer gels and biological cells, *Phys. Rev. Lett.*, **85** (2000), 880–883.

[MAc] Mahaffy, R.E., Park, S., Gerde, E., Käs, J., and Shih, C.K., Quantitative analysis of the viscoelastic properties of thin regions of fibroblasts using atomic force microscopy, *Biophys. J.*, **86** (2004), 1777–1793.

[MAd] Malvern, L.E., **Introduction to the Mechanics of a Continuum Medium**, Prentice-Hall, Englewood Cliffs, NJ (1969).

[MAe] Mason, T.G., Ganesan, K., Van Zanten, J.H., Wirtz, D. and Kuo, S.C., Particle tracking microrheology, *Phys. Rev. Lett.*, **79** (1997), 3282–3285.

[MAf] Mason, T.G. and Weitz, D.A., Optical measurements of frequency-dependent linear viscoelastic moduli of complex fluids, *Phys. Rev. Lett.*, **74** (1995), 1250–1253.

[MYa] Myers, B.S., MacElhaney, J.H., and Doherty, B.J., The viscoelastic responses of the human cervical spine in torsion: Experimental limitations of quasi-linear theory and a method for reducing these effects, *J. Biomech.*, **24** (1991), 811–817.

[NAa] Nasseri, S., Bilston, L.E. and Phan-Thien, N., Viscoelastic properties of pig kidney in shear, experimental results and modelling. *Rheol. Acta*, **41** (2002), 180–192.

[NDa] N'Dry, N.A., Shyy W., and Tran–Son–Tay, R., Computational modeling of cell adhesion and movement using a continuum–kinetics approach, *Biophys. J.*, **85** (2003), 2273–2286.

[PHa] Phan-Thien, N., Nasseri, S., and Bilston, L.E., Oscillatory squeezing flow of a biological material. *Rheol. Acta*, **39** (2000), 409–417.

[PIa] Pinto, and Fung, Y.C., Mechanical properties of the heart muscle in the passive state, *J. Biomech.*, **6** (1973), 597–606.

[PAa] Palececk, S.P., Loftus, J.C., Ginsberg, M.H., Lauffenburger, D.A., and Horwitz, A.F., Integrin-ligand binding properties govern cell migration speed through cell-substratum adhesiveness, *Nature*, **385** (1997), 537–540.

[RAa] Radmacher, M., Fritz, M., Kacher, C.M., Cleveland J.P., and Hansma, P.K., Measuring the viscoelastic properties of human platelets with the atomic force microscope, *Biophys. J.*, **70** (1996), 556–567.

[ROa] Rotsch, C., Jacobson, K., and Radmacher, M., Dimensional and mechanical dynamics of active and stable edges in motile fibroblasts investigated by using atomic force microscopy. *Proc. Natl. Acad. Sci. USA*, **96** (1999), 921–926.

[SAa] Sanjeevi, R., A viscoelastic model for the mechanical properties of biological materials, *J. Biomech.*, **15** (1982), 107–109.

[SCa] Schmidt, C., Hinner, B., and Sackmann, E., Microrheometry underestimates the viscoelastic moduli in measurements on F-actin solutions compared to macrorheometry, *Phys. Rev. Letters*, **61** (2000), 5646–5653.

[SHa] Shoemaker, P.A., Schneider, D., Lee, M.C., and Fung, Y.C., A constitutive model for two dimensional soft tissues and its application to experimental data, *J. Biomech.*, **19** (1986), 695–702.

[SKa]  Skalak, R., Modeling the mechanical behavior of blood cells, *Biorheology*, **10** (1973), 229–238.

[SMa]  Smith, B.A., Tolloczko, B., Martin, J.G., and Grütter, P., Probing the viscoelastic behavior of cultured airway smooth muscle cells with atomic force microscopy: Stiffening induced by contractile agonist, *Biophys. J.*, **88** (2005), 2994–3007.

[SOa]  Sollich, P., Lequeux, F., Hébraud, P., and Cates, M.E., Rheology of soft glassy materials, *Phys. Rev. Lett.*, **78** (1997), 2020–2023.

[STa]  Stossel, T., On the crawling of animal cells, *Science*, **260** (1993), 1086–1094.

[TSa]  Tseng, Y., Kole, T.P., and Wirtz, D., Micromechanical mapping of live cells by multiple-particle-tracking microrheology, *Biophys. J.*, **83** (2002), 3162–3176.

[VEa]  Verdier, C., Review: Rheological properties of living materials. From cells to tissues, *J. Theor. Medicine*, **5** (2003), 67–91.

[VEb]  Verdier, C., Jin, Q., Leyrat, A., Chotard-Ghodsnia, R., and Duperray, A., Modeling the rolling and deformation of a circulating cell adhering on an adhesive wall under flow, *Arch. Physiol. Biochem.*, **111** (2003), 14–14.

[WAa]  Wagner, M.H., A constitutive analysis of extensional flows of polyisobutylene, *J. Rheol.*, **34** (1990), 943–958.

[YAa]  Yamada, S., Wirtz, D., and Kuo, S.C., Mechanics of living cells measured by laser tracking microrheology, *Biophys. J.*, **78** (2000), 1736–1747.

[YEa]  Yeung, A. and Evans, E., Cortical shell-liquid core model for passive flow of liquid-like spherical cells in micropipettes, *Biophys. J.*, **56** (1989), 139–149.

# 2

# *Biochemical and Biomechanical Aspects of Blood Flow*

M. THIRIET

*REO team*
*Laboratoire Jacques-Louis Lions, UMR CNRS 7598,*
*Université Pierre et Marie Curie, F-75252 Paris cedex 05, and*
*INRIA, BP 105, F-78153 Le Chesnay Cedex.*

ABSTRACT. The blood vital functions are adaptative and strongly regulated. The various processes associated with the flowing blood involve multiple space and time scales. Biochemical and biomechanical aspects of the human blood circulation are indeed strongly coupled. The functioning of the heart, the transduction of mechanical stresses applied by the flowing blood on the endothelial and smooth muscle cells of the vessel wall, gives examples of the links between biochemistry and biomechanics in the physiology of the cardiovascular system and its regulation. The remodeling of the vessel of any site of the vasculature (blood vessels, heart) when the blood pressure increases, the angiogenesis, which occurs in tumors or which shunts a stenosed artery, illustrates pathophysiological processes. Moreover, focal wall pathologies, with the dysfunction of its biochemical machinery, such as lumen dilations (aneurisms) or narrowings (stenoses), are stress-dependent. This review is aimed at emphasizing the multidisciplinary aspects of investigations of multiple aspects of the blood flow.

## 2.1   Introduction

Biomechanics investigates the cardiovascular system by means of mechanical laws and principles. Biomechanical research related to the blood circulation is involved

1. In the motion of human beings, such as gait (blood supply, venous return in transiently compressed veins)

2. In organ rheology influenced by blood perfusion

3. In heat and mass transfer, especially in the context of mini-invasive therapy of tumors

4. Cell and tissue engineering

5. In the design of surgical repair and implantable medical devices

Macroscale biomechanical model of the cardiovascular system have been carried out with multiple goals:

1. Prediction

2. Development of pedagogical and medical tools

3. Computations of quantities inaccessible to measurements

4. Control

5. Optimization

In addition, macroscale simulations deal with subject-specific geometries, because of a high between-subject variability in anatomy, whatever the image-based approaches, either numerical and experimental methods, using stereolithography. The research indeed aims at developing computer-assisted medical and surgical tools in order to learn, to explore, to plan, to guide, and to train to perform the tasks during interventional medicine and mini-invasive surgery. However, this last topic is beyond the goal of the present review.

More and more studies deal with multiscale modeling in order to appropriately take into account the functioning of the blood circulation, from the molecular level (nanoscopic scale, nm), to the cell organelles associated with the biochemical machinery (microscopic scale, μm), to the whole cell connected to the adjoining cells and the extracellular medium, subdomain of the investigated tissue (mesoscopic scale, mm), and to the entire organ (macroscopic scale, cm). The genesis and the propagation of the electrochemical wave is the signal that commands the myocardium contraction, hence the blood propelling by the heart pump. It is characterized by ion motions across specialized membrane carriers. The myocardium contraction itself requires four-times nanomotors, associated with the actin–myosin binding and detachment. The actin and myosin filaments are assembled in contractile sarcomeres within the cardiomyocytes. The latter microscopic elements gather in muscular fibers, acting as a syncytium in the myocardium. The regulation of the vessel lumen caliber and of the wall structure is mediated by the blood flow (endothelial mechanotransduction). The endothelium is, indeed, permanently exposed to biochemical and biomechanical stimuli which are sensed and transduced, leading to responses that involve various pathways. In particular, the endothelium is subjected to hemodynamic forces (pressure, friction) that can vary both in magnitude and in direction during the cardiac cycle. The endothelium adapts to this mechanical environment using short- and long-term mechanisms. Among the quick reactions, it modulates the vasomotor tone by the release of vasoactive compounds. The endothelium actively participates in inflammation and healing. Chronic adaptation leads to wall remodeling and vascular growth, with the formation of functional collaterals and vessel regression.

## 2.2 Anatomy and Physiology Summary

Anatomy deals with the macroscopic scale and biochemistry with the nanoscopic level. The cardiovascular system is mainly composed of the cardiac pump and a circulatory network. The heart is made up of a couple of synchronized pumps in parallel, composed of two chambers. The left heart propels blood through the systemic circulation and the right heart through the pulmonary circulation.

The cardiovascular system provides adequate blood input to the different body organs, responding to sudden changes in demand of nutrient supplies. For a stroke volume of 80 ml and a cardiac frequency of 70 beats per mn, a blood volume of 5.6 l is propelled per mn. The travel time for oxygen delivery between the heart and the peripheral tissues has a magnitude $\mathcal{O}(s)$.

### 2.2.1 Heart

The heart is located within the mediastinum, usually behind and slightly to the left of the sternum (possible mirror-image configuration). The base of the heart is formed by vessels and atria, and the apex by the ventricles. The septum separates the left and right hearts. The left ventricle is the largest chamber and has the thickest wall. The pericardium surrounds the heart and the roots of the great blood vessels. The pericardium restricts excessive heart dilation, and thus limits the ventricular filling.

Four valves at the exit of each heart cavity, between the atria and the ventricules, the atrioventricular valves, and between the ventricules and the efferent arteries, the ventriculoarterial valves, regulate the blood flow through the heart and allow bulk unidirectional motion through the closed vascular circuit. The tricuspid valve, composed of three cusps, regulates blood flow between the right atrium and ventricle. The pulmonary valve controls the blood flow from the right ventricle into the pulmonary arteries, which carry the blood to the lungs to pick up oxygen. The mitral valve, which consists of two soft thin cups, lets oxygen-rich blood from pulmonary veins pass from the left atrium to the left ventricle. The aortic valve guards the exit of the left ventricle. Like the pulmonary valve, it consists of three semilunar cusps. Immediately downstream from the aortic orifice, the wall of the aorta root bulges to form the Valsalva sinuses. Papillary muscles protrude into both ventricular lumina and point toward the atrioventricular valves. They are connected to chordae tendineae, which are attached to the leaflets of the respective valve.

The heart is perfused by the right and left coronary arteries, originating from the aorta just above the aortic valve. These distribution coronary arteries lie on the outer layer of the heart wall. These superficial arteries branch into smaller arteries that dive into the wall. The heart is innervated by both components of the autonomic nervous system. The parasympathetic innervation originates in the cardiac inhibitory center and is conveyed to the heart by way of the vagus nerve. The sympathetic innervation comes from the cardiac accelerating center. Normally, the parasympathetic innervation represents the dominant neural influence on the heart.

Deoxygenated blood from the head and the upper body and from the lower limbs and the lower torso is brought to the right atrium by the superior (SVC) and by the inferior (IVC) venae cavae. When the pulmonary valves are open, the left ventricle ejects blood into the pulmonary artery. The pulmonary veins carry oxygen-rich blood from the lungs to the left atrium. The aorta receives blood ejected from the left ventricle. The right and left hearts, with their serial chambers, play the role of a lock between a low-pressure circulation and a high-pressure circuit. The atrioventricular and ventriculoarterial couplings set the ventricle for the filling and pressure adaptation and for the ejection, respectively.

The heart has an average oxygen requirement of 6–8 ml.min$^{-1}$ per 100 g at rest. Approximately 80% of oxygen consumption is related to its mechanical work (20% for basal metabolism). Myocardial blood flow must provide

| EDV | 70–130 ml |
|---|---|
| ESV | 20–50 ml |
| SEV | 60–100 ml |
| $f_c$ | 60–80 beats.mn$^{-1}$, $1 - 1.3$ Hz |
| $q$ | 4–7 l.mn$^{-1}$ (70–120 ml.s$^{-1}$) |
| Ejection fraction | 60–80% |

Table 2.1. Physiological quantities at rest in healthy subjects. $f_c$ decreases and then increases with aging; SV decreases with aging ($q \sim 6.5$ l.mn$^{-1}$ at 30 years old and $q \sim 4$ l.mn$^{-1}$ at 70 years old).

this energy demand. The myocardium also uses different substrates for its energy production, mostly fatty acid metabolism, which gives nearly 70% of energy requirements, and glucids.

The cardiac output is the amount of blood that crosses any point in the circulatory system and pumped by each ventricle per unit of time. In a healthy person at rest, CO $\sim 5$–6 l.mn$^{-1}$. The cardiac output is determined by multiplying the stroke volume (blood volume pumped by the ventricle during one beat) by the heart rate $f_c$. The stroke volume is the difference between the end-diastolic volume (EDV) and the end-systolic volume (ESV). Values of physiological quantities at rest in healthy subjects are given in Table. 2.1. Various factors determine the cardiac output. The preload is a stretching force exerted on the myocardium at the end of diastole, imposed by the blood volume. The afterload is the resistance force to contraction. The cardiac index is calculated as the ratio between the blood flow rate $q$ and the body surface area ($2.8 < CI < 4.2$ $l$.mn$^{-1}$.m$^{-2}$).

The stroke volume can be modified by changes in ventricular contractility (Frank–Starling effect) and in velocity of fiber shortening. Increased inotropy augments the ventricular pressure time gradient and therefore the ejection velocity. The left ventricle responds to an increase in arterial pressure by augmenting contractility and hence SV, whereas EDV may return to its original value (Anrep effect). An increase in heart rate creates a positive inotropic state (Bowditch or Treppe effect). Most of the signals that stimulate inotropy induce a rise in $Ca^{++}$ influx.

The total oxygen consumption is subject- and age-dependent (2–10 ml/ mn/100g). The heart has the highest arteriovenous O$_2$ difference. Contraction accounts for at least $\sim 75$% of myocardial oxygen consumption (MVO2). The coronary blood flow is equal to $\sim 5$% of the cardiac output. Whenever O$_2$ demand increases, various substances promote coronary vasodilatation: adenosine, $K^+$, lactate, nitric oxide (NO), and prostaglandins. Oxygen extraction in the capillary bed is more effective during diastole because capillaries, which cross the relaxed myocardium, are not collapsed. During systole, the myocardium contraction hinders the arterial perfusion (systolic compression) but more or less improves the venous drainage, such as the inferior limb venous return that is enhanced by the contraction of surrounding muscles which compress the valved veins. Activation of sympathetic nerves innervating the coronary arteries causes transient vasoconstriction mediated by $\alpha$1-adrenoceptors. The brief vasoconstrictor response is followed by vasodilation due to augmented vasodilator production and $\beta$1-adrenoceptor activation. Parasympathetic stimulation of the heart induces a slight coronary vasodilation.

In order to fit the body needs, the heart increases its frequency and/or the ejection volume. The afterload is determined by the arterial resistances, mainly controlled by the sympathetic innervation (the higher the resistances and the arterial pressure, the smaller the ejected volume). The preload affects the diastolic filling, and, consequently, the end-diastolic values of the ventricular volume and pressure. The blood circulation is controlled by a set of regulation mechanisms, which involve the central command (the nervous system), the peripheral organs via hormone releases, and the local phenomena (mechanotransduction).

The time scale of the short-term regulation of the circulation is $\mathcal{O}(\mathrm{s})$ – $\mathcal{O}(\mathrm{mn})$, whereas for the long-term one, it is $\mathcal{O}(\mathrm{h}) - \mathcal{O}(\mathrm{day})$. The short-term control includes several reflexes, which involve the following inputs and outputs: the arterial pressure, the heart rate, the stroke volume, and the peripheral resistance and compliance. So the autonomic nervous system can receive complementary information from the circulation and has several processing routes. Control of the peripheral resistance and compliance is slower than command of the heart period and the stroke volume. There are several types of mechanosensitive receptors in the circulation, such as the baroreceptors.

Delayed mechanisms involve circulating hormones such as catecholamines, endothelins (ET), prostaglandins, NO, angiotensin, and others. Late-adaptive mechanisms are provided by the kidneys, which control the volemia through the $Na^+$ and water reabsorption under action of the renin–angiotensin–aldosterone system (RAAS). Sympathetic stimulation via $\beta$1-receptors, renal artery hypotension, and decreased $Na^+$ delivery to the distal tubules stimulate the release of renin by the kidney. Renin cleaves angiotensinogen into angiotensin 1. Angiotensin-converting enzyme acts to produce angiotensin 2, which constricts arterioles, thereby raising peripheral resistance and arterial pressure. It also acts on the adrenal cortex to release aldosterone, which increases $Na^+$ and water retention by the kidneys. Angiotensin 2 stimulates the release of vasopressin (or antidiuretic hormone ADH) from the posterior pituitary, which also increases water retention by the kidneys. Angiotensin 2 favors noradrenalin release from sympathetic nerve endings and inhibits noradrenalin reuptake by nerve endings, hence enhancing the sympathetic function.

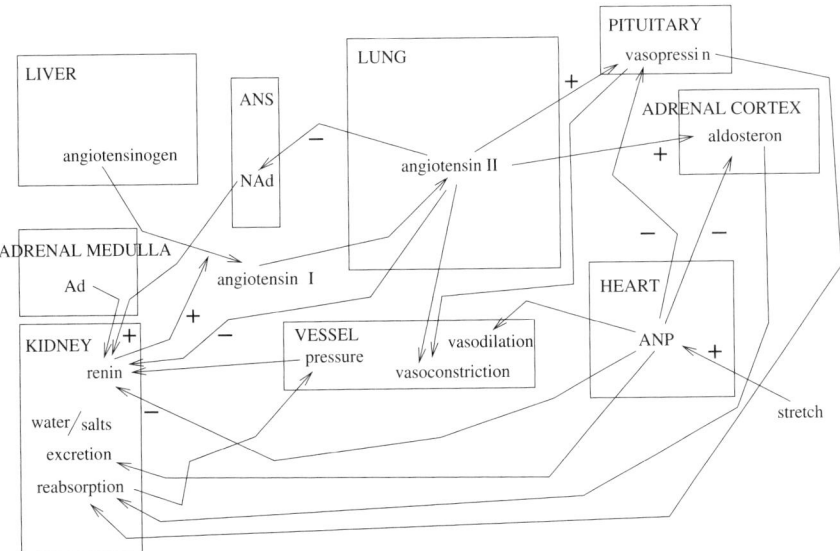

Figure 2.1. The atrial natriuretic peptide (ANP) and the renin–angiotensin system.

The endocrine heart acts as a modulator of the activity of the sympathetic nervous system and the renin–angiotensin–aldosterone system in particular [DEa] (Figure 2.1). The natriuretic peptides control the body fluid homeostasis and the blood volume and pressure. Both ANP (atrial natriuretic peptide or A-type) and BNP (B-type) are synthesized by the cardiac cells as preprohormones, which are processed to yield prohormones and, ultimately, hormones. They then are released into the circulation at a basal rate. Augmented secretion follows hemodynamical or neuroendocrine stimuli. They relax vascular smooth muscle cells and regulate their proliferation. They decrease the baroreflex activity. They inhibit renin release by the kidneys, augment the glomerular filtration rate and decrease the tubular sodium reabsorption. In the adrenal cortex, they inhibit aldosterone synthesis and release.

*Cardiac cycle.* The heartbeat is a two-stage pumping action over a period of about one second or less: a longer first diastole and a systole. More precisely, the heart rhythm focuses on the activity of the left venticle which consists of four main phases:

1. Isovolumetric relaxation (IR), with closed atrioventricular and ventriculoarterial valves

2. Ventricular filling (VF), subdivided into rapid and reduced filling phases, with open atrioventricular and closed ventriculoarterial valves

| Phase | Cycle Timing | Duration | Starting Event |
|-------|-------------|----------|----------------|
| IC    | 0–50        | 50       | Mitral valve closure |
|       |             |          | ECG R wave peak |
| SE    | 50–300      | 250      | Aortic valve opening |
| IR    | 300–400     | 100      | Aortic valve closure |
| VF    | 400–800     | 400      | Mitral valve opening |

Table 2.2. Duration (ms) of the four main phases of the cardiac (left ventricle) cycle ($f_c = 1,25$ Hz, i.e. 75 beats per mn).

3. Isovolumetric contraction (IC), with closed atrioventricular and ventriculoarterial valves

4. Systolic ejection (SE), subdivided into rapid and slow ejection, with closed atrioventricular and open ventriculoarterial valves

Duration of these four phases of the cardiac cycle is given in Table 2.2.

The systole and the diastole are dynamically related. Systolic contraction provides heart recoil and energy which is stored for active diastolic dilation and aspiration [ROb]. Moreover, heart motion during systole pulls the large blood vessels and surrounding mediastinal tissues which react by elastic recoil. Heart diastolic rebound can participate in ventricular filling.

The heart has chaotic behavior. Its irregular nonperiodic behavior characterizes a pump able to react quickly to any changes of the body environment. The normal heartbeat indeed exhibits complex nonlinear dynamics. At the opposite, stable, periodic cardiac dynamics give a bad prognosis. A decay in random variability over time, which is associated with a weaker form of chaos, is indicative of congestive heart failure [POb]. This feature, positive with respect to the heart function, is a handicap in signal and image processing, ensemble averaging being used to improve the signal-to-noise ratio.

### 2.2.2 Circulatory System

The blood is propelled under high pressure through a network of branching arteries of decreasing size, arterioles, and capillaries to the tissues where it delivers nutrients and removes catabolites. The blood is collected through merging venules and returns to the heart through veins under lower pressures. Each blood circuit (systemic and pulmonary circulation) is thus composed of three main components, arterial, capillary, and venous. The microcirculation starts with the arterioles ($10 < d < 250$ µm; $d$: vessel caliber) and ends with the venules.

Blood flows depend on the vasculature architecture and local geometry. The vasculature is characterized as a large between-subject variability in vessel origin, shape, path, and branching. Because the flow dynamics strongly depends on the vessel configuration, subject-specific models are required for improved diagnosis and treatment.

### 2.2.3  Hemodynamics

Hemodynamics differs at the different length scales of the circulatory circuit. In microcirculation, the non-Newtonian two-phase blood flows at low Reynolds number ($Re$). In the macrocirculation, the blood, supposed to be Newtonian in normal conditions, unsteadily flows at high $Re$.

The governing equations of a vessel unsteady flow of an incompressible fluid, of mass density $\rho$, dynamic viscosity $\mu$, and kinematic viscosity $\nu = \mu/\rho$, which is conveyed with a velocity $\mathbf{v}(\mathbf{x},t)$ ($\mathbf{x}$: Eulerian position; $t$: time), are derived from the mass and momentum conservation:

$$\nabla \cdot \mathbf{v} = 0,$$
$$\rho(\mathbf{v}_t + \mathbf{v} \cdot \nabla)\mathbf{v} = \mathbf{f} + \nabla \cdot \mathbf{C}, \tag{2.1}$$

where[1] $\mathbf{v}_t \equiv \partial\mathbf{v}/\partial t$, $\mathbf{f} = -\nabla\Phi$ is the body force density ($\Phi$: potential from which body force per unit volume are derived, which is, most often neglected), and $\mathbf{C}$ the stress tensor. The constitutive equation for an incompressible fluid is: $\mathbf{C} = -p_i'\mathbf{I} + \mathbf{T}$, where $p_i' = p_i + \Phi$ (when $\Phi$ is neglected, $p_i' = p_i$), $\mathbf{I}$ is the identity tensor, and $\mathbf{T}$ the extra-stress tensor. When the fluid is Newtonian, the stress tensor is a linear expression of the velocity gradient and the pressure; $\mathbf{T} = 2\mu\mathbf{D}$, where $\mu = \mu(T)$ ($T$: temperature), and $\mathbf{D} = (\nabla\mathbf{v} + \nabla\mathbf{v}^T)/2$ is the deformation rate tensor. The equation set (2.1) leads to the Navier–Stokes equations:

$$\rho(\mathbf{v}_t + (\mathbf{v} \cdot \nabla)\mathbf{v}) = -\nabla p_i' + \mu\nabla^2\mathbf{v}. \tag{2.2}$$

The formulation of the dimensionless equations depends on the choice of the variable scale. The dimensionless equations exhibit a set of dimensionless parameters. The Reynolds number $Re = V_q R/\nu$ ($V_q$: cross-sectional average velocity; $R$: vessel radius) is the ratio between convective inertia and viscous effects. When the flow depends on the time, both mean $\overline{Re} = Re(\overline{V_q})$ and peak Reynolds numbers $\widehat{Re} = Re(\widehat{V_q})$, proportional to mean and peak $V_q$, respectively, can be calculated. $Re = V_q\delta_S/R$ is used for flow stability study ($\delta_S$: Stokes boundary layer thickness). The Stokes

---

[1] $\nabla = (\partial/\partial x_1, \partial/\partial x_2, \partial/\partial x_3)$, $\nabla\cdot$, and $\nabla^2 = \sum_{i=1}^{3} \partial^2/\partial x_i^2$ are the gradient, divergence, and Laplace operators, respectively.

number $Sto = R(\omega/\nu)^{1/2} = R/\delta_S$ is the square root of the ratio between time inertia and viscous effects. The Strouhal number $St = \omega R/V_q$ is the ratio between time inertia and convective inertia ($St = Sto^2/Re$). The Dean number $De = (R_h/R_c)^{1/2}Re$, for laminar flow in curved vessels, is the product of the square root of the vessel curvature ratio by the Reynolds number. The modulation rate, or amplitude ratio, also plays a role in flow behavior.

The vasculature is made of successive bends and branchings. The embranchment can be, at a first approximation, supposed to be constituted of two juxtaposed bends, with a slip condition on the common wall within the stem. Bends present either gentle or strong curvature, and various curvature angles up to 180 degrees (aortic arch, intracranial segment of the internal carotid artery). The bend then represents the simplest basic unit of the circulatory system. Vessel curvature leads to helical blood motion. The vasculature does not present any symmetric planar bifurcation or junction. Bends, embranchments, and junctions induce 3-D developing flows [THb, THc]. Change in cross-section along the vessel length also generates 3-D flows. Several features affect the flow stability: the vessel curvature according to disturbance amplitude, the wall distensibility, the flow period, and the frequency content of the pressure signal. Laminar flow in blood vessel is a weak assumption.

The mechanical energy provided by the myocardium is converted into kinetic and potential energy associated with the elastic artery distensibility, as well as viscous dissipation. The aortic flow is a discontinuous flow characterized by a strong windkessel effect, with restitution of systolic-stored blood volume during diastole. In the arterial tree, the flow is pulsatile with a possible bidirectional flow period during the cardiac cycle, and back flow in certain arteries, such as in the femoral artery, whereas the flow rate is always positive in other, as in the common carotid artery.

### 2.2.4  Lymphatics

The circulatory system has a specialized open circuit conveying the lymph into the veins. The lymphatic vessels maintain the fluid balance and are involved in immunity. The lymph has a composition similar to plasma but with small protein concentration. The lymph flow is very slow.

Terminal lymphatics are composed of an endothelium with intercellular gaps surrounded by a highly permeable basement membrane. Larger lymphatic vessels have SMCs and are similar to veins, with thinner walls. Large lymphatic vessels have muscular walls. Lymphatic vessels have valves that prevent back lymph motion. The lymph is thus conveyed into the systemic circulation via the thoracic duct and subclavian veins. Spontaneous and stretch-activated vasomotion in terminal lymphatic vessels helps to convey lymph. Sympathetic nerves cause contraction.

## 2.2.5 Microcirculation

The microcirculation, with its four main duct components, arterioles, capillaries, venules, and terminal lymphatic vessels, regulates blood flow distribution within the organs, the transcapillary exchanges, and the removal of cell wastes. Arterioles are small precapillary resistance vessels. They are richly innervated by sympathetic adrenergic fibers and highly responsive to sympathetic vasoconstriction via both $\alpha1$ and $\alpha2$ postjunctional receptors. Venules are collecting vessels. Sympathetic innervation of larger venules can alter venular tone which plays a role in regulating capillary hydrostatic pressure.

The arteriolar flow is characterized not only by the important pressure loss but also by a decrease in inertia forces and an increase in viscous effects. Both the Reynolds number $Re$ and the Stokes number $Sto$ become much smaller than one. Centrifugal forces do not significantly affect the flow in the microcirculation, where the motion is quasi-independent of the vessel geometry. A two-phase flow appears in the arterioles of a few tens of $\mu m$ with a near-wall lubrification zone, or plasma layer, and a cell-seeded core. The arteriolar flow is characterized by the Fahraeus effect. The decrease in local $Ht$ associated with the vessel bore can be explained by a selection between the two phases of the blood, the plasma flowing more quickly than the blood cells. The Fahraeus–Lindqvist effect is related to the relative apparent blood viscosity dependence on the tube diameter and hematocrit in small pipes. The Fahraeus–Lindqvist effect can be explained by the interaction of the concentrated suspension of deformable erythrocytes with the vessel wall [FAa].

In the capillaries, the lumen size is smaller than the deformed flowing cell dimension and the plasma is trapped between the cells. The blood flow is then multiphasic. This vasculature compartment does not allow blood phase separation. The blood effective viscosity thus increases in capillaries with respect to its value in arterioles and venules. In this exchange region, where the blood velocity is low, the quantity of interest is the transit time of conveyed molecules and cells. The capillary circulation, characterized by (1) a low flow velocity and (2) a short distance between the capillary lumen and the tissue cells, is adapted to the molecule exchanges.

Fenestrated capillaries have a higher permeability than continuous capillaries. The solvent transport due to the transmural pressure drop $\Delta p$ is decreased by the difference in osmotic pressures $\Delta \Pi$ due to the presence of macromolecules in the capillary lumen that do not cross the wall. In most capillaries, there is a net filtration of fluid by the capillary endothelium (filtration exceeds reabsorption). A fraction of filtrated plasma is sucked back from the interstitial liquid into capillaries and the remaining part is drained by lymphatic circulation into the large veins. During inflammation, capillaries become leaky. VEGF, histamine, and thrombin disturb the endothelial barrier [RAa]. In most microvessels, the macromolecule transport is done by transcytosis and not through porous clefts. In microvessels with continuous endothelium, the main route for water and small solutes is the endothelium cleft. The estimated between-cell exchange area is of the order of 0.4% of the total capillary surface area. Transcapillary water flows and microvasculature transfer of solutes, from electrolytes to proteins, in both continuous and fenestrated endothelium, can be described in terms of these porous in-parallel routes.

1. A water pathway across the endothelial cells

2. A set of small pores (caliber of 4–5 nm)

3. A population of larger pores (bore of 20–30 nm) [MIa]

The fiber matrix model of the endothelium glycocalyx and the cleft entrance, associated with pore theory, of capillary permeability sieves solutes [WEa].

## 2.3  Blood

The blood performs several major functions:

1. Transport through the body

2. Regulation of bulk equilibria

3. Body immune defense against foreign bodies

The blood contains living cells and plasma. The plasma represents approximately 55% of the blood volume. The remaining is hematocrit (Ht), i.e. percent of packed cells. The electrolytes, cations, and anions, contribute to the osmotic pressure $\Pi$, which is mainly regulated by the kidneys. Plasma contains 92% water and 8% proteins and other substances. Glucids are energy sources. The fibrinogen acts on platelet and erythrocyte aggregation. Plasma proteins are composed of fibrinogen, albumin, and globulins. They participate in the blood colloidal osmotic pressure which keeps fluids within the vascular system. As does fibrinogen, globulins induce reversible erythrocyte aggregation. The four main types of circulating lipoproteins, which differ in size, density, and content, include chylomicrons, very-low-density lipoproteins (VLDL), low-density lipoproteins (LDL), and high-density lipoproteins (HDL). The lipoproteins convey cholesterol esters and triglycerides in blood. Triglycerides are delivered to muscles and adipose tissues for energy production and storage. Cholesterol is used to build cell membranes and is a precursor for the synthesis of the steroid hormones, vitamin D, and bile acids. A part is excreted in the biliary ducts as free cholesterol or as bile acids (partial cyclic travel) and another part is conveyed in blood. The lipoproteins are associated with apolipoproteins.

### 2.3.1 Blood Cells

There are three main kinds of flowing cells: erythrocytes, leukocytes, and platelets (Table 2.3).

*Erythrocytes.* The erythrocyte or red blood cell (RBC) is a hemoglobin solution bounded by a thin flexible membrane (nonnucleated cell). In its undeformed state, it has a biconcave disc shape with a greater thickness in its outer ring (diameter of $7.7 \pm 0.7$ µm, central and peripheral thicknesses of $1.4 \pm 0.5$ µm and $2.8 \pm 0.5$ µm). It is then susceptible to deformation, in particular with a parachute shape, when moving through tiny capillaries. The RBC shape represents an equilibrium configuration that minimizes the curvature energy of a closed surface for given surface area and volume with a geometrical asymmetry, the phospholipid outer layer of the RBC membrane having slightly more molecules than the inner one. Each RBC contains hemoglobin molecules, which consist of four globin chains $\alpha$ and $\beta$. The hemoglobin contains four iron atoms $Fe^{2+}$, in the center of hemes. It carries oxygen $O_2$ from lungs to tissues and helps to transport carbon dioxide $CO_2$ from tissues to lungs. The hemoglobin is also involved in pH regulation. *Reticulocytes* are slightly immature cells (0.5–2% of the total RBC count).

*Leukocytes.* The leukocyte or white blood cell (WBC) plays a role in immune defense. Five kinds of WBCs exist, three types of *granulocytes*, which have about the same size (8–15 µm), neutrophils, eosinophils, and basophils, and two kinds of agranular leukocytes, lymphocytes (8–15 µm)

| Blood Cells | Quantity $(mm^{-3})$ | Cell Volume(%) |
|---|---|---|
| RBC | $5.10^6$ | 97 |
| WBC | $5.10^3$ | 2 |
| thrombocyte | $3.10^5$ | 1 |

Table 2.3. Blood cell approximated geometry and relative concentration.

and monocytes (15–25 μm). The neutrophil is able to phagocytize foreign cells, toxins, and viruses. Scouting neutrophils look for possible invading agents. The eosinophil phagocytizes antigen–antibody complexes. The basophils release two mediator kinds, either from preformed granules or newly generated mediators. The lymphocyte plays an important role in immune response. There are *T-lymphocyte* subsets, inflammatory T cells that recruit macrophages and neutrophils to the site of tissue damage (CD4 + T-cell), cytotoxic T lymphocytes (CTL or CD8 + T-cell) that kill infected, allograft, and tumor cells, and helper T-cells that enhance the production of antibodies by B-cells. The B-lymphocyte produces antibodies. A small fraction of the circulating lymphocytes are natural killer (NK) cells. The monocytes leave the blood stream by diapedesis to become macrophages.

*Thrombocytes.* Thrombocytes, or platelets, of size 2–4 μm, are cell fragments involved in coagulation. Platelet activation is affected by hemodynamic forces. The two major secretory granule types include numerous α-granules and large dense granules. They contain substances involved in clotting, cell adhesion, and chemotaxis.

*Hematopoiesis.* All blood cell types are produced in the bone marrow from stem cells. Survival and proliferation are regulated by cytokines and hormones. Many kinds of hematopoeitins ensure a dynamical balance between cell differentiation and proliferation. Erythropoietin, produced by the kidney cortex stimulated by hypoxemia in the renal arteries and by the liver, activates the production of RBCs. The erythropoietin also prevents the destruction of viable tissue surrounding injuries, such as infarction. Thrombopoietin participates in hematopoiesis in general, and to thrombopoiesis in particular. Colony stimulating factors (CSF) stimulate the proliferation of stem cells of the bone marrow in adults. Granulocyte–monocyte colony-stimulating factor (GM-CSF) induces proliferation of multilineage progenitors and the growth of certain WBC colonies. Granulocyte colony-stimulating factor (G-CSF) stimulates proliferation and maturation of monopotent neutrophil progenitors which differentiate into neutrophils. Triggered by macrophage colony-stimulating factors (M-CSF) the granulocyte–macrophage progenitor cells differentiate into monocytes. Stem cell factor (SCF) promotes the production of NK cells. Interleukins (IL) are also involved in the hematopoiesis.

### 2.3.1.1  Clotting

In normal vessels, the cardiovascular endothelium prevents clotting, because of the presence at the wetted surface of substances such as thrombomodulin, protein C, lipoprotein-associated coagulation inhibitor, tissue factor pathway inhibitor, protease-nexin, and heparan sulfate. The endothelium

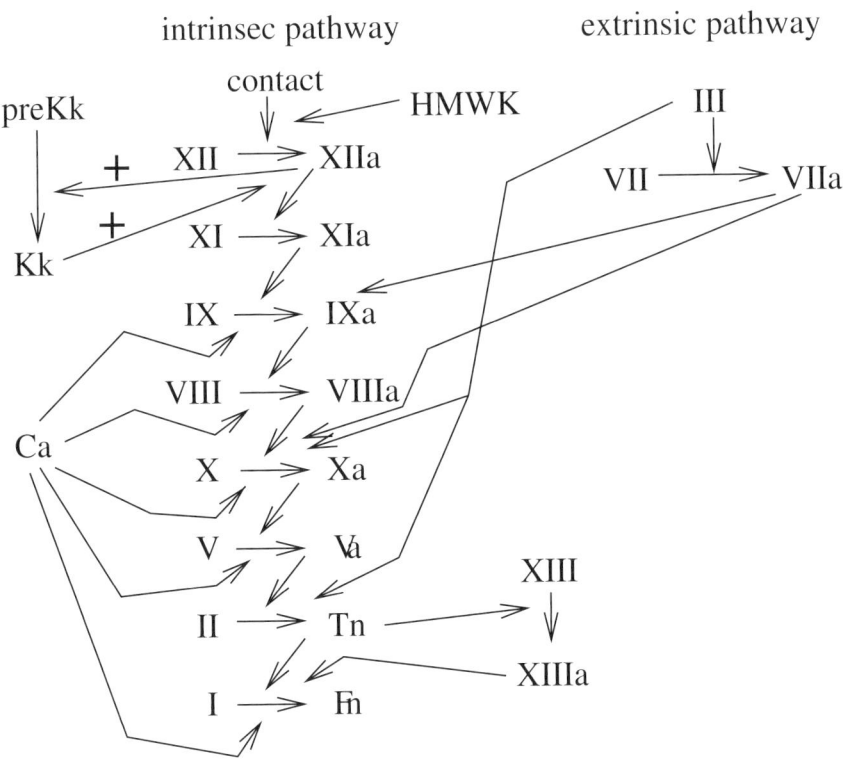

Figure 2.2. Coagulation factors and the reaction cascade. (I: fibrinogen, II: prothrombin, III: tissue thromboplastin, V: proaccelerin, VII: proconvertin, VIII: antihemophilic factor A, IX: antihemophilic factor B, X: Stuart–Prower factor, XI: plasma thromboplastin antecedent, XII: Hageman or contact factor, XIII: fibrin-stabilizing factor, Fn: fibrin, Tn: thrombin, Kk: kallikrein, HMWK: high molecular weight kininogen).

inhibits platelet aggregation by releasing prostacyclins (PGI2) and NO. Figure 2.2.

The endothelium synthesizes coagulation factors, the von Willebrand factor, and tissue factor. When the endothelium is damaged, blood must clot in order to prevent huge bleeding. The hemostatic process depends on the stable adhesion and aggregation of platelets with the subendothe- lial matrix molecules at the vessel injury site. The primary hemostasis refers to the plug formed by platelets. Various plasma clotting factors form fibrin strands that strengthen the platelet plug. The primary hemostasis involves a set of adhesion receptors and proteins (von Willebrand factor, collagen, fibrinogen, etc.). The secondary hemostasis has two pathways, the intrinsic and the extrinsic one, which join in a common pathway that leads to fibrin formation. The intrinsic pathway is characterized by the formation of the primary complex on collagen by high-molecular-weight kininogen, prekallikrein, and Hageman factor. The activation cascade transforms fibrinogen into fibrin to form the clot. In the extrinsic pathway, factor VII is activated by the tissue factor to activate factors IX and X. The common pathway begins with activation of factor X by activated factor IX and/or VII. Thrombin is then produced and converts fibrinogen into fibrin. It also activates factors VIII and V as well as platelets.

Mathematical multiscale models of either clotting on a breach of the vessel wall or thrombosis after a rupture of an atherosclerotic plaque have been developed in the presence of a flow of an incompressible viscous fluid [KUa, FOb]. Compounds and platelet transport by convection and diffusion are assumed to take place in a near-wall thin plasma layer. A competition occurs between the activation of the coagulation stages and the removal by the flowing fluid of the clotting factors and cells away from the reaction site. The thrombus growth and embolus shedding from the thrombus can be predicted by the stress field exerted by the moving fluid on the thrombus.

### 2.3.2 Blood Rheology

The blood can be considered to be a concentrated dispersed RBC suspension in a solution, the plasma, which is composed of ion and macromolecules, especially fibrinogen and globulins, interacting between them and bridging the RBCs. In large blood vessels (macroscale), the ratio between the vessel bore and the cell size ($\kappa_{vp} > \sim 50$) is such that the blood is considered as a continuous homogeneous medium. In capillaries (microscale), $\kappa_{vp} < 1$ and the blood is heterogeneous, transporting deformed cells in a Newtonian plasma. In the mesoscale ($1 < \kappa_{vp} < \sim 50$), the flow is annular diphasic with a core containing cells and a marginal plasma layer.

Shear-step experiments show that the blood has a shear-thinning behavior [CHc]. Blood exhibits creep and stress relaxation during stress formation and relaxation [JOb]. Blood exposures to sinusoidal oscillations of constant amplitude at various frequencies reveal a strain-independent loss modulus and a strain-dependent storage modulus. The blood is a viscoelastic material that experiences a loading history. Furthermore it is thixotropic. Its rheological properties are dictated by the flow-dependent evolution of the blood structural changes, thereby by the kinetics of both RBC aggregation and deformation. The changes in the blood internal structure are indeed due to the reversible RBC aggregation and deformation. The generalized Newtonian model, which is now commonly used when non-Newtonian behavior is considered, is not suitable because it does not take into account the loading history.

The input rheological data, which are provided by experimental results obtained in steady-state conditions, starting from a rouleau network, are far from the physiological ones in the arteries [THd]. In a large blood vessel, the flowing blood is characterized by a smaller convection time scale than the characteristic time of the cell bridging. Therefore, in the absence of stagnant blood regions, the blood, in large vessels, can then be considered to have a constant viscosity. Besides, shear-induced platelet activation has been modeled, using a complex viscoelastic model previously developed [RAc] and a threshold-dependent-triggered activation function [ANa].

## 2.4 Signaling and Cell Stress-Reacting Components

The cell has a nucleus and several organelles within its cytoplasm, wrapped in a membrane (0.1–10 μm), a phospholipid bilayer with adhesion molecules and other compounds acting in cell junctions in signaling and in substance transport. Biomechanical studies currently deal with larger elements, the cell membrane and the nucleus, as well as the cytoskeleton, which is composed of networks of various types of filaments. The continuum hypothesis is supposed to be valid because the problem length scale, although small with respect to the cell size, remains greater than the cell organelle size. However, the cytoskeleton elements are in general too small. Its behavior is then investigated in a domain that contains a solution of cytoskeleton components rather than the cell itself. The stripped cytoskeleton can also be considered as a discrete structure of stress-bearing components.

### 2.4.1 Cell Membrane

The plasma membrane, or plasmalemma, is a barrier between the extracellular (ECF) and the intracellular (ICF) fluids. The cell nucleus and organelles have their own membranes. The phospholipid bilayer of the plasma membrane embeds proteins and glucids. The membrane has specialized sites for exchange of information, energy, and nutrients, essentially made from transmembrane proteins.

Phosphoinositides are involved in numerous cell-life events, such as smooth muscle cell contraction and endothelial cell production of vasoactive molecules. Phosphoinositides can specifically interact with proteins having lipid-binding domains. Lipid rafts compartmentalize the membrane into domains, forming microdomains of the cell membrane with phosphoinositides, which are involved in cell endocytosis.

Membrane glucidic copulae of the external membrane layer contribute to the membrane asymmetry. Membrane glucids also participate in the protein structure and stability of the membrane. Furthermore, membrane glucids modulate the function of membrane proteins. Surface polysaccharides are used directly for cell recognition and adhesion. For example, the laminin, a glycoprotein, allows adhesion of collagen to endothelial cells.

Ion pumps and channels, and gap junctions on the other hand, coordinate the electrical activity and the molecular exchanges. Signal transduction allows the cell to adapt to the changing environment, using various procedures at the molecular scale. The transducers can involve

1. Mechano-sensitive ion channels

2. Conformational changes of molecules

3. Molecular switches in the cell membrane or the cytosol

Many different kinds of receptors, mostly integral membrane proteins, of multiple messengers are localized in the cell membrane. These communication receptors are transduction molecules. Attachment of the ligand on its corresponding receptor induces a reaction cascade. The ligand fixation first triggers synthesis of second messengers, such as cyclic nucleotides, cyclic adenosine monophosphate (cAMP), and cyclic guanosine monophosphate (cGMP), phosphoinositids, and so on, responsible for cell responses of the extracellular ligand (first messenger). Both agonists and antagonists can fix on receptors, generally specific, with or without effect, respectively. Antagonists block agonist fixation. A given messenger can have several receptors. For example, adrenalin has $\alpha 1$, $\alpha 2$, $\beta 1$, and $\beta 2$ receptors, acetylcholine has nicotinic and muscarinic receptors. The receptor types include

1. Receptors coupled to G proteins

2. Receptors whose cytoplasmic domain is activated when the receptor is linked to its ligand and activates one or more specific enzymes to simultaneously stimulate multiple signaling pathways

3. Receptors linked directly or indirectly to ion channels

Protein tyrosine kinases (PTK) modulate multiple cellular events, such as differentiation, growth, metabolism, and apoptosis. PTKs include not only transmembrane receptor tyrosine kinases (RTK) but also cytoplasmic nonreceptor tyrosine kinases (NRTK). Growth factors are major ligands of RTKs. Signaling proteins that bind to the intracellular domain of RTKs include RasGAP, PI3K, phospholipase C $\gamma$, phosphotyrosine phosphatase (PTP) SHP, and adaptor proteins involved in the construction of the clathrin coat. The Eph family, the largest family of RTKs, interacts with ephrins. Signaling mediated by ephrins and Eph RTKs regulates a variety of processes including cell shape, cell adhesion and separation, and

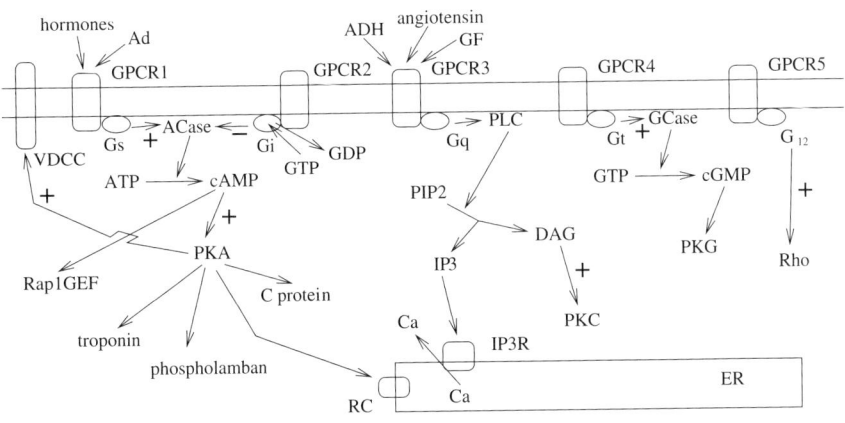

Figure 2.3. G-protein-coupled receptors (GPCR) are associated with G proteins with its three subunits Gα, which binds GDP (inactive state) or GTP, after ligand binding and stimulation, Gβ, and Gγ. Activated Gα activates an effector. Several types of Gα include Gαs (stimulatory), Gαq, Gαi (inhibitory), Gαt, and Gα12. Gαs stimulates adenylyl cyclase (ACase), which produces a second messenger, cyclic AMP (cAMP). Gαq activates phospholipase C (PLC) which generates second messengers, inositol trisphosphate (IP3) and diacylglycerol (DAG). Gαi inhibits adenylyl cyclase. Gαt stimulates guanylyl cyclase (GCase), which forms cyclic GMP (cGMP). Gα12 activates RhoA GTP-binding proteins.

cell motion (attraction and repulsion), modulating the activity of the actin cytoskeleton.

Receptor activation can involve guanosinetriphosphatases (GTPases). These protein switches cycle between two conformations induced by the binding of either guanosine diphosphate (GDP) or guanosine triphosphate (GTP). They then are flicked off (inactive GDP-bound state) and on (active GTP-bound state). The two major kinds of GTPases include the large guanine nucleotide-binding proteins, the G proteins, and the small GTPases.

The G proteins activate (Gs) or inactivate (Gi) adenylate cyclase to regulate the intracellular cAMP level (second messanger) (Figure 2.3). The activation of G proteins is induced by ligand-bound G protein-coupled receptors (GPCR). They then regulate the production or the influx of second messengers. The small Rho GTPases regulate the assembly of filamentous actin (F-actin) in response to signaling. Their effectors induce the assembly of contractile actin–myosin filaments (stress fibers in particular) and of integrin-containing focal adhesions. Thereby, the small Rho GTPases act in vascular processes, such as smooth muscle cell contraction, cell adhesion, endothelial permeability, platelet activation, leukocyte extravasation, and migration of smooth muscle cells (SMC) and endothelial cells (EC) involved in wall remodeling and angiogenesis [VAa]. They are also required in vascular disorders associated with pathological remodeling and altered cell contractility. The Rho-kinase, an effector of the small Rho GTPase, is involved in atherosclerosis as well as in poststenting restenosis.

Small Rho GTPases can be activated via GPCRs, RTKs, and cytokine receptors. The activation of GTPases into GTP-bound conformations is controlled by specific guanine nucleotide exchange factors (GEF), which activate the Rho GTPases. GTP is hydrolyzed to GDP by the GTPase in combination with GTPase-activating proteins (GAP). In the absence of signaling, the major fraction of small GDP-bound Rho GTPases is sequestrated in the cytosol, bound to guanine dissociation inhibitors (GDI). GDIs slow the rate of GDP dissociation from the Rho GTPases, which remain inactive.

Membrane-associated GTPases Ras and Rho/Rac activate intracellular pathways in response to extracellular signals. The Ras GTPases include Ras, Rap, Ral, and others [BOc]. Both Rap and Ras can bind the same effectors in order to regulate intracellular signaling events. Ras activates effectors, members of the Raf kinase family, the phosphatidylinositol 3 kinase (PI3K) and members of the RalGEF family. Raf activation includes translocation to the plasma membrane, induction of a conformational change by Ras, and phosphorylation. Activated Raf1 activates extracellular signal-regulated kinase (ERK). The Ras-Raf-ERK signal transduction pathway controls proliferation, differentiation, and apoptosis. Cross-actions between the Ras-Raf-ERK and the Ras-Raf-PI3K-protein kinase B (PKB) pathways modulate cell-life modes [ZIa]. ERK belongs to mitogen-activated protein kinases (MAPK), involved in signal transduction. The other main groups of MAPKs include c-Jun N-terminal kinase (JNK) and p38. Mitogen-activated protein kinase p38 is activated by osmotic pressure changes and cytokines. It acts in a cascade that involves the MAPK kinase kinases (MAPKKK) and MAPK kinases (MAPKK) [CHa]. Rho, Rac, and Cdc42 are the three most known classe of the Rho protein family. Each Rho classe has its specific effects on the actin cytoskeleton.

There are many diacylglycerol receptors (DAGR), protein kinase C (PKC) and D (PKD), chimaerins, which target Rho GTPases, and others as translocation activators or inhibitors. PKC$\delta$ and PKC$\epsilon$ are implicated in the evolution of the cardiac function after myocardial infarction [RAd]. PKC$\delta$ and PKC$\epsilon$ are also implicated in vasculogenesis. PKC$\alpha$ and PKC$\epsilon$ control integrin signaling to ERK [KEb].

Signaling networks are also associated with focal points of enzyme activity. A-kinase anchoring proteins (AKAP) contribute to spatial regulation of signaling events, signal-organizing molecules, targeting protein kinases and phosphatases to specific sites where the enzymes control the phosphorylation state of neighboring substrates [CAe]. Within a site, a given AKAP can link to diverse substrates. Different AKAPs within a given site can assemble distinct signaling complexes. The displacement of enzymes into and out of these complexes contributes to the temporal regulation of signaling events.

### 2.4.2 Endocytosis

The endocytosis is the internalization of molecules from the cell surface into membrane compartments, and then into vesicles for cellular trafficking. It starts with the binding between a molecule and its surface receptor. Ligand–receptor interactions often need aggregation of numerous ligand–receptor complexes in a site where the membrane bulges to form a vesicle. The two major paths include clathrin- and caveolae/lipid-raft-mediated endocytosis.

The clathrin-dependent route is responsible for the internalization of nutrients, growth factors, and receptors, as well as antigens and pathogens. Adaptor-proteins stimulate the formation of the clathrin coat. Once inside the cytosol, clathrin is rapidly released for subsequent use by exocytosis (recycling). The naked vesicles fuse with an endosome and the ligands are separated from their receptors. The caveolins form caveolae, types of lipid rafts. The cytoplasmic motion of caveosomes depends on the microtubule network. The caveosome route is regulated by NRTKs and PKC [LEa]. Both clathrin- and caveolae/lipid-raft-mediated endocytosis are modulated by groups of specific kinases. Within each group, some kinases act directly whereas other kinases modulate the endocytic path. Certain kinases exert opposite effects on the two main endocytic kinds for coordination between endocytic routes.

Molecule internalization can also be done by via structures that contain glycosylphosphatidylinositol-anchored proteins (GPI-AP) and fluid-phase markers (caveolae- and clathrin-independent pathway) [KIa]. A transient burst of actin polymerization accompanies endocytic internalization.

### 2.4.3 Cell Cytoskeleton

The cell deformation and motility is due to the cytoskeleton, a fibril network with articulation nodes from which the cytoskeleton can reorganize itself. In particular, it undergoes stresses and responds to minimize local stresses. The cytoskeleton contracts and forms stiffer bundles to make the cell rigid. Its anchorage on adjacent tissue elements allows an ensemble deformation. Manifold molecules and fibers form this dynamic cell framework which also determines cytosol organization and intracellular displacements [DEb]. There are three classes of cytoskeleton filaments: the microfilaments, the microtubules, and the intermediary filaments. The spanning network, which fills the whole cytosol, is a fourth element. It can determine the sites of protein synthesis and the assembling locations of filaments and of microtubules. It acts on cell organelle motility.

*Microfilaments.* The microfilaments contain several proteins. The myosin filaments are localized along the actin filaments. The actin filaments are involved in cell configuration, adhesion, and motility. Actin filaments are anchored on the plasma membrane, using mooring proteins (talin, vinculin) [PAa]. The microfilaments can be used as mooring and transmission lines in a stress field, as towlines during motion. The actin cytoskeleton dynamics is maintained by the balance between actin-binding proteins and actin-severing proteins. Actin aggregation is induced by filamin and cortactin, whereas profilin can inhibit actin polymerization (but stimulates assembly of actin filaments). Cofilin has a depolymerizing activity. $\alpha$-Actinin favors formation of actin stress fibers.

*Microtubules.* The microtubules are long polymers of $\alpha$ and $\beta$ tubulins. The tubulin polymerizes in the presence of guanosine triphosphate and calcium. The microtubule-associated protein facilitates microtubule assembling. The microtubules are thicker and less stable than microfilaments. They are the stiffer element of the cytoskeleton. The microtubules are organized as a scaffolding within the cytoplasm. They control the cell–organelle distribution. Mitochondria and the endoplasmic reticulum (ER) are located along the microtubule network. The microtubules are also necessary for vesicles traveling across the cytosol. Intracellular transport of organelles involves dynein and kinesin. Kinesin is the motor protein for tubulin, which moves along tubulin, as actin does along myosin, having ATPase sites.

The centrosome is a cell body from which the microtubules radiate [GLc]. The centrosome contains two centrioles, each one composed of nine cylindrical elements like a paddle wheel and constituted of three microtubules.

*Intermediate filaments.* The intermediate filaments cross the cytoplasm either as bundles or isolated elements often in parallel to the microtubules. They are composed of vimentin, desmin, keratin, among others.

*Cell motility.* The cells display a set of internal motions, change their shape in a stress field, and migrate. Cell motility results from actin polymerization into filaments and depolymerization. The Rho GTPases regulate the actin cytoskeletal dynamics during cell motility and cell shape changes [NOa]. Paxillin (Pax) is a cytoskeletal and focal adhesion docking protein that regulates cell adhesion and migration [TUb]. Pax is also implicated in the regulation of integrin and growth factor signaling. Pax binds to focal adhesion kinase (FAK), vinculin, and interacts with Rho GTPases [SCa].

*Modeling of the cytoskeleton mechanics.* A 2-D model of the cytoskeleton dynamics has been developed to describe stress-induced interactions between actin filaments and anchoring proteins [CIb]. A small shear induces rearrangment of the four-filament population toward an orientation parallel to the streamwise direction.

Reactive flow model of contractile networks of dissociated cytoplasm under an effective stress in a square domain is associated with a system of non-linear partial derivative equations with boundary conditions [DEd]. Crucial dynamical factors of the cytoskeleton mechanics are

1. The viscosity of the contractile network associated with an automatic gelation as the network density enlarges, without undergoing large deformation

2. A cycle of polymerization-depolymerization

3. A control of network contractibility and of cell-surface adhesion

Tensegrity models consider deformable cells as a set of beams and cables that sustain tension and compression [INa]. The stiffness depends on the prestress level, and for a given prestress state, to the applied stretch, in agreement with experimental findings [WAb].

*Extravasation.* Circulating blood cells have adhesion receptors that enable the cells subjected to flow forces to adhere to the vessel wall. Flowing cells undergo a sequential-step extravasation, the kinetics of which is shear-dependent. The steps include tethering, rolling, activation, firm adherence, locomotion, diapedesis, and finally transendothelial migration [SCb]. The endothelium can either favor or inhibit flowing cell adhesion on its wetted surface. Adhesion molecules attract the flowing cells toward intercellular spaces for transmigration. Conversely, the endothelium produce 13-hydroxyoctadecadienoic acid (13-HODE) which confers a resistance to platelet or monocyte adherence.

### 2.4.4  Adhesion Molecules

There are different types of adhesion molecules.

*Cadherins.* The cadherins, which contain calcium binding sites, connect cells together, one cadherin binding to another in the extracellular space. The cadherin interacts via catenins with actinin and vinculin to link the cadherin–catenin complex to the actin cytoskeleton. Vascular endothelial cadherins (VE-cadherin) anchor the adherens junctions between endothelial cells. p120 Catenin, $\beta$-catenin, and $\alpha$-catenin link VE-cadherin to the actin cytoskeleton.

*Selectins.* The selectins are expressed in endothelial cells and blood cells for binding two cell surfaces in the presence of $Ca^{++}$. They slow intravascular leukocytes before transendothelial migration. Three selectin kinds are defined according to the cell in which they were discovered. L-Selectin is expressed on leukocytes for targeting activated endothelial cells. E-Selectin is produced by endothelial cells after cytokine activation. P-Selectin is stored for rapid release in platelet granules or Weibel–Palade bodies of endothelial cells [WAa].

*Integrins.* The integrins connect actin filaments of the cell cytoskeleton to proteins of the extracellular matrix. Various integrins combine different kinds of two subunits $\alpha$ and $\beta$ [SMb]. They mediate signaling to or from the environment. They form complexes with cytoskeletal proteins, adaptor proteins, and protein tyrosine kinases, which initiate signaling cascades. They are involved in the regulation of vascular tone and vascular permeability. Various proteins link the integrins to the cytoskeleton, such as tensin and laminin. Among these proteins, certain ones have several binding sites, therefore, cross-linking actin filaments. They include $\alpha$-actinin, fimbrin, and ezrin-radixin-moesin (ERM).

*Ig cell adhesion molecules.* Certain members of the immunoglobulin (Ig) superfamily, the Ig cell adhesion molecules (IgCAM), are involved in calcium-independent cell-to-cell binding. Among them, intercellular adhesion molecules (ICAM) are expressed on activated endothelial cells, being the ligand for integrins expressed by WBCs. Platelet–endothelial cell adhesion molecule 1 (PECAM-1) belongs to WBCs, platelets, and intercellular junctions of endothelial cells. Vascular cell adhesion molecule 1 (VCAM-1), once binds to $\alpha_4\beta_1$ integrin, and induces firm adhesion of leukocytes on endothelium.

### 2.4.5 Intercellular Junctions

Intercellular junctions are tiny specialized regions of the plasma membrane. Several functional categories include

1. Impermeable junctions, which maintain an internal area chemically distinct from surroundings

2. Adhering junctions, which reinforce tissue integrity

3. Communicating junctions for exchange of nutrients and signals with the environment

Within the junctions, membrane proteins have specific configurations. Different histological kinds of intercellular junctions exist.

*Desmosomes.* The desmosomes contain two classes of desmosomal cadherins, the desmocollins and the desmogleins, each having several subtypes, which are specific to the differentiation status and to the cell type. The intercellular space is filled with filaments that bridge not only membranes but also cytoskeletons of adjacent cells. Two main desmosome types exist. Belt desmosomes contain actin filament susceptible to contract in the presence of ATP, $Ca^{++}$, and $Mg^{++}$, in order to close the gap during cell apoptosis. Spot desmosomes contain filaments and transmembrane linkers that connect cytoplasmic networks of tonofilament bundles for mechanical coupling between adjacent cells. Hemidesmosomes allow adhesion of cells to basement membrane. Cells subjected to mechanical stresses have numerous spot desmosomes and hemidesmosomes, which limit cell distensibility and distribute stresses among layer cells and to the underlying tissues to minimize disruptive effects.

*Zonula adherens.* The zonula adherens is a cell-to-cell adhesion via cadherin–calcium dependent bridging [YAb]. These cadherin-based adhesive contacts link the cytoskeletal proteins of a given cell not only to the cytoskeleton of its neighboring cells, but also to the proteins of the extracellular matrix. Actin filaments are associated with the adherens junctions through catenins.

*Tight junctions.* Tight junctions leave tiny between-cell space. They selectively modulate paracellular permeability and act as a boundary between the apical and basolateral plasma membranes. Several proteins are involved: cingulin, claudin, occludin, junctional adhesion molecules (JAM), symplekin, zonula occludens proteins (ZO), and so on. E-cadherin is specifically required for tight junction formation and is involved in signaling rather than cell contact [TUa]. The RhoA GTPase regulates the tight-junction assembly and the cell polarity [OZa].

*Gap junctions.* Gap junctions build between-cell channels, which bridge adjacent membranes. These intercellular protein channels allow low-molecular-weight molecules, small signaling molecules, and ions to diffuse between neighboring cells. Various connexins are involved in gap junctions.

*Focal adhesions.* Focal adhesions are complexes of clustered integrins and associated proteins that link fibronectin, collagen, laminin, and vitronectin with the cytoskeleton of cultured cells and mediate cell adhesion [ZAb]. Focal adhesion proteins (FAP) include talin, vinculin, tensin, paxillin, and focal adhesion kinase (FAK), among others [CLa]. The focal adhesion kinase mediates several integrin signaling pathways. FAK signaling to Rho GTPases regulates changes in actin and microtubules in cell protrusions of migrating cells.

### 2.4.6   Extracellular Matrix

The extracellular matrix (ECM) supports cell functions during growth, division, development, differentiation, and apoptosis, as well as tissue formation and remodeling. Cells interact and communicate with other cells and with ECM. ECM is composed of several molecule classes:

1. Structural proteins, collagen, and elastin

2. Specialized proteins, such as fibrillin, laminin, and fibronectin

3. Proteoglycans, or mucopolysaccharides, such as chondroitin sulfate, heparan sulfate, keratan sulfate, and hyaluronic acid

Proteoglycans are composed of a protein to which are attached glycosaminoglycans (GAG). They have an important water-binding capacity, which amplifies the volume occupied by the macromolecules.

Cell anchorage and migration are due to glycoproteins, in particular to fibronectins. These connecting elements are fixed to collagen and elastin of ECM and to the cell membrane on the other hand [HYa]. The fibronectins act on clotting and healing. They also promote chemotaxis [CLb] and activate integrin signaling [GIa].

Proteolytic degradation and remodeling of ECM is controlled by matrix metalloproteinases (MMP) and their inhibitors, the tissue inhibitors of metalloproteinases (TIMP). MMPs are involved in the evolution of atherosclerosis and aneurisms. TIMPs also have mitogenic and cell growth promoting activity.

*Basement membrane.* The basement membrane (BM) is a specialized extracellular matrix sheet at the interface between the connective tissue and the endothelium. It gives an anchorage surface for endothelium which protects against shearing and detachment. Moreover, it acts as a selective barrier for macromolecular diffusion. It also influences the functions of contacting cells (regulation of cell shape, gene expression, proliferation, migration, and apoptosis) [AUa]. The basement membranes contain laminins, type IV collagen, and proteoglycans. The laminin and type IV collagen networks are linked by nidogens [YUa]. The basement membrane binds a variety of growth factors [GOa].

*Interstitial matrix.* The interstitial matrix affects the functions of contacting cells. The interstitial matrix has a fibrillar structure, with a large amount of collagen. The structure of the interstitial matrix depends mainly on the type of fibrils and on the type and amount of proteoglycans.

*Elastin.* A first type of major fibers is given by elastin fibers. The desmosine cross-links elastin to form elastin fibers [ROa]. Elastin binds to cells via elastonectine. Elastin fibers are the most elastic biomaterials, at least up to a stretch ratio of 1.6 [FUa], the loading and unloading cycles being nearly superimposed. A value of the elastic modulus of elastin fibers of 0.4–0.6 MPa is often considered [AAa, CAd, VIa].

*Collagens.* The collagens, the second type of major fibers, are structural proteins that form fibrils, characterized by a triplet of helical chains and stabilized by covalent cross-links. The triple helix domain is common for all collagens. The heterogeneity resides in the assembling mode and in the resulting structure. The collagen is surrounded by extensible glycoproteins and proteoglycans. The rheological properties of pure collagen are thus difficult to assess.

The mechanical properties of the blood vessels depend on the interaction between elastic and collagenous elements. Elastin and collagen not only intervene in the vessel wall rheology, via their mechanical properties, their density, and their spatial organization, but also control the function of the smooth muscle cells.

### 2.4.7 Microrheology

The cell is a complex body that is commonly decomposed into three major rheological components: the plasma membrane, the cytosol, and the viscoelastic nucleus. The membrane and the cytosol can be assumed to be a poroviscoelastic and a poroplastoviscoelastic material, respectively [VEa]. With such a decomposition, macroscopic laws, in particular constitutive laws, are supposed to be valid because the cell size is much greater than the size of its microscopic components.

Rheological sensors must have a suitable size, greater than the size of the cell organelles, as demonstrated by micro- and macrorheometric measurements of the storage and loss moduli [SCc]. Several rheological techniques have been recently developed to explore cell rheology, which include among the usual methods, micropipette technique [EVa], twisting magnetocytometry [LAb], and optical tweezer [HEd]. Micropipette aspiration allows us to study continuous deformation and penetration of a cell into a calibrated micropipette (bore $< 10$ μm) at various suction pressures ($[10^{-1} - 10^4$ Pa]) in order to determine a cell's apparent viscosity by measuring the rate of cell deformation and the pressure. The leading edge of the cell is tracked in a microscope to an accuracy of $\pm 25$ nm. Associated basic continuum models, which assume that the cell is a viscous fluid contained in a cortical shell, yield apparent viscosity, shear modulus, and surface tension [YEa]. The results depend on the ratio of the cell size to the micropipette caliber. Soft cells, such as neutrophils and red blood cells, develop about 16 times smaller surface tension than more rigid cells, such as endothelial cells [HOa].

The atomic force microscope (AFM), a combination of the principles of the scanning tunneling microscope and the stylus profilometer, provides a force range of [10pN to 100nN] [BIa]. Twisting magnetocytometry (TMC) uses ligand-coated ferromagnetic beads to apply controlled mechanical stress to cells via specific surface receptors. The sampled cell is subjected to a magnetic field and the bead position is recorded using videomicroscopy. The torque resulting from the shear is measured to determine the viscosity and the elastic modulus using a Kelvin model. The bead size affects the results. Optical tweezers trap dielectric bodies by a focused laser beam through the microscope objective. Optical traps can be used to make quantitative measurements of displacements ($\mathcal{O}(1)$ nm) and forces ($\mathcal{O}(1)$ pN) with time resolution ($\mathcal{O}(1)$ ms).

Measurements have been carried out on round and spread endothelial cells, as well as on isolated nuclei [CAb]. The nonlinear force-deformation curves have been found to be affected by the cell morphology, the nucleus influence being much greater in spread cells, the most common in vivo shape. Cell adhesion affects the cell rheological properties. Due to the cell adaptation to its environment associated with cytoskeleton structural changes, material parameters depend in particular on the cytoskeleton polymerization state (thixotropy). Last but not least, the results of the rheological tests depend not only on the cell state, but also on the techniques, and, for a given technique, on the experimental procedure (cell environment, loading conditions, impacted region size, etc.).

## 2.5 Heart Wall

The heart wall is composed of several layers:

1. The internal thin endocardium

2. The thick muscular myocardium

3. The external thin epicardium

The double-layered pericardium is composed of the outer parietal pericardium and the epicardium. The pericardial cavity, which contains a lubrificating fluid, separates the two pericardium layers. The heart has a fibrous skeleton with its central fibrous body, which prevents early propagation of action potential. The central fibrous body provides extensions:

1. The valve rings, into which are inserted the cardiac valves

2. The membranous interventricular septum

*Mathematical histology.* Structure–function features of the heart have been mathematically investigated. Two fiber networks have been particularly studied: the network of collagen fibers of the aortic valve cusp, and the myofibers of the left ventricle wall, using a simple model of mechanically-loaded fibers.

The structure of the aortic leaflet has been derived from its function, which is assumed to consist of supporting a uniform pressure load, undergone by a single family of fibers under tension [PEb]. The equation of equilibrium for the fiber structure is solved to determine its architecture. The computed fiber architecture resembles the real one. Assuming constant myofiber cross-sectional area, symmetry with respect to the ventricle axis, small wall thickness with respect to the other dimensions, and a stress tensor resulting from hydrostatic pressure and myofiber stress, the bundles of myofibers have been shown to be located on approximate geodesics on a nested set of toroidal surfaces centered on a degenerate torus in the equatorial plane of the cylindrical part of the left ventricle [PEa].

*Heart valves.* The cardiac valves are sheets of connective tissue, attached to the wall at the insertion line. Like the heart wall, the valves are covered by the endothelium. They contain many collagen and elastic fibers and some smooth muscle cells. The cusp is a multilayer structure, with the fibrosa, the spongiosa, and the ventricularis, which is absent in the coaptation region.

In vitro uniaxial traction tests-sort the valve strips in increasing order of stiffness:

1. Axial strips of pulmonary valves

2. Axial strips of aortic valves

3. Circumferential strips of aortic valves

4. Circumferential strips of pulmonary valves [STd]

However, axial and azimuthal strips of porcine aortic valve leaflets are stiffer than the corresponding strips of the pulmonary ones, the circumferential strips being stiffer than the axial ones [JEa].

Heart structure provides the three properties of contractibility, automatism, and conduction due to two kinds of cardiac muscular cells: cardiomyocytes and nodal myocytes.

## 2.5.1 Cardiomyocyte

Cardiomyocytes (CMC) are striated nucleated cells that are electrically excited in order to contract and relax rhythmically. The striated appearance of the muscle fiber is created by arrays of parallel filaments, the thick filaments of myosin and the thin filaments of actin. The sarcomere is the anatomical unit of muscular contraction, and the hemisarcomere its functional unit. Myofibrils are held in position by scaffolds of desmin filaments, anchored by costameres enriched in vinculin along the plasma membrane surface. The costameres maintain the spatial structure of sarcomeres and couple CMCs to ECM. The membrane skeleton is made from spectrin and dystrophin, adapting the membrane to CMC functioning and contributing to the force transmission. The costameres, the membrane skeleton, and the cytoskeleton are linked to ECM by membrane protein such as integrins, dystrophin–glycoprotein complexes, and $\beta$ dystroglycan–laminin bonds. The sarcoplasmic reticulum (SR) broadens out at multiple sites to form junctional SR cisternae tightly coupled to the sarcolemna and its repeated invaginations, the T-tubules. CMCs are joined by intercalated discs which contain gap junctions in order to allow electrochemical impulses, or action potentials, to spread rapidly and orderly so that the cell contraction is almost synchronized.

The contraction is induced by the four-time nanomotor composed of interdigited myosin and actin filaments.

1. The myosin head detaches from actin and fixes ATP.

2. ATP is hydrolized and the myosin head binds to actin.

3. The myosin head releases the phosphate and undergoes a conformational change [RAe].

4. The myosin releases ADP and remains anchored to actin.

CMC interdigitated actin and myosin filaments slide over each other to shorten the sarcomere during contraction (sliding-filament model). The myosin contains myosin heads, which are binding sites for actin and ATP. The troponin and tropomyosin allow actin to interact with myosin heads in the presence of $Ca^{++}$. When links break, bonds reform farther along actin to repeat the process.

The collagen and elastin fibers form a network forming a surrounding trellis and between-CMC struts, which avoid excessive CMC stretching. Atrial CMCs contain small granules, especially in the right atrium. These granules secrete natriuretic peptides.

*Modeling and simulations.* Homogenization considers objects of length scale $L_o$, which have a relative periodicity, and thus are made from repeatable basic units of length scale $L_u$. In the myocardium, the basic unit contains a limited number of CMCs, inside which the electric field is computed. A constitutive law for the myocardium has been derived from discrete homogenization. A CMC set has been modeled by a quasiperiodic discrete lattice of elastic bars [CAb].

A first stage of heart modeling deals with the mechanical behavior of the myocardium, with its constitutive law. Both systolic and diastolic deformations of the left ventricle are heterogeneous [AZa,FOa,BOa]. Diffusion tensor imaging has been used to characterize cardiac myofiber orientations, with the reduced encoding imaging (REI) methodology [HSa]. The cardiac myofiber direction in each computational mesh element is the mean value of the noisy information contained in the voxels enclosed in the mesh element.

The contraction is more synchronous than the depolarization. The myofibers are differently stretched whether they are early or lately depolarized, without consequences due to their spatial arrangment. Moreover, even if cross-bridging in different cardiomyocytes is simultaneous, the cardiomyocytes can contract differently depending on the force applied to each cardiomyocyte by its own environment. Most of the biomechanical studies have been focused on the mechanical behavior of the left ventricle wall. The material is composite and infiltrated by liquids. Its muscular fibers, with various orientations, are embedded in a matrix with small blood vessels. The myocardial fibers are reinforced and connected by collagen fibers.

The constitutive law takes into account two main phases, active and passive, of the time-dependent heart cyclic behavior. The direct problem computes the stress and strain fields in the given geometry, using given constitutive law and loadings. Finite element models of the left ventricle which undergoes large displacements most often neglect the heterogeneity in wall properties, the myocardial fiber orientation, and the wall thickness variation in the axial and in the azimuthal directions, and assume an uniform transmural pressure. The first simulations were performed in idealized geometries. With the development of medical imaging, imaging data were used to determine the computational domain, reconstructing ventricle cross-section models and, later, heart cavities [HEc]. Numerical results differ from the findings obtained in idealized geometries.

Models of the behavior of the myocardium subjected to large deformations have been developed [ODa]. The Cauchy stress tensor $\mathbf{C}$ is given by $\mathbf{C} = \mathbf{F}(\partial W/\partial \mathbf{G})\mathbf{F}^T - p\mathbf{I}$, where $\mathbf{F}$ is the transformation gradient tensor and $\mathbf{G}$ the Green–Lagrange strain tensor, $W$ the deformation energy function, and $p$ is a Lagrange multiplier. When the material is incompressible, as are most of the biological materials which are rich in water, $det(\mathbf{F}) = 1$. The first Piola–Kirchhoff stress tensor $\mathbf{P}$ is expressed with respect to the deformed configuration: $\mathbf{P} = det(\mathbf{F})\mathbf{F}^{-1}\dot{\mathbf{C}}$. The passive state is defined by the configuration in the absence of internal and external stresses. The active state means myocardium contraction associated with a new configuration but without applied stress (free contraction without environmental constraint). The loaded state corresponds to an active state that undergoes internal and external stresses. The resulting deformation energy function during the cardiac cycle is the sum of three terms, which simulate the passive ground matrix, the passive elastic, and the active components of the muscular fibers [YAa]. A good agreement is found with experimental data [VIb]. The radial stress is the strain component that has the greatest magnitude during the cardiac cycle. The Cauchy stress reaches its highest value at the beginning of the systolic ejection, between the mid-wall and the endothelium.

In order to derive a constitutive law for the myocardium during the whole cardiac cycle, the heart wall has been modeled by a homogeneous, incompressible, transversally isotropic material [BOd]. The wall behavior during the cardiac cycle is continuously described using several states,

1. A passive unstressed state

2. A virtual state defined by a constant geometry but a rheology change

3. An active state of contraction without rheology change

The time-dependent strain energy function is composed of two terms, a passive and an active strain energy function (SEF) associated with the passive fibers and with the cardiomyocytes, respectively:

$$W(\mathbf{G}, t) = W_{pas}(\mathbf{G}) + \beta(t) W_{act}(\mathbf{G})$$

($\beta(t)$: activation function).

### 2.5.2 Nodal Cells

Nodal cells are small muscular cells with few myofibrils, which create or quickly spread the depolarization wave in the myocardium. The electrochemical signal starts with the spontaneous depolarization of the nodal cells of the sino-atrial node (SAN), the "natural pacemaker". The action potential spreads through the atria to reach the atrioventricular node (AVN) and to produce atrial contraction. This node imposes a short delay in impulse transmission to the ventricles. The action potential then runs in the His bundle and the Purkinje fibers, which penetrate into the myocardium to end on CMCs.

Ion pumps and exchangers of the CMC membrane maintain steep ion concentration and electrical gradients across the membrane. The resting membrane potential is $\sim -88$ mV (inside negative). The action potential is initiated by depolarization of the sarcolemna. The $Na^+$ channels open first and then rapidly inactivate. The quick cellwards $Na^+$ motion increases the transmembrane potential to $\sim +30$ mV (phase 0). Phase 1 corresponds to the first rapid repolarization associated with a transient outward motion of $K^+$. Phase 2 is associated with a slow inward $Ca^{++}$ current (plateau), involving L-type $Ca^{++}$ channels. During phase 3, delayed rectifier $K^+$ channels induce rapid repolarization. Throughout phase 4, the resting membrane potential is regulated by a background $K^+$ current. The refractory period is a protective mechanism in order to maintain efficient successive blood fillings and ejections.

### 2.5.3   Excitation–Contraction Coupling

The intracellular calcium content is an important factor that triggers the contraction, determining the inotropy. Its removal kinetics from the cytosol characterizes the heart lusitropy. Voltage-dependent $Ca^{++}$ channels (VDCC) are located at sarcolemmal–sarcoplasmic reticulum junctions close to the ryanodine channels (Figure 2.4). They induce a $Ca^{++}$ influx and elicit $Ca^{++}$ release from the ryanodine channels with a negative feedback [SIa]. Moreover, the $Ca^{++}$ influx is limited by $Ca^{++}$-dependent inactivation of the channels due to calmodulin [ZUa]. The protein S100A1 increases $Ca^{++}$ release from the sarcoplasmic reticulum by interacting with ryanodine channels. $Ca^{++}$ release from the sarcoplasmic reticulum into the cytoplasm sufficiently increases the cytosolic $[Ca^{++}]_i$ to induce contraction. $Na^+ - Ca^{++}$ porters, which exchange three $Na^+$ for one $Ca^{++}$, operate during phase 2, stabilizing $[Ca^{++}]_i$. Relaxation requires $Ca^{++}$ removal from the cytosol, by sarco(endo)plasmic reticulum $Ca^{++}$-ATPase (SERCA) pumps, by $Na^+ - Ca^{++}$ exchangers (NCX), by mitochondrial $Ca^{++}$ uniporters, and by plasma membrane $Ca^{++}$-ATPase pumps (PMCA). Phospholamban, which is associated with SERCA, inhibits this pump. Protein kinase A (PKA) phosphorylates phospholamban, and thus has a lusitropic effect. Most $Ca^{++}$ returns to SR where it is stored by bonds with calsequestrin.

The second stage of heart modeling refers to the genesis and the propagation of the action potential. The epicardial depolarization and deformation of the ventricular wall have been simultaneously measured using electrode brushes and videorecording of optical markers [DEc] or multielectrode socks

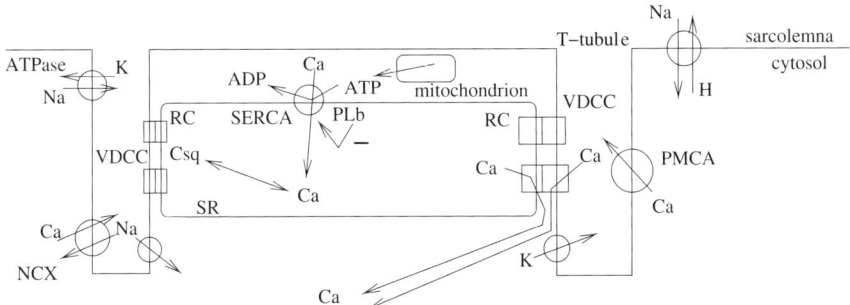

Figure 2.4. The cardiomyocyte, its ion carriers and calcium fluxes.

and MRI tagging [FAb]. The electrochemical wave propagation model provides the arrival time of the depolarization in the various parts of the myocardium, the local myofiber orientations affecting the ventricular depolarization timing due to the anisotropic myocardium conductivity. However, the nodal tissue remains difficult to locate.

The basic tractable phenomenological monodomain model of depolarization and repolarization consists of two variables $u$ and $v$ [FIa, Naa], in which fast dynamics are coupled to slow ones. The Aliev–Panfilov model gives an example of dimensionless FitzHugh–Nagumo system [ALa].

Bidomain models have been proposed to simulate electrophysiological waves in the myocardium [GEa, BOe]. Bidomain models take into account the intracellular and the extracellular spaces, separated by the CMC syncytium membrane. Both domains have their own volume-averaged properties, especially the conductivity of the extra- and intracellular spaces. The problem to numerically solve is very complex. Cardiac fibers have anisotropic conduction properties, the impulse propagation being faster in the axial direction than transversally. A conductivity tensor $\mathbf{M}$ is then introduced, assuming that the conductivity values are identical in all directions perpendicular to the muscular fiber direction [COa]. The collection of CMCs, end-to-end or side-to-side connected by specialized junctions, immersed in the extracellular fluid and ground matrix, is modeled as a periodic array that leads to a homogenization procedure, with homogenized conductivity tensors $\mathbf{M}_i$ and $\mathbf{M}_e$. The membrane current density $J_m$ is then given by:

$$J_m = -\nabla \cdot \mathbf{i}_i = \nabla \cdot \mathbf{M}_i \nabla u_i = \nabla \cdot \mathbf{i}_e = -\nabla \cdot \mathbf{M}_e \nabla u_e,$$

where $u_i$ and $u_e$ are the electric potentials of the intra- and extracellular spaces and $\mathbf{i}_i$ and $\mathbf{i}_e$ the currents $(\nabla \cdot (\mathbf{M}_i \nabla u_i + \mathbf{M}_e \nabla u_e) = 0)$. In its general form, the bidomain model is defined by [BOe]:

$$\kappa_{av}(C_m u_t + 1/R_m \, f(u, v)) = \nabla \cdot (\mathbf{M}_i \nabla u_i),$$

where $u = u_i - u_e$ is the action potential, $v$ the recovery variable, $\kappa_{av}$ the surface area-to-volume ratio of the cardiac myofibers, and $C_m$ and $R_m$ the cell membrane capacitance and resistance.

### 2.5.3.1 *Electromechanical Coupling*

The electromechanical ICEMA model of the electrochemical wave propagation is based on the FitzHugh–Nagumo equations and the myocardium functioning model on the Hill–Maxwell rheological model, associated with the muscular fiber direction and the dynamics equations [BEb, CHb, SEa]. The ICEMA heart model thus connects the microscopic and macroscopic enumerate levels, via a mesoscopic step. The constitutive equations require a small number of state variables, so that the solution of direct problems can be quickly obtained and inverse problems can be solved.

The Huxley nanoscopic-scale sliding filament theory states that the binding $f$ and rupture $g$ frequencies of the actin–myosin bridges are functions of the elongation $x$ [HUd]. The cross-bridge proportion $n$ with elongation $\tilde{x} = x/\ell$ ($\ell$: maximum bridge length) with the kinetics defined by $f$ and $g$ is given by:

$$\dot{n} = dn/dt = (1-n)f - ng. \tag{5.1}$$

The sarcomere contraction generates a shortening $s = s_0(1 + \epsilon_c)$ ($\epsilon_c$ : sarcomere deformation), with a velocity $s_0\dot{\epsilon}_c$ assuming a synchronized motion of the set of bridgings (subscript $c$: contractile component). The theory of actin–myosin cross-bridge dynamics and the moments [ZAa] applied to the cross-bridge model describes the contraction at the sarcomere and at the myofiber scale, respectively.

The functioning of the nanomotors, i.e. the actin and myosin molecules, is controlled by $Ca^{++}$ and adenosine triphosphate. $[Ca^{++}]_i$ has been given by a relatively simple function of time [HUc]. The amount of calcium linked to troponin C is commonly assumed to be equal to the cytosolic concentration, i.e. to the extent of the calcium influx driven by the action potential. The myocardium contraction results from a conformational change of the actin–myosin bridge, coupled with ATP hydrolysis. A unique command signal ($u > 0$ during contraction and $u < 0$ during active relaxation) has been defined, which involves two parameters, $k_{ATP}$, mainly the ATP hydrolysis rate, which is regulated by the actin–troponin–tropomyosin complex and $k_{Ca}$, the calcium extraction rate by the sarcoplasmic reticulum [BEb]. Such a model does not take into account either the oxygen consumption or the link between the ATP and $Ca^{++}$. Heart mitochondria use more oxygen, but produce ATP at a faster rate than liver ones [CAc]. $Ca^{++}$ overload can induce mitochondrial dysfunction in disease in the presence of a pathological stimulus. Calcium, ATP, and reactive oxygen species (ROS) are indeed in close connection [BRa]. The mitochondrion produces ATP, which synthesis is stimulated by $Ca^{++}$. The dysregulation of mitochondrial $Ca^{++}$ can lead to elevated concentration of ROS and apoptosis.

The collective behavior of the sarcomere fibers is governed by the relationship between the stress $\sigma$ and the strain $\epsilon$ in the myofiber direction (viscoelastoplastic behavior). The evolution of the stiffness $k_c$ and the active stress $\sigma_c$ of the contractile component, knowing the strain rate $\dot{\epsilon}_c$ and the command $u$, are given by the following set of ordinary differential equations [BEa].

$$\begin{cases} \dot{k}_c = -(\alpha|\dot{\epsilon}_c| + |u|)k_c + k_0|u|_+ \\ \dot{\sigma}_c = k_c\dot{\epsilon}_c - (\alpha|\dot{\epsilon}_c| + |u|)\sigma_c + \sigma_0|u|_+ \\ \sigma = d(\epsilon_c)(\sigma_c + k_c\tilde{x}_0) + \mu_c\dot{\epsilon}_c \end{cases} \tag{5.2}$$

The parameters $k_0$ and $\sigma_0$ are related to the maximum available actin–myosin cross-bridges in the sarcomere. $d(\epsilon_c) \in (0,1)$ is the modulation function of the active stress which accounts for the length-tension curve[2] and $\mu_c$ the viscosity ($k_c(0) = \sigma_c(0) = 0$). The cross-bridge detachment rate is given by $|u| + \alpha|\dot{\epsilon}_c|$.

The action potential u is modeled by the two-variable Aliev–Panfilov equation system [ALa]:

$$\begin{cases} u_t - \nabla \cdot (\kappa_e \nabla u) = f(u) - v \\ \qquad\qquad v_t = \varepsilon(\beta u - v) \end{cases},\qquad(5.3)$$

where $\varepsilon \ll 1$, $f(u) = u(u - \alpha)(u - 1)$ ($\alpha \in [0, 1/2]$), and $\kappa_e$ is an electrical conductivity.

In the macroscopic scale, the cardiac fibers are embedded in a collagen sheath, connected by collagen struts and supported by a ground matrix with elastin fibers. The mesoscopic myofiber constitutive law is incorporated in a Hill–Maxwell rheological model [CHb]. The sarcomere set of a myofiber is represented by a single contractile element (CE). The activity of CE depends on the action potential u. The electrochemically driven contraction and relaxation obey Eq. (5.2). The isometric deformations are modeled by an elastic serial element (ESE) in series with CE. ESE lengthens when CE shortens at a constant myofiber length. A third viscoelastic element (EPE) in parallel to the CE-ESE branch is introduced. EPE educes the force developed from a certain myofiber length in the absence of stimulation. These three components are not related to the muscle constituents.

The cardiac tissue is supposed to be not purely incompressible [SAa].

$$\mathbf{P} = -p\,\det(\mathbf{F})\mathbf{S}_r^{-1} + \sigma_p^e(\mathbf{G}) + \sigma_p^v(\mathbf{G}, \dot{\mathbf{G}}) + \sigma_{1D}\,\hat{\mathbf{f}} \otimes \hat{\mathbf{f}},\qquad(5.4)$$

where $\mathbf{P}$ denotes the second Piola–Kirchhoff stress tensor, $p = -B(\det(\mathbf{F}) - 1)$ ($B$: bulk modulus, used as coefficient of incompressibility penalization), $\mathbf{F}$ the deformation gradient, $\mathbf{G}$ the Green–Lagrange strain tensor, and $\mathbf{S}_r$ the right Cauchy–Green deformation tensor, $\sigma_p^e(\mathbf{G}) \propto \rho_0 \partial W^e/\partial \mathbf{G}$ (elastic part of EPE, $\rho_0$: density of the reference state, $W^e$: elastic strain energy density), and $\sigma_p^v(\mathbf{G}, \dot{\mathbf{G}}) = \partial W^v/\partial \dot{\mathbf{G}}$ (viscoelastic part of EPE, $W^v$: viscous strain energy pseudo-density) the stresses in the passive materials, $\sigma_{1D}$ the stress generated by the active element CE, and $\hat{\mathbf{f}}$ the local unit direction vector of the myofiber.

---

[2] The troponin C sensitivity for $Ca^{++}$ and the cross-bridge availability depend on the sarcomere length.

In the preliminary stage, the valves are not incorporated in the model. Constraints on the volume variations of the ventricle are then added:

$$\begin{cases} q \geq 0 \text{ when } p_V = p_a & \text{(ejection)} \\ q = 0 \text{ when } p_A < p_V < p_a & \text{(isovolumic phases)}, \\ q \leq 0 \text{ when } p_V = p_A & \text{(filling)} \end{cases} \tag{5.5}$$

where $q = -\dot{V}_V$ is the blood flow ejected from the ventricle ($V_V$: ventricular volume), $p_V$ the ventricular blood pressure, $p_A$ the atrial pressure, and $p_a$ the arterial pressure. A regularization is used to overcome numerical failures on flow rate computations. A windkessel or a 1-D model of the blood flow provides $p_a$.

Using Eqs. (5.2) and (5.4), and incorporating the multiple elements of the simplified system

1. The myofibers

2. The passive components

3. The valve model

4. The upstream atrium and the exiting artery

the following equation set is obtained [CHb].

$$\begin{cases} \rho \ddot{\mathbf{u}} - \nabla \cdot (\mathbf{F} \cdot \sigma) = 0 \\ \sigma = \sigma_{1D} \, \hat{\mathbf{f}} \otimes \hat{\mathbf{f}} + E_p(\mathbf{G}) \\ \sigma_{1D} = \sigma_{\mathbf{c}}/(1 + \epsilon_s) = \sigma_s/(1 + \epsilon_{\mathbf{c}}) \\ \sigma_{\mathbf{c}} = E_{\mathbf{c}}(\epsilon_{\mathbf{c}}, \mathbf{u}) \\ \sigma_s = E_s((\epsilon_{1D} - \epsilon_{\mathbf{c}})/(1 + \epsilon_{\mathbf{c}})) \\ \epsilon_{1D} = \sum_{i,j} G_{ij} f_i f_j \\ \dot{\mathbf{g}} = \mathcal{G}(\mathbf{g}, t) \\ \text{initial + boundary conditions} \end{cases} \tag{5.6}$$

where $E_p(\mathbf{G}) = -p \det(\mathbf{F}) \mathbf{S}_r^{-1} + \sigma_p(\mathbf{G}, \dot{\mathbf{G}})$, $E_{\mathbf{c}}$ is a function expressing system (5.2), and $\epsilon_{1D}$ is the deformation in the myofiber direction. From thermomechanical considerations, $1 + \epsilon_{1D} = (1 + \epsilon_{\mathbf{c}})(1 + \epsilon_s)$. $\mathbf{g}$ stands for $V_V$, or $p_V$, or $p_A$, or $p_a$, the last equation accounting for the set of ordinary differential equations modeling the valve opening and closure and the arterial pressure changes.

The action potential is initiated at the ends of the Purkinje network localized according to the literature data [DUb]. The model parameters are calibrated according to the available data, which provide

1. The arterial parameters [STb, WEb]

2. $k_c$ and $\tau_c$ [WUa], from which are computed the respective asymptotic values $k_0$ and $\sigma_0$

3. The estimation of the passive behavior from conflicting literature data [VEb, PIb]

### 2.5.4 Vessel Wall

As blood pulses in an artery, its wall alternatively stretches and rebounds. Wall expansion and relaxation are due to the rheological properties of the vessel wall, and, thus to its composition and structure.

*Main structural elements.* The structure components exist in every kind of blood vessel, except capillaries, although the element amount and the structure vary between the vessel types. Endothelial cells line the blood–wall interface. The specific cells of connective tissues are fibroblasts, which produce the ground matrix and fibers, and fibrocytes. Elastin provides vessel distensibility and collagen tensile strength. Smooth muscle cells (SMC) are responsible for the lumen size.

*Wall structure.* The wall structure is circumferentially layered. The tunica number and tunica structure vary according to the vessel type and size. The wall of large blood vessels has three main layers. The internal intima is composed of the inner endothelium and a subendothelial connective tissue. The internal elastic lamina (IEL) delimits the intima from the media. The media is formed by layers of circumferential smooth muscle cells and connective tissue with fibers. The external elastic lamina (EEL) is located between the media and the adventitia. The adventitia mainly consists of connective tissue with some SMCs, nerves, vasa vasorum, and lymphatic vessels.

The media is the main site of histological specializations of artery walls. Vessels proximal to the heart are elastic arteries, involved in the windkessel effect. The thick media contains thin concentric fenestrated lamellae of elastin. EEL is not very well defined and the adventitia is thinner than in distal muscular arteries. Muscular arteries have thinner intima and a media which is characterized by numerous concentric layers of SMCs. EEL can be clearly observed.

The vein walls are thinner than artery walls, and the caliber is larger. The intima is very thin. IEL and EEL are either absent or very thin. The media is thinner than the adventitia. Medium-sized veins are characterized by the presence of valves in order to prevent the transient blood returning to upstream segments during muscular compression. The largest veins have very thick adventitia, with bundles of longitudinal SMCs and vasa vasorum. Valves are absent.

Arterioles are composed of an endothelium surrounded by one or a few concentric layers of SMCs, which regulate blood flow. Capillaries are small exchange vessels composed of endothelium surrounded by basement membrane with three structural types: continuous capillaries have tight intercellular clefts; fenestrated capillaries are characterized by perforations in endothelium; and discontinuous capillaries are defined by large intercellular and basement membrane gaps. Venules are composed of endothelium surrounded by basement membrane for the postcapillary venules and smooth muscle for the larger venules.

### 2.5.4.1 *Vascular Smooth Muscle Cell*

The vascular smooth muscle cell, which contains $\alpha$-smooth muscle actin, carries out slow and sustained contractions. Actin and myosin in SMCs are not arranged into distinct bands. SMC activity is regulated by $[Ca^{++}]_i$ due to $Ca^{++}$ entry via VDCCs, to $Ca^{++}$ release from endoplasmic reticulum, and to receptor-dependent $Ca^{++}$ channels. Vasodilator and vasoconstrictor influences are exerted upon a basal vascular tone. A vasomotor tone is indeed spontaneously developed in most arterioles [DUa]. In isolated arterioles and arteries, the basal tone is developed for vessel physiological pressure [DAa].

Myosin light chain (MLC) phosphorylation, which leads to the contraction, is tightly controlled by the relative activities of the counterregulatory enzymes myosin light chain kinase (MLCK) and myosin phosphatase. MLCK is activated by calcium and calmodulin. Caldesmon is a calmodulin-binding protein implicated in the regulation of actomyosin interactions. Calponin, a $Ca^{++}$ and calmodulin-binding troponin T-like protein, binds to tropomyosin and to calmodulin. MLC phosphorylation leads to crossbridge formation between the myosin heads and the actin filaments, and hence, to smooth muscle contraction. Dephosphorylation of myosin light chains by PKC leads to relaxation. The cGMP-dependent protein kinase 1$\alpha$ mediates SMC relaxation [SUa]. The RhoA pathway inhibits the myosin phosphatase.

### 2.5.4.2 *Pericytes*

The pericytes, which surround the capillaries, can also encircle precapillary arterioles and postcapillary venules. A basal lamina separates the endothelial cells and the pericytes. However, tight and gap junctions can develop between the endothelial cells and the pericytes. A basement membrane can also be found along the outer surface of the pericytes. The pericytes regulate the capillary bore and, then, the tissue perfusion, as well as the transport from the blood via pericytic processes at interendothelial clefts. The pericytes secrete vasoactive autoregulating substances and release structural constituents of the basement membrane and interstitial matrix. Pericyte coverage is required during angiogenesis.

### 2.5.4.3 Endothelial Cells

The endothelium is involved

1. In blood-wall exchange control

2. In vasomotor tone modulation

3. In coagulation regulation

4. In vessel wall growth and remodeling

5. In inflammation and immune pathways, driving the leukocyte adhesion

The wetted cell membrane is covered by the glycocalyx, made of proteoglycans, glycosaminoglycans, glycoproteins, and glycolipids. The glycocalyx is the first barrier to molecular transport from the flowing blood.

### 2.5.4.4 Mechanotransduction

The shear stress and the pressure exerted on the wall by the blood generate a basal tone of SMCs in the absence of neurogenic and hormonal influences. The hemodynamic stresses act on the smooth muscle cells via stress transmission or via compound release by the endothelial cells. The endothelial membrane is the first wall component to bear stresses from the circulating blood. Stresses can act on mechanosensitive ion channels, on cell-membrane receptors, on adhesion molecules, on proteins associated with cytoskeleton proteins, on elements of cell junctions, and so on. These changes lead to biochemical responses.

In vitro effects of flow over a cultured EC layer and cyclic stretch of the culture support have been investigated to study the responses of endothelial cells in well-defined mechanical conditions. Support and perfusion media used in flow chambers can introduce substances that can interfere with the cell response to the investigated stimulus. Consequently, experimental testing and result interpretation must be carefully handled.

Stresses applied on EC wetted or on abluminal surface affect

1. The cell shape and orientation [DEe], as well as the cell ultrastructure [NOc]

2. The cell rheology [THa, SAc], the endothelial cell becoming stiffer

3. Cell proliferation

4. Cell metabolism and transport

5. Cell adhesion to its support and the matrix content

Endothelial cells respond, in particular, to space and time changes in wall shear stress (WSS). Studying the responses of the endothelial cells to step flows, impulse flows, ramp flows, inverse ramp flows, and pulsatile flows, it has been shown that the time derivative of WSS, and not the shear stress itself, is directly responsible for the EC reactions [BAa].

*Nitric oxide.* NO is a vasodilator and inhibits vasoconstrictor influences. NO also inhibits platelet and leukocyte adhesion to the endothelium. NO has an antiproliferative effect on SMCs. It is produced from L-arginine by nitric oxide synthase (NOS, Figure 2.5). There are two isoforms of NOS: constitutive (cNOS) and inducible (iNOS). NO is continuously produced. Its release is enhanced by multiple stimuli. NO acts via cGMP, after binding to guanylyl cyclase. cGMP decreases $Ca^{++}/Cam$ stimulation of myosin light chain kinase. NO can be released from the endothelium by $\alpha$ 2-adrenoceptor activation, serotonin, aggregating platelets, leukotrienes, adenosine diphosphate, and bradykinin [VAc]. Hypoxemia also yields vasodilatation [POa]. The time gradients of the wall shear stress induce transient high-concentration burst of NO release via G proteins [BAa].

Endogenous NO contributes to CMC "hibernation" by reducing oxygen consumption and preserving calcium sensitivity and contractile function without an energy cost during ischemia [HEe]. The endothelium and the myocardium then are able to adapt to ischemia. The hypoxemia also induces the production of prostaglandins PGE2 and leukotriens LktC4 by CMCs.

*Endothelin.* The endothelin (ET) is a potent vasoconstrictor. It also regulates the extracellular matrix synthesis by stimulated vascular smooth muscle cells (Figure 2.6). In human myocardium in vitro, endothelin exerts a positive inotropic effect via sensitization of cardiac myofilaments to calcium and activation of the sodium exchanger [PIa]. However, the generated coronary vasoconstriction balances the positive inotropic and chronotropic effects. Endothelin is also a growth factor for cardiomyocytes [ITa]. The ET-1 release is shear-dependent [MOa].

*Other vasoactive substances.* Prostacyclin (PGI2) is another endothelium-derived vasodilator. Endothelium-derived hyperpolarizing factor (EDHF) can also be a prominent vasodilator hampered by NO. The endothelial

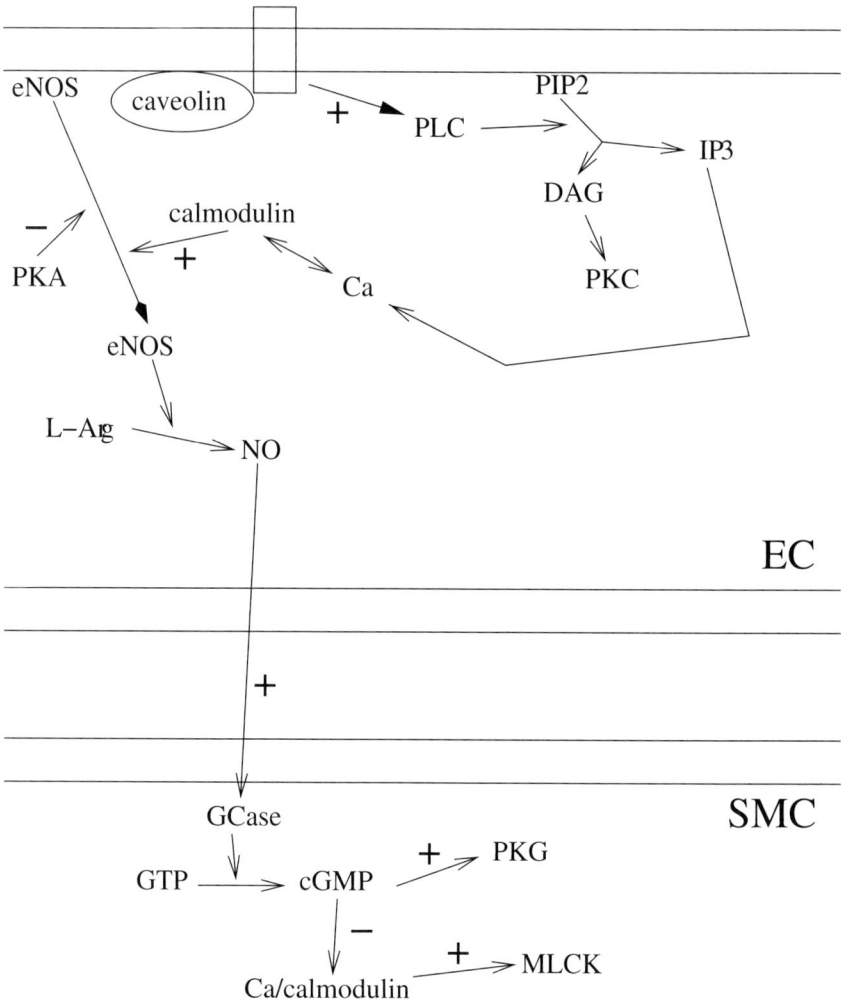

Figure 2.5. Nitric oxide (NO) produced in the endothelial cells by nitric oxide synthase (eNOS). eNOS is bound to caveolin in the cell membrane (inactive state). Activation of phospholipase C (PLC) by the ligand-bound receptor produces inositol triphosphate (IP3) and diacylglycerol (DAG) from phosphatidylinositol biphosphate (PIP2). IP3 increase in intracellular calcium content, which activates calmodulin. The latter dissociates eNOS from caveolin (cytosolic translocation). Protein kinase A (PKA) inactivates eNOS, which then relocates to the membrane caveolin (source: www.sigmaaldrich.com). NO stimulates guanylyl cyclase (GCase). cGMP decreases the stimulation by the Ca/calmodulin complex of myosin light chain kinase (MLCK) and activates protein kinase G (PKG).

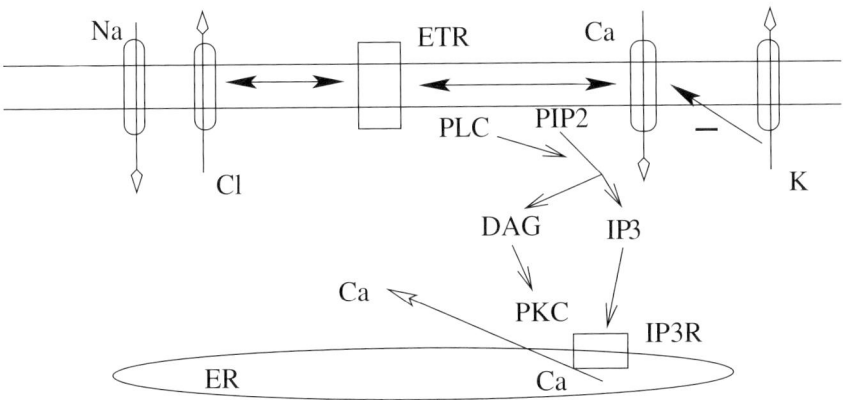

Figure 2.6. Binding of endothelin 1 to its receptor (ETR) induces activation of phospholipase C (PLC), which degrades phosphatidylinositol bisphosphate (PIP2) into diacylglycerol (DAG) and inositol trisphosphate (IP3). IP3 links to its receptor (IP3R) on the endoplasmic reticulum (ER) in order to release calcium (Ca) from ER. DAG activates protein kinase C (PKC). ETR is associated with a calcium channel of the cell membrane, which opens in response to ETR binding, hence further increasing the cytosolic calcium content. ET also opens chloride (Cl, Cl⁻ efflux), sodium (Na, Na⁺ influx) and potassium (K, K⁺ efflux) channels. K⁺ efflux inhibits the Ca channel (Source: www.sigmaaldrich.com).

cells also produce endothelium-derived contracting factors, which include superoxide anions [KAa], endoperoxides, and thromboxane A2.

The uridine adenosine tetraphosphate (Up4A) vasoconstricts the blood vessel.

*Myogenic response.* The myogenic response, independent of the vascular endothelium, couples SMC contraction or relaxation to SMC deformation [JOa]. Stresses can act

1. On exchangers and transporters

2. On membrane ion channels

3. On membrane bound enzymes to modulate activity of contractile proteins

$Ca^{++}$ influx leads to phospholipase C activation and release of inositol triphosphate and diacylglycerol [SHa]. Protein kinase C is involved in the myogenic response in the microcirculation [MEb]. The phosphatidylinositol metabolism stimulated by mechanical factors enhances $[Ca_i^{++}]$ and causes a translocation of PKC from cytosol to membrane [NOc].

### 2.5.5   Vessel Wall Rheology

The vessel walls exhibit longitudinal and circumferential prestresses. In vitro rheology measurements on excised or isolated vessels need suitable vessel conservation and preconditioning. Various experimental methods have been developed; they are summarized in a literature review [HAa]. Uniaxial loading is widely used because carefully controlled 2-D/3-D experiments are difficult to carry out on biological tissues. Rheological properties differ whether tests are performed on isolated segments of the vasculature or in vivo. In in vitro experiments, connections and interactions between regions of the anatomical system and between the wall and its neighborhood are removed, although they affect the wall rheology. Furthermore, excised tissues are more or less dried and not perfused. Conversely, in vivo measurements are carried out in a noisy environment due to blood circulation and respiration on targeted regions of limited surface areas, most often without preconditioning and without control of influence factors.

Imaging velocimetry (US, MRV) can be used as indirect methods. Two exploring stations are not sufficient because a single value of the wave speed does not take into account the nonlinear pressure-dependency of the wave speed. It has been proposed to use three stations, two giving the BCs, and one the pressure-dependency, assuming a 1-D flow. It is also possible to process the pressure signals from two stations in three identifiable wave points (foot, peak, notch) with different time delay between the two waves, the absence of reflexion being assumed [STb]. With aging, the wall becomes stiffer and the wave speed increases. Besides, MRI shows that the wall deformation is not uniform as well as the circumferential variation in wall strains during the cardiac cycle [DRa]. Moreover, the muscular tone affects the vessel compliance, especially in muscular arteries and arterioles.

Uniaxial loadings exhibit nonlinear force-deformation relationships. Stress–strain relationships have been mainly explained by the wall microstructure. The ability to bear a load is mostly done by elastin and collagen fibers. The nonlinearity is commonly understood as an initial response of elastin fibers and a progressive recruitment of collagen fibers. At low strain, collagen fibers are not fully stretched and elastin fibers play a dominant role. At high strain, the higher stiffness of the stretched collagen fibers affects the elastic properties [FUb]. When the elastin content is higher than the collagen one, the elastic modulus decreases and distensibility increases and vice versa.

Constitutive equations are based on the strain energy function (SEF), which links stresses to strains via a differentiation. Logarithmic and exponential formulation have been proposed in the literature, but they are not fully appropriate for numerical simulations. Polynomial expressions are prefered. The vessel wall can be considered as made of three elements, elastine, collagen (the response of which depends on the fiber stretch level), and SMCs (the response of which depends on deformation-dependent tone level) [ZUb]. Usually, the collagen fibers, embedded in the ground matrix and undulated in the rest configuration, are supposed to be gradually recruited. The vessel wall has been modeled as an isotropic elastic material containing an anisotropic helical network of stiff collagen fibers with a given orientation with respect to the circumferential direction [HOb].

Constrained mixture models, which meld classical mixture and homogenization theories, consider the specific turnover rates and configurations of the main constituents to study stress-dependent wall growth and remodeling [HUa, GLa]. But, appropriate knowledge of constituent material properties is still lacking.

## 2.5.6 Growth, Repair, and Remodeling

The growth and the remodeling of the wall of the vasculature (heart, blood vessels) are controlled. The processes, modulated by biomechanical quantities, are coordinated by biochemical mechanisms requiring factors and signals.

### 2.5.6.1 *Growth Factors*

Tissue growth needs mechanical, electrical, structural, and chemical signals to grow into functional 3-D tissue. Interactions of cells with ECM provide structural cues for normal cellular activity. Cell responses to various environmental signals are mediated by growth factors (GF). GFs promote not only cell meiosis, maturation, and functioning, but also tissue growth and remodeling. Cell growth is controlled by a balance between growth-promoting and growth-inhibiting factors. GFs can have autocrine, paracrine, juxtacrine, or endocrine effect (Figure 2.7). The epidermal growth factor (EGF) has proliferative effects especially on fibroblasts. The platelet-derived growth factor (PDGF), fibroblast growth factor (FGF), and vascular endothelial growth factor (VEGF) are involved in angiogenesis (Figure 2.8). The transforming growth factor-$\beta$ is a growth inhibitor for endothelial cells and fibroblasts. Cytokines are growth factors that modulate activities of immune cells. Interleukins (IL) are growth factors targeted to hematopoietic cells. Interferons (IFN) are cytokines produced by the cells of the immune systems in response to foreign agents. Sphingosine 1-phosphate (S1P), a lipid growth factor, mediates locomotion and maturation of endothelial cells. Endothelial cells have an intracellular reserve of functional S1P1 in caveolae. S1P acts on various pathways via

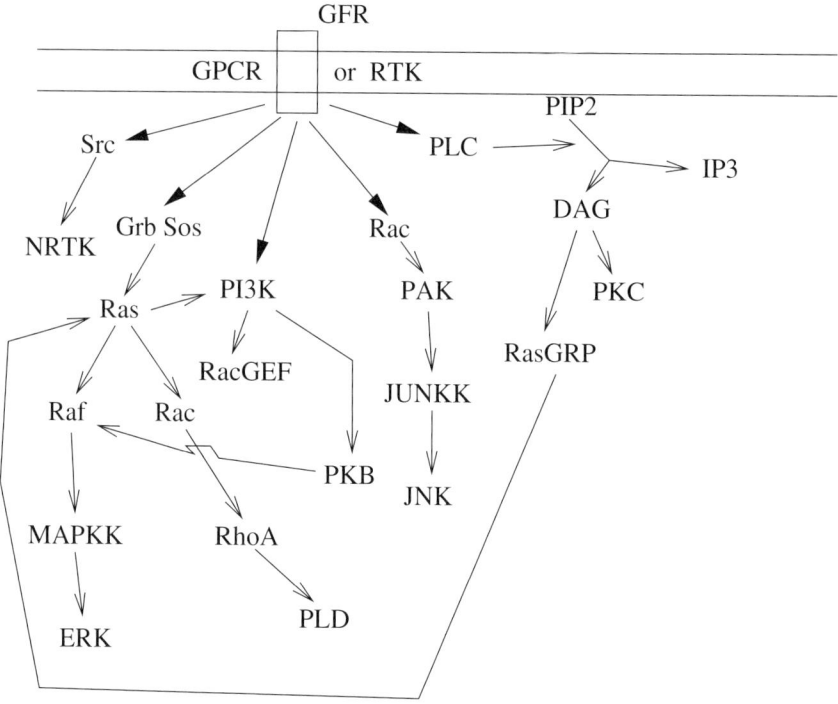

Figure 2.7. Growth factor-bound receptor activates adaptor protein Grb coupled to the guanine nucleotide releasing factor Sos (Grb–Sos complex), and, subsequently Ras, Raf, mitogen-activated protein kinase (MAPK), and extracellular regulated kinase (ERK) and Rac, RhoA, and PLD on the other hand. Ligand-bound receptors also activate (1) phosphatidylinositol 3-kinase (PI3K), protein kinase B (PKB), (2) Src, (3) Rac and JNK, and (4) phospholipase C (PLC) (Source: www.sigmaaldrich.com).

G-protein-coupled receptors (GPCR, Figure 2.9). S1P tightens adherens junctions between endothelial cells, characterized by VE cadherins.

### 2.5.6.2  *Chemotaxis*

Chemotaxis requires several main processes:

1. Cell alignment along the chemoattractant gradient

2. Cell polarization

3. Protrusion at the leading edge (cell front) and retraction at the trailing edge (cell back) of cytoskeletal elements, which all implicate small GTPases [MEa]

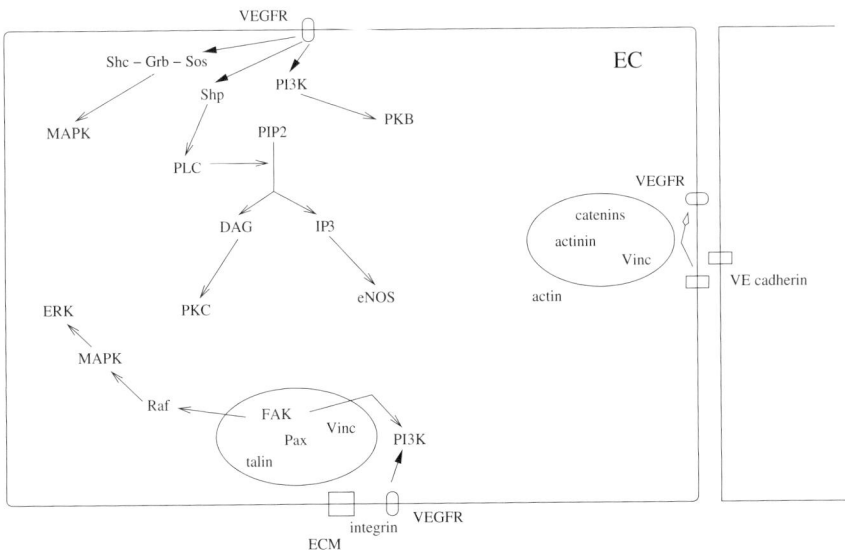

Figure 2.8. The vascular endothelial growth factor (VEGF) stimulates angiogenesis, particularly proliferation of endothelial cells (EC). VEGF binds to its receptors (VEGFR-1, VEGFR-2, VEGFR-3). It then activates a cell type-dependent signaling cascade via Shc and Grb, tyrosine phosphatases Shp and PLC-$\gamma$, phosphatidylinositol 3-kinase (PI3K), and so on. Vascular endothelial (VE)-cadherins are involved in the adherens junctions between neighboring ECs. VE-cadherins interact with catenins, and, subsequently, with $\alpha$-actinin and vinculin (Vinc), and with the actin cytoskeleton. VE-cadherin acts with VEGFR-2 to control the PI3K/PKB pathway. ECs are linked to the extracellular matrix (ECM) via integrins, such as $\alpha_v\beta_3$ and focal adhesion molecules, such as focal adhesion kinase (FAK), talin, paxillin (Pax), and Vinc (Source: www.sigmaaldrich.com).

The chemotactic flux depends

1. On a chemotactic response function of the available cell number and of the chemoattractant concentration

2. On chemoattractant concentration gradient [MUa]

The time gradient of the chemoattractant concentration depends on the production and destruction rates as well as its diffusion flux. The Keller–Segel model is widely used for the chemical control of cell movement [KEa]. A new formulation of the system of partial differential equations has been obtained by the introduction of a new variable and is approximated via a mixed finite element technique [MAb]. Chemotaxis is used in vasculogenesis and angiogenesis [AMa].

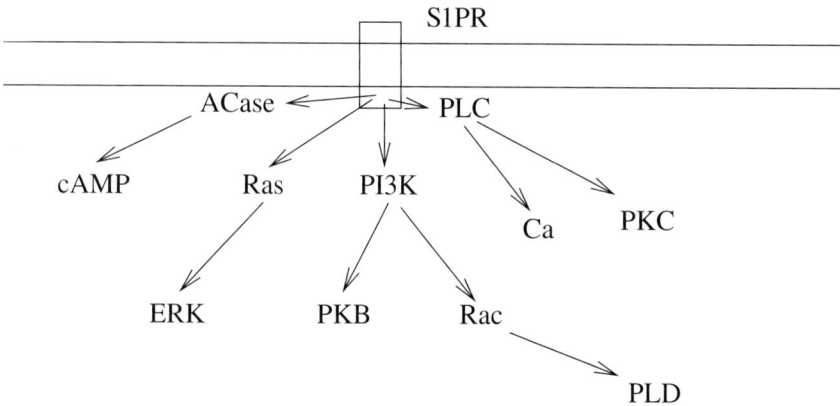

Figure 2.9. Sphingosine 1-phosphate (S1P) is produced intracellularly and is then secreted. It acts on adenylyl cyclase (ACase), Ras, phosphatidylinositol 3-kinase (PI3K), and phospholipase C (PLC).

### 2.5.6.3   Growth and Repair

Vasculogenesis defines formation of capillary plexus from endothelial stem cells (embryological process). A primitive vascular network is formed during embryogenesis through the assembly of angioblasts. Angiogenesis is characterized by maturation of or generation from a primary vascular network. Arteriogenesis deals with formation of mature arterioles and arteries with SMCs, for example, for collateral development in order to bypass an obstructed artery. Blood and the vessel wall are also involved in defense and repair processes. Both chemotaxis (directed response of cells according to a chemoattractant concentration gradient) and haptotaxis (adhesion gradient associated with the concentration of the constituent of the support medium, i.e. gradient of extracellular matrix density) are involved during the tissue development.

*Vasculogenesis.* During vasculogenesis, angioblasts differentiate and determine "blood pockets" which lengthen to form irregular capillaries. These pipes connect to each other in a nonhierarchical inhomogeneous network of primitive vessels. Once associated with the heart pump, this network, which conveys moving blood, remodels with branching. Vessels through which blood flows with high and/or quick flow rates widen and narrow. The network progressively matures with arteries, capillaries, and veins [LEb].

   Optimal design of vessel branching is based on cost functions that are the sum of the rate at which work is done on blood and the rate at which energy is used, supposedly proportional to the vessel volume for each vessel segment [MUa]. Other cost functions have been proposed, based on the minimal total surface area of blood vessels, the minimal total volume, or the minimal total wall shear force on the vessel wall or the minimal power of the blood flow.

*Angiogenesis.* Angiogenesis is a process leading to the generation of blood vessels through sprouting from existing blood vessels, because it involves migration and proliferation of endothelial cells from preexisting vessels. Localized production of growth factors promotes tissue expansion and de-termination of the position of branching nodes, with possible adaptation. The formation of new blood vessels or the remodeling of existing blood vessels requires the controlled growth of various cell types by different factors. Development of vascular trees also includes wall stress adapta-tion mechanisms. Limitation in angiogenesis is provided by angiostatin and endostatin. Antiangiogenic compounds are useful in cancers, such as inhibitors

1. Of ECM remodeling

2. Of adhesion molecules

3. Of activated endothelial cells

4. Of angiogenic mediators or receptors

Angioinhibins and other factors negatively influence angiogenesis, either by inhibiting endothelial cell proliferation or by preventing cell migration.

*Arteriogenesis.* Once the lumen of a main artery narrows too much, the lumen of a small artery increases to form a collateral in order to maintain the blood flow. Arteriogenesis is initiated by the monocyte chemoattractant protein-1 (MCP1) [VAb]. Various substances are also required at different stages of arteriogenesis; among these, TGF$\beta$, PDGF, FGF2, GM-CSF, and TNF-$\alpha$. Attracted monocytes produce fibronectin and proteoglycans as well as proteases in order to remodel the extracellular matrix. These inflammatory cells then produce growth factors to stimulate EC and SMC proliferation.

*Stem cells.* Stem-cell revascularization for tissue oxygenation and myocar-dium regeneration for pump functioning are proposed therapies. However, cell therapy can be ineffective or even hazardous in certain clinical settings and in specific subgroups of patients [HIa, HEb]. In the heart, resident stem cells can lead to endothelial cells, smooth muscle cells, and cardiomy-ocytes [BEa].

*Tissue engineering.* Bioreactors are devices used for the growth of tissues in an artificial environment that mimics the physiological conditions. In vivo biomechanical and chemicophysical conditions are created for in vitro cell conditioning and for construction of blood vessel, heart valves, and so on, with similar features to the native tissue ones. Histological niches, which define the cell status, its living site, its functioning, and its interactions with the environment, must then be replicated in bioreactors.

In planar cultures on matrices rich in collagen IV and laminin, the endothelial cells form clusters and pull on the matrix, generating tension lines that can extend between the cell aggregates. The matrix eventually condenses along the tension lines, along which the cells elongate and migrate, building cellular rods. The rate of change in cell density is equal to the balance between the convection and the strain-dependent motion. The inertia being negligible, the forces implicated in the vasculogenesis model include the traction exerted by the cells on the ECM, the cell anchoring forces, and the recoil forces of the matrix [MUb]. In order to study the role of the mechanical and chemical forces in blood vessel formation, a mathematical model has been developed using a finite difference scheme to simulate the formation of vascular networks in a plane [MAa]. The numerical model assumes

1. Traction forces exerted by the cells onto the extracellular matrix

2. A linear viscoelastic matrix

3. Chemotaxis

Spontaneous formation of networks can be explained via a purely mechanical interaction between the cells and the extracellular matrix.

*Myocardium remodeling.* Acute myocardial infarction leads to necrosis of cardiomyocytes and other cells. The heart contains rare cardioblasts susceptible to division, for self-regeneration and maturation [LAa]. CMC proliferation, from CMCs, resident stem cells, endothelial cells, fibroblasts, or migrated hematopoietic stem cells, in an area adjacent to the infarcted zone can be a regeneration source. Adequate input in growth factors is necessary to stimulate myocardial regeneration with needed angiogenesis and ECM formation in order to avoid maladaptive remodeling of the myocardium. Cardiac hypertrophy is induced by sustained pressure overload. Multiple hypertrophy signaling pathways are triggered by the pressure: calcineurin, phosphoinositide-3 kinase/protein kinase B, and ERK1/2.

*Wall remodeling.* Blood vessels are subjected to mechanical forces that regulate vascular development, adaptation, and genesis of vascular diseases. A chronic increase in arterial blood flow leads to vessel enlargement and reduction in WSS to physiological values. Wall remodeling is characterized by SMC proliferation and migration. Wall remodeling implies changes in rheological properties, and, consequently, the material constants of the constitutive law must be updated. Moreover, the constitutive equation must include not only the stress history but also material history due to wall restructuralization.

Flow-induced changes involved in long-term vascular tissue growth and remodeling have been studied using the continuum approach and motion decomposition [HUa, HUb]. The proposed homogenized, constrained mixture theory is used to develop a 3-D constitutive law that takes into account the three primary load-bearing constituents (SMC, collagen, and elastin) with time-varying mass fractions due to the turnover of cells and extracellular matrix fibers during the wall remodeling under a varying stress field. The turnover of constituent $i$ is described by its total mass evolution, introducing two evolution functions for production and for degradation rates. Besides, axial extension quickly increases the length of a carotid artery and the rate of turnover of cells and matrix, the turnover rates correlating with the stress magnitude. Numerical simulations show that moderate (15%) increases in axial extension generate much greater axial stress than circumferential stress augmentation induced by marked (50%) rise in blood pressure [GLb]. A 2-D constrained mixture model, based on different constitutive relations, shows that the turnover of cells and matrix in altered configurations is effective in restoring nearly normal wall mechanics.

## 2.6 Cardiovascular Diseases

Cardiovascular diseases can develop because of a favorable genetic ground and of exposure to risk factors. They are primarily located either in the blood vessels, mainly the arteries, or in the heart. Due to the lack of space, this review is only focused on two major arterial pathologies, the aneurisms and atherosclerosis. Both wall diseases are targeted by physical and mathematical modeling and by mini-invasive treatments, coiling and stenting.

### 2.6.1 Atheroma

Atherosclerosis is defined by a deposition of fatty materials and then fibrous elements in the intima, beneath the endothelium. The artery wall thickens and the lesion secondarily protudes into the lumen. Atheroma may be scattered throughout the large and medium thick-walled systemic arteries, especially in branching regions. The inflammation is composed of four main stages characterized by

1. Foam cells

2. Fatty streaks

3. Intermediate lesions

4. Uncomplicated, then complicated, plaques

When atherosclerosis begins, LDL, which have crossed the endothelium, are oxidized in the intimal connective layer. These modified LDL induce an immune response. Attracted monocytes and T-lymphocytes migrate from the blood stream into the artery wall. Migrated monocytes multiply and are transformed into macrophages with scavenger receptors at the cytoplasmic membrane. Oxidized LDL (oxLDL) are bound to the scavenger receptors and then ingested by the macrophages. Ingestion of oxLDL is unsaturatable and leads to foam cell formation. OxLDL and proinflammatory cytokines induce expression of adhesion molecules, and cell chemotaxis. Secreted chemokines, such as macrophage chemoattractant protein (MCP), and growth factors stimulate the migration of smooth muscle cells from the media to the intima. SMCs can dedifferentiate, losing their contractile properties. Dysfunctional SMCs contribute to lipid accumulation and calcification in the atherosclerotic plaque. Agglomeration of foam cells, of T-lymphocytes, of SMCs, with extracellular matrix synthesis forms fatty streaks. The streaks gradually become larger, covered by a fibrous cap. The fibrous cap increases and a necrotic core can occur. Episodic fibrous cap ruptures initiate thrombus formations.

Mediators of immunity are involved at various stages of atherosclerosis. Ceramide is implicated in atherogenesis. Following uptake by the endothelial cells (EC), a part of the LDL-derived ceramide is converted into sphingosine, whereas another part accumulates inside the cells, with an increased rate of apoptosis [BOf]. Oxidative stress, a consequence of hypoxemia, is implicated in atherosclerosis. Statins are regulators of NO synthesis by ECs. They inhibit leukocyte transmigration [SAb]. Apolipoprotein E regulates the cholesterol metabolism. Its level decay increases atherosclerosis risk.

Various works were performed to discover which blood dynamic factors participate in atherogenesis. Strong correlation has been found between low wall shear stress (WSS) region and atherosclerotic plaque localization [CAf]. The atherosclerosis mainly occurs not only where WSS is low, but also where WSS strongly changes both in time and in space. Intimal thickening may correspond to a remodeling response. Secondarily, a lesion can develop, caused by disturbed transmural fluxes of cells and lipoproteins. The stresses applied by the blood flow on the vessel wall are involved in the pathogenesis as well as in complications, such as damage of the fibrous cap and cracking of the plaque. Investigations have been mostly performed in idealized geometry, assuming rigid walls due to atheroma-induced hardening and passive homogeneous plaques. In any case, accumulation of lipidic molecules is not only secondary to increased influx with changes in endothelial permeability, but is also the consequence of decreased wall efflux, especially by the microcirculation of the outer half of the vessel wall.

The stenosis induces ischemia and tissue infarction, most often by a distal flow blockage by emboli. Flow perturbations induced by a severe lumen narrowing can be detected by ultrasound techniques. The systolic jet through the stenosis is associated with recirculation zones. Flow separations can also be produced during the diastole. The blood stagnation or, at least, low-speed transport can more easily trigger clotting on the more or less damaged endothelium of the constricted segment.

### 2.6.2 Aneurism

The aneurism is the product of a multifactorial process that leads a gradual dilation of an arterial segment over years. The aneurism wall stretches and becomes thinner and weaker than normal artery walls. Consequently, an untreated aneurism can rupture. Two main kinds of aneurisms exist, fusiform wall dilation and saccular bulging of the artery wall. Fusiform aneurisms are often complications of atheroma. Saccular aneurisms can occur after an infection or a traumatizing of the wall, whereas congenital aneurisms are located at the branching sites of the brain arterial network.

A mechanically induced degeneration of the wall internal elastic lamina has thus been proposed as the initiating cause of congenital aneurisms with genetic predisposition. Imbalance between matrix metalloproteinases (MMP) and their inhibitors, the tissue inhibitors of metalloproteinases (TIMP), are involved in formation of abdominal aortic aneurisms (AAA). AAAs are characterized by chronic inflammation, destructive remodeling of the extracellular matrix, and increased activity of MMPs [IRa]. In AAAs, the volume fraction of elastin and of SMCs decreases and the combined content of collagen and ground substance increases [HEa]. The percentage of chondroitin sulphate increases and that of heparan sulphate decreases [SOa].

In clinical practice, the rupture risk must be estimated. Stress distribution in the aneurism is investigated to find regions subjected to high stresses and to plan the treatment accordingly. The geometry of the aneurism can be reconstructed from patient 3-D images. Experiments and numerical simulations have been mostly carried out in AAAs and intracranial saccular aneurisms (ISA), focused on the stress field either in the lesion wall [RAb, DIa] or in the blood cavity [STa, BOb]. A saccular aneurism illustrates the help of numerical simulations for the choice of the treatment (Figure 2.10). The high-pressure zone in the neck gives an argument in favor of surgical clipping because the aneurism is superficial with easy surgical access. It is not possible, indeed, to protect the neck efficiently with coiling because coils in this location will always induce emboli. However, if the endovascular

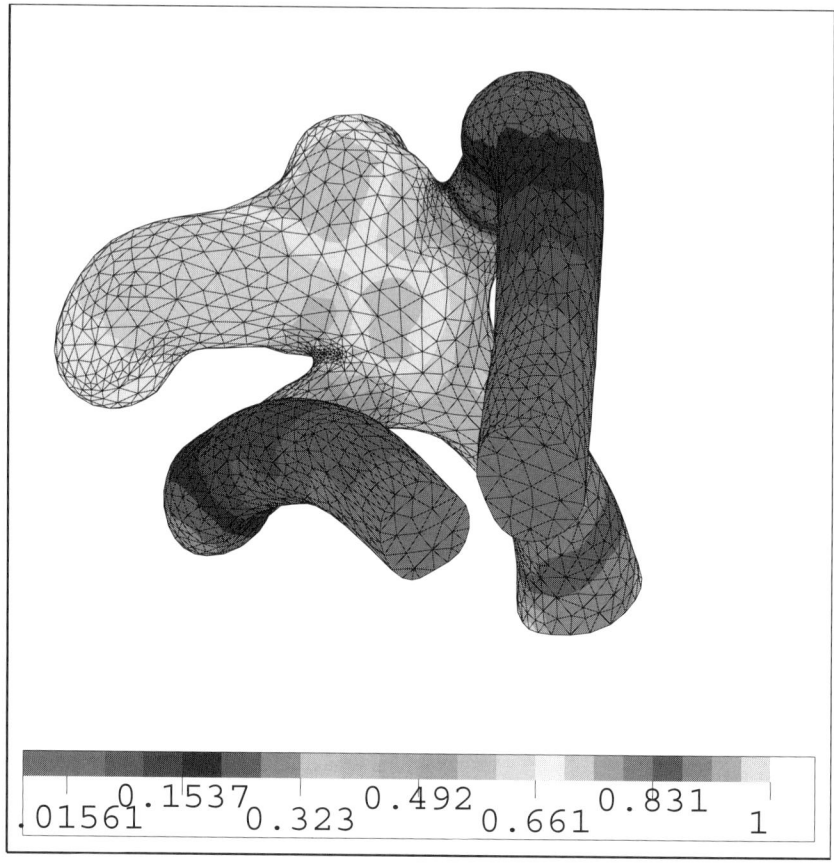

Figure 2.10. Pressure field in a terminal branching aneurism at peak flow.

treatment is chosen with respect to heavily invasive surgery, frequent angiography control must be done because of the high risk of recanalization. Model improvements are needed in order to take into account the aneurism wall responses to the stress field applied by the blood, at the tissue level as well as at the cell scale, both being associated with synthesis or degradation of multiple compounds.

## 2.7   Conclusion

Three-dimensional numerical modeling that completely describes the myocardium activity must couple models at various length scales in order to take into account the biochemical machinery that triggers and is responsible for heart contractions. Such a coupling indeed involves

1. Electrochemical wave propagation

2. Myocardium contraction–relaxation

3. Blood systolic ejection, which is generated by the contraction of the whole left ventricle tuned by the action potential.

Future investigations are then aimed at developing patient-specific tools, which combine not only medical images but also physiological signals to the heart model. The complete heart model takes into account the metabolic (perfusion), the electrochemical, and the mechanical activities. Such a computer tool can be used to solve inverse problems in order to estimate parameters and state variables from observations of the cardiac function (data assimilation). The fluid dynamics within the coronary network, which irrigates the myocardium, can be based on a hierarchical approach, taking into account both the large coronary arteries and the intramural vessels. The three-dimensional blood flow model in distensible right and left coronary arteries (proximal part of the heart perfusion network, located on the heart surface) can be coupled with the one-dimensional flow model for wave propagation, which can correspond to the first six generations of branches [SMa], and a poroelastic model of the small arteries and the microcirculation which cross the heart wall, using homogenization [CIa, HUe], the wall permeability depending on the wall deformation. This hierarchical flow enumerate is coupled with oxygen transport.

Mechanotransduction is another example of intricate biomechanical reactions interlinked to biomechanical phenomena. The continuum level remains used at the cell scale to estimate the stress distribution in the wall layers of the vasculature and the interactions among the large cell components, the nucleus, the cytosol, and the membrane, the small cell organelles being neglected. Such interactions can affect the local flow and, consequently, mass transport and mechanotransduction. The mechanotransduction is investigated to clarify the manifold stress-induced processes from sensing to processing, and to better define the niches of the vascular cells. The better the niche, the more efficient is cell and tissue engineering for tissue replacement. Genes involved in mechanotransduction, coding for ion channels, or responsive substances are targets for additional studies.

Nowadays, biomechanical models are beginning to incorporate involved microscale events. Biomechanics also contributes to the development of new diagnosis methods, of new measurement techniques, from signal acquisition to processing, of new surgical or medical implantable devices, and of new therapeutic strategies.

## 2.8   References

[AAa]  Aaron, B.B. and Gosline, J.M., Elastin as a random-network elastomer: a mechanical and optical analysis of single elastin fibers, *Biopolymers*, **20** (1981), 1247–1260.

[ALa]  Aliev, R.R. and Panfilov, A.V., A simple two-variable model of cardiac excitation, *Chaos Solitons Fractals*, **7** (1996), 293–301.

[AMa]  Ambrosi, D., Bussolino, F., Preziosi, L., A review of vasculogenesis models *Journal of Theoretical Medicine*, **6** (2005), 1–19.

[ANa]  Anand, M. and Rajagopal, K.R., A mathematical model to describe the change in the constitutive character of blood due to platelet activation, *C.R. Acad. Sci. Paris*, Mécanique, **330** (2002), 557–562.

[AUa]  Aumailley, M. and Gayraud, B., Structure and biological activity of the extracellular matrix, *J. Mol. Med.*, **76** (1998), 253–65.

[AZa]  Azhari, H., Weiss, J.L., Rogers, W.J., Siu, C.O., Zerhouni, E.A., and Shapiro, E.P., Noninvasive quantification of principal strains in normal canine hearts using tagged MRI images in 3-D, *Am. J. Physiol.*, **264** (1993), H205–216.

[BAa]  Bao, X., Lu, C., and Frangos, J.A., Mechanism of temporal gradients in shear-induced ERK1/2 activation and proliferation in endothelial cells, *Am. J. Physiol., Heart Circ. Physiol.*, **281** (2001), H22–29.

[BEa]  Beltrami, A.P., Barlucchi, L., Torella, D., Baker, M., Limana, F., Chimenti, S., Kasahara, H., Rota, M., Musso, E., Urbanek, K., Leri, A., Kajstura, J., Nadal-Ginard, B., and Anversa, P., Adult cardiac stem cells are multipotent and support myocardial regeneration, *Cell*, **114** (2003), 763–776.

[BEb]  Bestel, J., Clément, F., and Sorine, M., A biomechanical model of muscle contraction, In **Medical Image Computing and Computer-Assisted Intervention** (MICCAI'01), Lectures Notes in Computer Science, Niessen, W.J., and Viergever, M.A. eds, Springer-Verlag, New York (2001), 2208, pp. 1159–1161.

[BIa]  Binning, G., Quate, C.F., and Gerber, C., Atomic force microscope, *Phys. Rev. Lett.*, **56** (1986), 930–933.

[BOa]  Bogaert, J. and Rademakers, F.E., Regional nonuniformity of normal adult human left ventricle, *Am. J. Physiol., Heart Circ. Physiol.*, **280** (2001), H610–620.

[BOb]  Boissonnat, J.D., Chaine, R., Frey, P., Malandain, G., Salmon, S., Saltel, E., and Thiriet, M., From arteriographies to computational flow in saccular aneurisms: The INRIA experience, *Med. Image Anal.*, **9** (2005), 133–143.

[BOc] Bos, J.L., de Rooij, J., and Reedquist, K.A., Rap1 signalling: Adhering to new models. *Nat. Rev. Mol. Cell. Biol.*, **2** (2001), 369–77.

[BOd] Bourdarias, C., Gerbi, S., and Ohayon, J., A three dimensional finite element method for biological active soft tissue, *Math. Model. Numer. Anal.*, **37** (2003), 725–739.

[BOe] Bourgault, Y., Ethier, M., and LeBlanc, V.G., Simulation of electrophysiological waves with an unstructured finite element method, *Math. Model. Numer. Anal.*, **37** (2003), 649–661.

[BOf] Boyanovsky, B., Karakashian, A., King, K., Giltyay, N.V., and Nikolova-Karakashian, M., Ceramide-enriched low-density lipoproteins induce apoptosis in human microvascular endothelial cells, *J. Biol. Chem.*, **278** (2003), 26992–26999.

[BRa] Brookes, P.S., Yoon, Y., Robotham, J.L., Anders, MW., and Sheu, SS. Calcium, ATP, and ROS: A mitochondrial love-hate triangle, *Am. J. Physiol., Cell Physiol.*, **287** (2004), C817–C833.

[CAa] Caille, N., Thoumine, O., Tardy, Y., and Meister, J.J., Contribution of the nucleus to the mechanical properties of endothelial cells. *J Biomech.*, **35** (2002), 177–187.

[CAb] Caillerie, D., Mourad, A., and Raoult, A., Cell-to-muscle homogenization. Application to a constitutive law for the myocardium, *Math. Model. Numer. Anal.*, **37** (2003), 681–698.

[CAc] Cairns, C.B., Walther, J., Harken, A.H., and Banerjee, A., Mitochondrial oxidative phosphorylation thermodynamic efficiencies reflect physiological organ roles, *Am. J. Physiol., Regul. Integr. Comp. Physiol.*, **274** (1998), R1376–R1383.

[CAd] Canfield, T.R. and Dobrin, P.B., Static elastic properties of blood vessels, In: **Handbook of Bioengineering**, Skalak, R. and Chien, S. Eds., McGraw-Hill, New-York (1987), pp. 16.1–16.28.

[CAe] Carnegie, G.K. and Scott, J.D., A-kinase anchoring proteins and neuronal signaling mechanisms, *Genes Dev.*, **17** (2003), 1557–1568.

[CAf] Caro, C.G., Fitzgerald, J.M., and Schroter, R.C., Atherosclerosis and arterial wall shear: observations, correlation and proposal of a shear dependent mass transfer mechanism for atherogenesis, *Proc. Royal Soc. London B*, **177** (1971), 109–159.

[CHa] Chang, L., and Karin, M., Mammalian MAP kinase signalling cascades, *Nature*, **410** (2001), 37–40.

[CHb] Chapelle, D., Clément, F., Génot, F., Le Tallec, P., Sorine, M., and Urquiza, J., A physiologically-based model for the active cardiac muscle, Lectures Notes in Computer Science, Katila, T., Magnin, I.E., Clarysse, P., Montagnat, J., and Nenonen, J. Eds, Springer-Verlag New York (2001), pp. 2230,

[CHc] Chien, S., Shear dependence of effective cell volume as a determinant of blood viscosity, *Science*, **168** (1970), 977–978.

[CIa] Cimrman, R. and Rohan, E., Modelling heart tissue using a composite muscle model with blood perfusion, in: **Computational Fluid and Solid Mechanics**, Bathe, K.J., Ed., Elsevier (2003).

[CIb] Civelekoglu, G. and Edelstein-Keshet, L., Modelling the dynamics of F-actin in the cell, *Bull. Math. Biol.*, **56** (1994), 587–616.

[CLa] Clark, E.A. and Brugge, J.S., Integrins and signal transduction pathways: The road taken, *Science*, **268** (1995), 233–239.

[CLb] Clark, R.A., Wikner, N.E., Doherty, D.E., and Norris, D.A., Cryptic chemotactic activity of fibronectin for human monocytes resides in the 120-kDa fibroblastic cell-binding fragment, *J. Biol. Chem.*, **263** (1988), 12115–12123.

[COa] Colli-Franzone, P., Guerri, L., and Tentoni, S., Mathematical modeling of the excitation process in myocardial tissue: Influence of fiber rotation on the wavefront propagation and potential field, *Math. Biosci.*, **101** (1990), 155–235.

[DAa] Davis, M.J., Kuo, L., Chilian, W.M., and Muller, J.M., Isolated, perfused microvessels, In **Clinically Applied Microcirculation Research**, Barker, J.H., Anderson, G.L. and Menger, M.D. eds, CRC, Boca Raton, 1995, pp. 435–456.

[DEa] de Bold AJ., Atrial natriuretic factor: A hormone produced by the heart. *Science*, **230** (1985), 767–770.

[DEb] de Brabander, M.J., Le cytosquelette et la vie cellulaire, *La Recherche*, **145** (1993), 810–820.

[DEc] Delhaas, T., Arts, T., Prinzen, F.W., and Reneman, R.S., Relation between regional electrical activation time and subepicardial fiber strain in the canine left ventricle, *Pflugers Arch.*, **423** (1993), 78–87.

[DEd] Dembo, M., The mechanics of motility in dissociated cytoplasm. *Biophys. J.*, **50** (1986), 1165–1183.

[DEe] Dewey, C.F., Bussolari, S.R., Gimbrone, M.A., and Davies, P.F., The dynamic response of vascular endothelial cells to fluid shear stress, *J. Biomech. Eng.*, **103** (1981), 177–185.

[DIa] Di Martino, E.S., Guadagni, G., Fumero, A., Ballerini, G., Spirito, R., Biglioli, P., and Redaelli, A., Fluid-structure interaction within realistic three-dimensional models of the aneurysmatic aorta as a guidance to assess the risk of rupture of the aneurysm, *Med. Eng. Phys.*, **23** (2001), 647–655.

[DRa] Draney, M.T., Herfkens, R.J., Hughes, T.J., Pelc, N.J., Wedding, K.L., Zarins, C.K., and Taylor, C.A., Quantification of vessel wall cyclic strain using cine phase contrast magnetic resonance imaging, *Ann. Biomed. Eng.*, **30** (2002), 1033–1045.

[DUa] Duling, B.R., Gore, R.W., Dacey, R. G., and Damon, D.N., Methods for isolation, cannulation, and in vitro study of single microvessels, *Am. J. Physiol.*, *Heart Circ. Physiol.*, **241** (1981), H108–H116.

[DUb] Durrer, D., van Dam, R.T., Freud, G.E., Janse, M.J., Meijler, F.L., and Arzbaecher, R.C., Total excitation of the isolated human heart, *Circulation*, **41** (1970), 899–912.

[EVa] Evans, E.A., New membrane concept applied to the analysis of fluid shear- and micropipette-deformed red blood cells, *Biophys. J.*, **13** (1973), 941–54.

[FAa] Fahraeus, R. and Lindqvist, T., The viscosity of the blood in narrow capillary tubes. *Am. J. Physiol.*, **96** (1931), 562–568.

[FAb] Faris, O.P., Evans, F.J., Ennis, D.B., Helm, P.A., Taylor, J.L., Chesnick, A.S., Guttman, M.A., Ozturk, C., and McVeigh, E.R., Novel technique for cardiac electromechanical mapping with magnetic resonance imaging tagging and an epicardial electrode sock, *Ann. Biomed. Eng.*, **31** (2003), 430–440.

[FIa] FitzHugh, R., Impulses and physiological states in theoretical models of nerve membrane, *Biophys. J.*, **1** (1961), 445–466.

[FOa] Fogel, M.A., Weinberg, P.M., Hubbard, A., and Haselgrove, J., Diastolic biomechanics in normal infants utilizing MRI tissue tagging, *Circulation*, **102** (2000), 218–224.

[FOb] Fogelson, A.L. and Guy, R.D., Platelet-wall interactions in continuum models of platelet thrombosis: Formulation and numerical solution, *Math. Med. Biol.*, **21** (2004), 293–334.

[FUa] Fung, Y.C., **Biomechanics**, Springer-Verlag, New York (1981).

[FUb] Fung, Y.C., **Biomechanics: Mechanical Properties of Living Tissues** Springer-Verlag, New York (1993).

[GEa] Geselowitz, D.B. and Miller, W.T., A bidomain model for anisotropic cardiac muscle, *Ann. Biomed. Eng.*, **11** (1983), 191–206.

[GIa] Giancotti, F.G. and Ruoslahti, E., Integrin signaling, *Science*, **285** (1999), 1028–1032.

[GLa] Gleason, R.L. and Humphrey, J.D., A mixture model of arterial growth and remodeling in hypertension: Altered muscle tone and tissue turnover, *J. Vasc. Res.*, **41** (2004), 352–363.

[GLb] Gleason, R.L. and Humphrey, J.D., Effects of a sustained extension on arterial growth and remodeling: A theoretical study, *J. Biomech.*, **38** (2005), 1255–1261.

[GLc] Glover, D.M., Gonzalez, C., and Raff, J.W., The centrosome, *Sci. Amer.*, **268** (1993), 62–68.

[GOa] Gohring, W., Sasaki, T., Heldin, C.H., and Timpl, R., Mapping of the binding of platelet-derived growth factor to distinct domains of the basement membrane proteins BM-40 and perlecan and distinction from the BM-40 collagen-binding epitope, *Eur. J. Biochem.*, **255** (1998), 60–66.

[HAa] Hayashi, K., Experimental approaches on measuring the mechanical properties and constitutive laws of arterial walls, *J. Biomech. Eng.*, **115** (1993), 481–488.

[HEa] He, C.M. and Roach, M.R., The composition and mechanical properties of abdominal aortic aneurysms, *J. Vasc. Surg.*, **20** (1994), 6–13.

[HEb] Heeschen, C., Lehmann, R., Honold, J., Assmus, B., Aicher, A., Walter, D.H., Martin, H., Zeiher, A.M., and Dimmeler, S., Profoundly reduced neovascularization capacity of bone marrow mononuclear cells derived from patients with chronic ischemic heart disease, *Circulation*, **109** (2004), 1615–1622.

[HEc] Heethaar, R.M, Pao, Y.C., and Ritman, E.L., Computer aspects of three-dimensional finite element analysis of stresses and strains in the intact heart, *Comput. Biomed. Res.*, **10** (1977), 271–285.

[HEd] Hénon, S., Lenormand, G., Richert, A., and Gallet, F., A new determination of the shear modulus of the human erythrocyte membrane using optical tweezers, *Biophys. J.*, **76** (1999), 1145–1151.

[HEe] Heusch, G., Post, H., Michel, M.C., Kelm, M., and Schulz, R., Endogenous nitric oxide and myocardial adaptation to ischemia, *Circ. Res.*, **87** (2000), 146–152.

[HIa] Hill, J.M., Zalos, G., Halcox, J.P.J., Schenke, W.H., Waclawiw, M.A., Quyyumi, A.A., and Finkel, T., Circulating endothelial progenitor cells, vascular function, and cardiovascular risk, *N. Engl. J. Med.* **348** (2003), 593–600.

[HOa] Hochmuth, R.M., Micropipette aspiration of living cells, *J. Biomech.*, **33** (2000), 15–22.

[HOb] Holzapfel, G.A. and Gasser, T.C., A new constitutive framework for arterial wall mechanics and a comparative study of material models, *J. Elasticity*, **61** (2000), 1–48.

[HSa] Hsu, E.W. and Henriquez, C.S., Myocardial fiber orientation mapping using reduced encoding diffusion tensor imaging, *J. Cardiovasc. Magn. Reson.*, **3** (2001), 339–347.

[HUa] Humphrey, J.D. and Rajagopal, K.R., A constrained mixture model for growth and remodeling of soft tissues, *Math. Model. Meth. Appl. Sci.*, **12** (2002), 407–430.

[HUb] Humphrey, J.D., Continuum biomechanics of soft biological tissues, *Proc. R. Soc. Lond. A*, **459** (2003), 3–46.

[HUc] Hunter, P.J., McCulloch, A.D., and ter Keurs, H.E., Modelling the mechanical properties of cardiac muscle, *Prog. Biophys. Mol. Biol.*, **69** (1998), 289–331.

[HUd] Huxley, A.F., Muscle structure and theories of contraction, *Prog. Biophys. Chem.*, **7** (1957), 255–318.

[HUe] Huyghe, J.M., Arts, T., and van Campen, D.H., Porous medium finite element model of the beating left ventricle. *Am. J. Physiol.*, **262** (1992), H1256-H1267.

[HYa] Hynes, R.O., Fibronectins, *Sci. Am.*, **254** (1986), 42–51.

[INa] Ingber, D.E., Madri, J.A., and Jamieson, J.D., Role of basal lamina in neoplastic disorganization of tissue architecture, *Proc. Natl. Acad. Sci. USA*, **78** (1981), 3901–3905.

[IRa] Irizarry, E., Newman, K.M., Gandhi, R.H., Nackman, G.B., Halpern, V., Wishner, S., Scholes, J.V., and Tilson, M.D., Demonstration of interstitial collagenase in abdominal aortic aneurysm disease, *J. Surg. Res.*, **54** (1993), 571–574.

[ITa] Ito, H., Hirata, Y., Hiroe, M., Tsujino, M., Adachi, S., Takamoto, T., Nitta, M., Taniguchi, K., and Marumo, F., Endothelin-1 induces hypertrophy with enhanced expression of muscle-specific genes in cultured neonatal rat cardiomyocytes, *Circ. Res.*, **69** (1991), 209–215.

[JEa] Jennings, L.M., Butterfield, M., Booth, C., Watterson K.G., and Fisher, J., The pulmonary bioprosthetic heart valve: its unsuitability for use as an aortic valve replacement. *J. Heart Valve Dis.*, **11** (2002), 668–678.

[JOa] Johnson, P.C., The myogenic response, In **Handbook of Physiology. The Cardiovascular System. Vascular Smooth Muscle**, sect. 2, vol. II, chapt. 15, Am. Physiol. Soc., Bethesda, MD, (1981), pp. 409–442.

[JOb] Joly, M., Lacombe, C., and Quemada, D., Application of the transient flow rheology to the study of abnormal human bloods, *Biorheology*, **18** (1981), 445–452.

[KAa] Katusic, Z.S., Superoxide anion and endothelial regulation of arterial tone, *Free Radic. Biol. Med.*, **20** (1996), 443–448.

[KEa] Keller, E.F. and Segel, L.A., Model for chemotaxis, *J. Theor. Biol.*, **30** (1971), 225–234.

[KEb] Kermorgant, S., Zicha, D., and Parker, P.J., PKC controls HGF-dependent c-Met traffic, signalling and cell migration, *EMBO J.*, **23** (2004), 3721–3734.

[KIa] Kirkham, M., Fujita, A., Chadda, R., Nixon, S.J., Kurzchalia, T.V., Sharma, D.K., Pagano, R.E., Hancock, J.F., Mayor, S., Parton, R.G., and Richards, A.A., Ultrastructural identification of uncoated caveolin-independent early endocytic vehicles, *J. Cell Biol.*, **168** (2005), 465–476.

[KUa] Kuharsky, A.L. and Fogelson, A.L., Surface-mediated control of blood coagulation: the role of binding site densities and platelet deposition, *Biophys. J.*, **80** (2001), 1050–1074.

[LAa] Laugwitz, K.L., Moretti, A., Lam, J., Gruber, P., Chen, Y., Woodard, S., Lin, L.Z., Cai, C.L., Lu, M.M., Reth, M., Platoshyn, O., Yuan, J.X., Evans, S., and Chien, K.R., Postnatal isl1+ cardioblasts enter fully differentiated cardiomyocyte lineages, *Nature*, **433** (2005), 647–653.

[LAb] Laurent, V., Planus, E., Isabey, D., Lacombe, C., and Bucherer, C., Propriétés mécaniques de cellules endothéliales évaluées par micromanipulation cellulaire et magnétocytométrie, **MecanoTransduction 2000**, Ribreau, C. et al. Eds., Tec&Doc, Paris (2000), pp. 373–380.

[LEa] Le, P.U. and Nabi, I.R., Distinct caveolae-mediated endocytic pathways target the Golgi apparatus and the endoplasmic reticulum, *J. Cell Sci.*, **116** (2003), 1059–1071.

[LEb] le Noble, F., Moyon, D., Pardanaud, L., Yuan, L., Djonov, V., Matthijsen, R., Breant, C., Fleury, V., and Eichmann, A., Flow regulates arterial-venous differentiation in the chick embryo yolk sac. *Development*, **131** (2004), 361–375.

[MAa] Manoussaki, D., A mechanochemical model of angiogenesis and vasculogenesis, *Math. Model. Numer. Anal.*, **37** (2003), 581–599.

[MAb] Marrocco, A., Numerical simulation of chemotactic bacteria aggregation via mixed finite elements, *Math. Model. Numer. Anal.*, **37** (2003), 617–630.

[MEa] Meili, R., Sasaki, A.T., and Firtel, R.A., Rho Rocks PTEN, *Nature Cell Biology*, **7** (2005), 334–335.

[MEb] Meininger, G.A. and Davis, M.J., Cellular mechanisms involved in the vascular myogenic response, *Am. J. Physiol., Heart Circ. Physiol.*, **263** (1992), H647–H659.

[MIa] Michel, C.C. and Curry, F.E., Microvascular permeability, *Physiol. Rev.*, **79** (1999), 703–761.

[MOa] Morawietz, H., Talanow, R., Szibor, M., Rueckschloss, U., Schubert, A., Bartling, B., Darmer, D., and Holtz, J., Regulation of the endothelin system by shear stress in human endothelial cells. *J. Physiol.*, **525** (2000), 761–770.

[MUa] Murray, C.D., The physiological principle of minimum work. I. The vascular system and the cost of blood volume, *Proc. Nat. Acad. Sci. (USA)*, **12** (1926), 207–214.

[MUb] Murray, J., Osher, G., and Harris, A., A mechanical model for mesenchymal morphogenesis, *J. Math. Biol.*, **17** (1983), 125–129.

[MUc] Murray, J.D., **Mathematical Biology**, Springer-Verlag, New York (2002).

[NAa] Nagumo, J., Arimoto, S., and Yoshizawa, S., An active pulse transmission line simulating nerve axons, *Proc. IRE*, **50** (1962), 2061–2070.

[NOa] Nobes, C.D. and Hall, A., Rho, rac, and cdc42 GTPases regulate the assembly of multimolecular focal complexes associated with actin stress fibers, lamellipodia, and filopodia. *Cell*, **81** (1995), 53–62.

[NOb] Nollert, M.U., Eskin, S.G., and McIntire, L.V., Shear stress increases inositol trisphosphate levels in human endothelial cells, *Biochem. Biophys. Res. Commun.*, **170** (1990), 281–287.

[NOc] Nollert, M.U., Diamond, S.L., and McIntire, L.V., Hydrodynamic shear stress and mass transport modulation of endothelial cell metabolism, *Biotechnol. Bioeng.*, **38** (1991), 588–602.

[ODa] Oddou, C. and Ohayon, J., Mécanique de la structure cardiaque, In **Biomécanique des fluides et des tissus**, Jaffrin, M.Y., and Goubel, F. Eds., Masson, Paris (1998), pp. 247–292.

[OZa] Ozdamar, B., Bose, R., Barrios-Rodiles, M., Wang, H.R., Zhang, Y., and Wrana, J.L., Regulation of the polarity protein Par6 by TGF-beta receptors controls epithelial cell plasticity. *Science*, **307** (2005), 1603–1609.

[PAa] Pavalko, F.M. and Otey, C.A., Role of adhesion molecule cytoplasmic domains in mediating interactions with the cytoskeleton, *Proc. Soc. Exp. Biol. Med.*, **205** (1994), 282–293.

[PEa] Peskin, C.S., Fiber architecture of the left ventricular wall: An asymptotic analysis, *Commun. Pure Appl. Math.*, **42** (1989), 79–113.

[PEb]  Peskin, C.S. and McQueen, D.M., Mechanical equilibrium determines the fractal fiber architecture of aortic heart valve leaflets, *Am. J. Physiol.*, **266** (1994), H319-H328.

[PIa]  Pieske, B., Beyermann, B., Breu, V., Loffler, B.M., Schlotthauer, K., Maier, L.S., Schmidt-Schweda, S., Just, H., and Hasenfuss, G., Functional effects of endothelin and regulation of endothelin receptors in isolated human nonfailing and failing myocardium, *Circulation*, **99** (1999), 1802–1809.

[PIb]  Pioletti, D.P. and Rakotomanana, L.R., Non-linear viscoelastic laws for soft biological tissues, *Eur. J. Mech. A/Solids*, **19** (2000), 749–759.

[POa]  Pohl, U. and Busse, R., Hypoxia stimulates release of endothelium-derived relaxant factor, *Am. J. Physiol.*, **256** (1989), H1595–H1600.

[POb]  Poon, C.S. and Merrill, C.K., Decrease of cardiac chaos in congestive heart failure, *Nature*, **389** (1997), 492–495.

[RAa]  Rabiet, M.J., Plantier, J.L., Rival, Y., Genoux, Y., Lampugnani, M.G., and Dejana, E., Thrombin-induced increase in endothelial permeability is associated with changes in cell-to-cell junction organization, *Arterioscler. Thromb. Vasc. Biol.*, **16** (1996), 488–496.

[RAb]  Raghavan, M.L., Vorp, D.A., Federle, M.P., Makaroun, M.S., and Webster, M.W., Wall stress distribution on three-dimensionally reconstructed models of human abdominal aortic aneurysm, *J. Vascular Surgery*, **31** (2000), 760–769.

[RAc]  Rajagopal, K.R. and Srinivasa, A.R., A thermodynamic framework for rate-type fluid models, *J. Non-Newtonian Fluid Mech.*, **88** (2000), 207–227.

[RAd]  Raval, A.P., Dave, K.R., Prado, R., Katz, L.M., Busto, R., Sick, T.J., Ginsberg, M.D., Mochly-Rosen, D., and Perez-Pinzon, M.A., Protein kinase C delta cleavage initiates an aberrant signal transduction pathway after cardiac arrest and oxygen glucose deprivation, *J. Cereb. Blood Flow Metab.*, **25** (2005), 730–741.

[RAe]  Rayment, I., Holden, H.M., Whittaker, M., Yohn, C.B., Lorenz, M., Holmes, K.C., and Milligan, R.A., Structure of the actin-myosin complex and its implications for muscle contraction, *Science*, **261** (1993), 58–65.

[ROa]  Robert, L., Elasticité des tissus et vieillissement, *Pour la science*, **201** (1994), 56–62.

[ROb]  Robinson, T.F., Factor, S.M., and Sonnenblick, E.H., The heart as a suction pump, *Sci. Amer.*, **6** (1986), 62–69.

[SAa]  Sainte-Marie, J., Chapelle, D., and Sorine, M., Data assimilation for an electro-mechanical model of the myocardium, In **Second M.I.T.**

**Conference on Computational Fluid and Solid Mechanics**, Bathe, K.J. Ed. (2003), pp. 1801–1804.

[SAb] Sata, M., Saiura, A., Kunisato, A., Tojo, A., Okada, S., Tokuhisa, T., Hirai, H., Makuuchi, M., Hirata, Y., and Nagai, R., Hematopoietic stem cells differentiate into vascular cells that participate in the pathogenesis of atherosclerosis, *Nat. Med.*, **8** (2002), 403–409.

[SAc] Sato, M., Theret, D.P., Wheeler, L.T., Ohshima, N., and Nerem, R.M., Application of the micropipette technique to the measurement of cultured porcine aortic endothelial cells viscoelastic properties, *Trans. ASME J. Biomech. Eng.*, **112** (1990), 263–268.

[SCa] Schaller, M.D. and Parsons, J.T., pp125FAK-dependent tyrosine phosphorylation of paxillin creates a high-affinity binding site for Crk, *Mol. Cell. Biol.*, **15** (1995), 2635–2645.

[SCb] Schenkel, A.R., Mamdouh, Z., and Muller W.A., Locomotion of monocytes on endothelium is a critical step during extravasation, *Nature Immunology*, **5** (2004), 393–400.

[SCc] Schmidt, F.G., Hinner, B., Sackmann, E., and Tang, J.X., Viscoelastic properties of semiflexible filamentous bacteriophage fd, *Phys. Rev. E Stat. Phys. Plasmas Fluids Relat. Interdiscip. Topics*, **62** (2000), 5509–5517.

[SEa] Sermesant, M., Forest, C., Pennec, X., Delingette, H., and Ayache, N., Deformable biomechanical models: application to 4D cardiac image analysis, *Med. Image. Anal.*, **7** (2003), 475–488.

[SHa] Sheng, M., McFadden, G., and Greenberg, M.E., Membrane depolarization and calcium induce c-fos transcription via phosphorylation of transcription factor CREB, *Neuron*, **4** (1990), 571–82.

[SIa] Sipido, K.R., Callewaert, G., and Carmeliet, E., Inhibition and rapid recovery of Ca2+ current during Ca2+ release from sarcoplasmic reticulum in guinea pig ventricular myocytes, *Circ. Res.*, **76** (1995), 102–109.

[SMa] Smith, N.P., Pullan, A.J., and Hunter, P.J., An anatomically based model of transient coronary blood flow in the heart, *SIAM J. Appl. Math.*, **62** (2002), 990–1018.

[SMb] Smyth, S.S., Joneckis, C.C., and Parise L.V., Regulation of vascular integrins, *Blood*, **81** (1993), 2827–2843.

[SOa] Sobolewski, K., Wolanska, M., Bankowski, E., Gacko, M., and Glowinski, S., Collagen, elastin and glycosaminoglycans in aortic aneurysms, *Acta Biochim. Pol.*, **42** (1995), 301–307.

[STa] Steinman, D.A., Milner, J.S., Norley, C.J., Lownie, S.P., and Holdsworth, D.W., Image-based computational simulation of flow dynamics in a giant intracranial aneurysm, *Am. J. Neuroradiol.*, **24** (2003), 559–566.

[STb] Stergiopulos, N., Tardy, Y., and Meister, J.J., Nonlinear separation of forward and backward running waves in elastic conduits, *J. Biomech.*, **26** (1993), 201–209.

[STc] Stergiopulos, N., Westerhof, B.E., and Westerhof, N., Total arterial inertance as the fourth element of the windkessel model *Am. J. Physiol., Heart Circ. Physiol.*, **276** (1999), H81–H88.

[STd] Stradins, P., Lacis, R., Ozolanta, I., Purina, B., Ose, V., Feldmane, L., and Kasyanov, V., Comparison of biomechanical and structural properties between human aortic and pulmonary valve, *Eur. J. Cardiothorac. Surg.*, **26** (2004), 634–639.

[SUa] Surks, H.K., Mochizuki, N., Kasai, Y., Georgescu, S.P., Tang, K.M., Ito, M., Lincoln, T.M., and Mendelsohn, M.E., Regulation of myosin phosphatase by a specific interaction with cGMP-dependent protein kinase I$\alpha$, *Science*, **286** (1999), 1583–1587.

[THa] Theret, D.P., Levesque, M.J., Sato, M., Nerem, R.M., and Wheeler, L.T., The application of a homogeneous half-space model in the analysis of endothelial cell micropipette measurements, *Trans. ASME J. Biomech. Eng.*, **110** (1988), 190–199.

[THb] Thiriet, M., Issa, R., and Graham, J.M.R., A pulsatile developing flow in a bend, *J. Phys. III* (1992), 995–1013.

[THc] Thiriet, M., Pares, C., Saltel, E., and Hecht, F., Numerical model of steady flow in a model of the aortic bifurcation, *ASME J. Biomech. Eng.*, 114 (1992), 40–49.

[THd] Thiriet, M., Martin-Borret G., and Hecht, F., Ecoulement rhéofluidifiant dans un coude et une bifurcation plane symétrique. Application à l'écoulement sanguin dans la grande circulation, *J. Phys. III*, **6** (1996), 529–542.

[TUa] Tunggal, J.A., Helfrich, I., Schmitz, A., Schwarz, H., Ganzel, D., Fromm, M., Kemler, R., Krieg, T., and Niessen, C.M., E-cadherin is essential for in vivo epidermal barrier function by regulating tight junctions, *EMBO J.*, **24** (2005), 1146–1156.

[TUb] Turner, C.E., Paxillin and focal adhesion signalling, *Nat. Cell Biol.*, **2** (2000), E231–236.

[VAa] van Nieuw Amerongen G.P. and van Hinsbergh, V.W.M., Cytoskeletal effects of Rho-like small guanine nucleotide-binding proteins in the vascular system, *Arterioscler. Thromb. Vasc. Biol.*, **21** (2001), 300–311.

[VAb] van Royen, N., Piek, J.J., Buschmann, I., Hoefer, I., Voskuil, M., and Schaper, W., Stimulation of arteriogenesis: a new concept for the treatment of arterial occlusive disease, *Cardiovasc. Res.*, **49** (2001), 543–553.

[VAc] Vanhoutte, P.M., Endothelial dysfunction and atherosclerosis, *Eur. Heart J.*, **18** (1997), E19–29.

[VEa] Verdier, C., Rheological properties of living materials : from cells to tissues, *J. Theor. Med.*, **5** (2003), 67–91.

[VEb] Veronda, D.R. and Westmann, R.A., Mechanical characterization of skin. Finite deformation, *J. Biomech.*, **3** (1970), 114–124.

[VIa] Viidik, A., Properties of tendons and ligaments, In **Handbook of Bioengineering**, Skalak, R., and Chien, S. Eds., McGraw-Hill, New York (1987), 6.1–6.19.

[VIb] Villarreal, F.J., Lew, W.Y., Waldman, L.K., and Covell, J.W., Transmural myocardial deformation in the ischemic canine left ventricle, *Circ. Res.*, **68** (1991), 368–381.

[WAa] Wagner, D.D., The Weibel-Palade body: The storage granule for von Willebrand factor and P-selectin, *Thromb. Haemost.* **70** (1993), 105–110.

[WAb] Wang, N., Butler, J.P., and Ingber, D.E., Mechanotransduction across the cell surface and through the cytoskeleton, *Science*, **260** (1993), 1124–1127.

[WEa] Weinbaum, S. and Curry, F.E., Modelling the structural pathways for transcapillary exchange *Symp. Soc. Exp. Biol.*, **49** (1995), 323–345.

[WEb] Westerhof, N., Bosman, F., De Vries, C.J., and Noordergraaf, A., Analog studies of the human systemic arterial tree, *J. Biomechanics*, **2** (1969), 121–143.

[WUa] Wu, J.Z. and Herzog, W., Modelling concentric contraction of muscle using an improved cross-bridge model, *J. Biomech.*, **32** (1999), 837–848.

[YAa] Yang, M., Taber, L.A., and Clark, E.B., A nonliner poroelastic model for the trabecular embryonic heart, *J. Biomech. Eng.*, **116** (1994), 213–223.

[YAb] Yap, A.S., Brieher, W.M., and Gumbiner, B.M., Molecular and functional analysis of cadherin-based adherens junctions, *Annu. Rev. Cell. Dev. Biol.*, **13** (1997), 119–146.

[YEa] Yeung, A. and Evans, E., Cortical shell-liquid core model for passive flow of liquid-like spherical cells into micropipets, *Biophys. J.*, **56** (1989), 139–149.

[YUa] Yurchenco, P.D. and O'Rear, J.J., Basal lamina assembly, *Curr. Opin. Cell Biol.*, **6** (1994), 674–681.

[ZAa] Zahalak, G.I., A distribution-moment approximation for kinetic theories of muscular contraction, *Math. Biosci.*, **114** (1981), 55–89.

[ZAb] Zamir, E. and Geiger, B. Components of cell-matrix adhesions, *J. Cell Sci.*, **114** (2001), 3577–3579.

[ZIa] Zimmermann, S., and Moelling, K., Phosphorylation and regulation of Raf by Akt (protein kinase B). *Science*, **286** (1999), 1741–1744.

[ZUa] Zühlke, R.D., Pitt, G.S., Deisseroth, K., Tsien, R.W., and Reuter, H., Calmodulin supports both inactivation and facilitation of L-type calcium channels, *Nature*, **399** (1999), 159–162.

[ZUb] Zulliger, M.A., Rachev, A., and Stergiopulos, N., A constitutive formulation of arterial mechanics including vascular smooth muscle tone, *Am. J. Physiol. Heart Circ. Physiol.*, **287** (2004), H1335–H1343.

# 3

# Theoretical Modeling of Enlarging Intracranial Aneurysms

S. BAEK, K. R. RAJAGOPAL, AND J. D. HUMPHREY

*Texas A&M University*
*Departments of Biomedical and Mechanical Engineering*
*College Station, USA*

ABSTRACT.  Rupture of intracranial aneurysms is the leading cause of spontaneous subarachnoid hemorrhage, which results in significant morbidity and mortality. The mechanisms by which intracranial aneurysms develop, enlarge, and rupture are unknown, and it remains difficult to collect the longitudinal patient-based information needed to improve our understanding. We suggest, therefore, that mathematical models hold considerable promise by allowing us to propose and test competing hypotheses on potential mechanisms of aneurysmal enlargement and to compare predicted outcomes with limited clinical information; in this way, we may begin to narrow the possible mechanisms and thereby focus experimental studies. Toward this end, we develop a constrained mixture model for evolving thin-walled, saccular, and fusiform aneurysms and illustrate its efficacy via computer simulations of lesions having idealized geometries. We also present a method to estimate linearized material properties over the cardiac cycle, which can be exploited when solving coupled fluid–solid interactions in a lesion.

## 3.1  Introduction

Intracranial aneurysms are focal dilatations of the arterial wall that usually
occur in or near the circle of Willis, the primary network of vessels that
supplies blood to the brain [HUb]. In general, these aneurysms have one
of two forms: fusiform lesions, which are elongated dilations of an artery,
and saccular lesions, which are local saclike out-pouchings. Despite significant accomplishments in molecular and cell biology as well as clinical
advances, intracranial aneurysms remain an enigma: how do they begin,
how do they enlarge, and how do they rupture? Rupture of intracranial
aneurysms is the leading cause of spontaneous subarachnoid hemorrhage
(SAH), which results in high morbidity and mortality rates. Although it
has been long thought that material instabilities are responsible for the
enlargement of aneurysms, recent nonlinear analyses cast doubt that such
instabilities play any role in the natural history (e.g. [KYa]). Rather, recent histopathological data and modeling suggest that aneurysms enlarge
due to growth and remodeling of collagen, the primary load-bearing constituent within the wall [HUa]. Note, therefore, that the natural history
of intracranial aneurysms consists of at least three phases: pathogenesis,
enlargement, and rupture (Figure 3.1). Albeit not well understood, some
initial insult to the cerebral artery causes a small out-pouching or dilation
of the arterial wall. We suggest that a stress-mediated process of growth

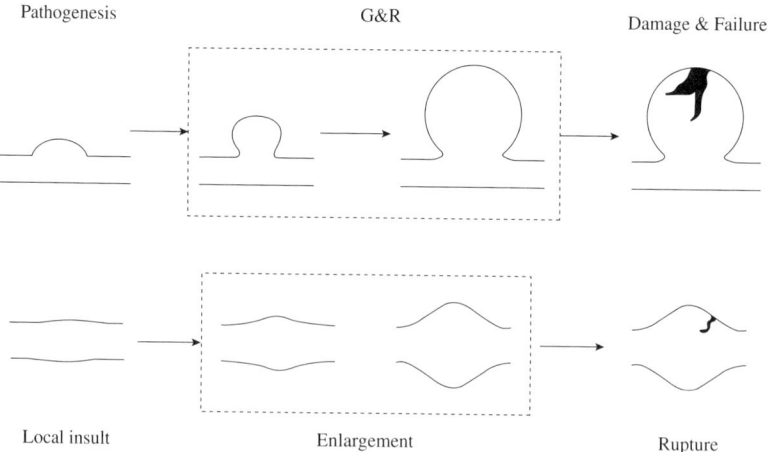

Figure 3.1. Schematic view of the natural history of intracranial saccular
(top) and fusiform (bottom) aneurysms.

and remodeling (G&R) is responsible for the subsequent enlargement and possible stabilization of the lesions.

In the early 1980s, Skalak [SKa] first attempted to model growth in soft tissue within the context of finite strain elasticity. His seminal work has been extended by Rodriguez et al. [ROa] and others (e.g. Taber [TAa], Rachev et al. [RAa]), and has served as the primary models of arterial G&R. Briefly, they suggested that shape changes of an unloaded tissue during growth can be decomposed into two fictitious deformations: first, independent growth of stress-free elements of tissue, which need not result in compatible elements, and second, an elastic deformation. The volumetric growth model of Skalak, Rodriguez et al., and others provides a mathematical method to model certain consequences of growth, but it does not model processes by which G&R occur. Recently, Humphrey and Rajagopal [HUc] presented an approach that is conceptually different, one that is based on a fundamental process by which growth and remodeling occur: the continual production and removal of constituents in potentially different stressed configurations. Because the kinetics of turnover and the way each constituent is deposited can differ markedly, they employ ideas from the theory of mixtures to account for the separate contributions of each constituent. Furthermore, because of inherent difficulties in prescribing traction and other boundary conditions in the theory of mixtures, they suggested a constrained mixture model wherein all solid constituents are assumed to have the same motion as that of the mixture despite different natural (stress-free) configurations. Moreover, part of the focus was geometrical alterations due to removal and new production of solid structural constituents as the main mechanism for G&R of soft tissue (see [HUc] for more details).

In this work, we adapt the constrained mixture approach to study the enlargement of intracranial aneurysms and postulate a new potential mechanism of aneurysmal enlargement. In particular, because the medial layer of the aneurysmal wall is degraded during the early development of a lesion and the remaining wall consists primarily of thin layers of collagen, we formulate a constrained mixture model for intracranial aneurysms within the context of a membrane theory. For numerical simulations, we employ initially ellipsoidal and cylindrical membranes for saccular and fusiform aneurysms, respectively. Although soft tissues are dissipative and a proper resolution of any process requires an appropriate thermodynamic framework, at this stage we assume that the body is purely elastic and solve inflation problems with these ideal geometries using the principle of virtual work. We also compare multiple competing hypotheses with regard to the production, removal, and alignment of the collagen fibers. Finally, we recognize that throughout G&R of the aneurysmal wall, hemodynamic loads play key roles, thus fluid–solid interactions should be taken into account. However, a full computation for coupled fluid–solid problems with a complex geometry and evolving nonlinear properties is even more challenging, requiring considerable computing time and cost. So, at the end of this chapter, we suggest how the theory of small deformations superimposed on large can be exploited when solving coupled fluid–solid interaction problems.

## 3.2 Theoretical Framework

### 3.2.1 Kinematics

Let the aneurysmal wall consist of a homogenized[1] mixture of collagen layers having different preferred fiber directions: that is, we treat collagen having different preferred fiber directions as different co-existing constituents (e.g. [CAa,CAb]). The multiple constituents are allowed to have continuous turnover during G&R, but they may have different rates of production and removal. When a new ($k$th) family of collagen fibers is produced at time $\tau \in (-\infty, t]$, it has a preferred fiber direction that is measured by the in-plane angle $\alpha^k(\tau)$ from the direction of an orthonormal vector in the tangent plane. In general, the in-plane angle for the preferred direction in each constituent can change over time and result in changes in material anisotropy.

In this work, we introduce a fixed configuration $\kappa_R$ as a computational domain. However, the fixed configuration $\kappa_R$ is different from the traditional reference configuration in that particles in $\kappa_R$ can be produced and removed so that the current configuration and the fixed configuration contain the same particles at each time. Now, let the positions of a particle of a lesion (mixture) be $\mathbf{X}$ and $\mathbf{x}$ in the fixed and current configurations of the lesion, and let the mapping $\chi_{\kappa_R}$ assign particles from the fixed configuration to the current configuration at time $t$, that is,

$$\mathbf{x} = \chi_{\kappa_R}(\mathbf{X}, t). \tag{2.1}$$

The deformation gradient $\mathbf{F}(t)$ is defined through

$$\mathbf{F}(t) := \frac{\partial \chi_{\kappa_R}}{\partial \mathbf{X}}. \tag{2.2}$$

Although all solid constituents are constrained to deform together, we imagine that each constituent has individual natural (i.e. stress-free)

---

[1] Even within the context of a single constituent inhomogeneous nonlinear elastic body, current procedures that lead to an homogenized model are fraught with serious difficulties (see [SAa])

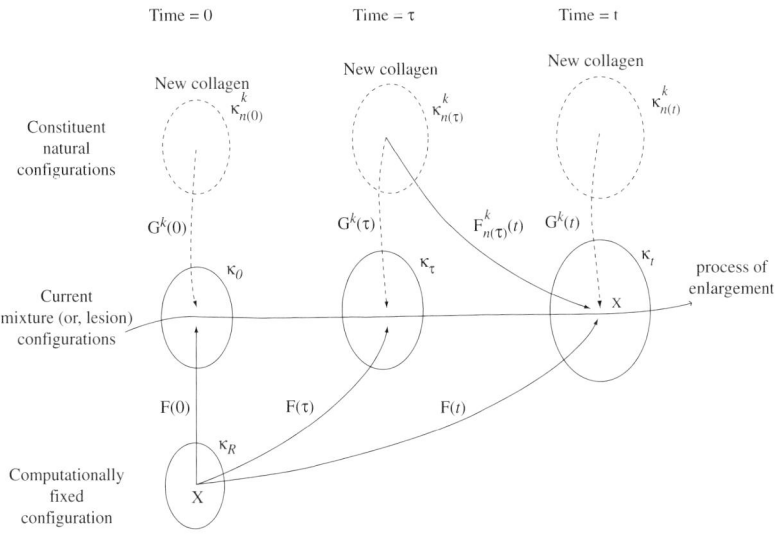

Figure 3.2. Schema of important configurations. The current mixture configurations $\kappa_\tau$ with $\tau \in (0, t]$ track the evolution of a lesion under the physiological condition. For computational purposes, we chose a fixed configuration $\kappa_R$ where particles are produced and removed so that the current configuration and the fixed configuration contain the same particles at time $\tau \in (0, t]$. Finally, although the newly produced collagen is incorporated into the wall under stress, we imagine the existence of individual natural (stress-free) configurations $\kappa^k_{n(\tau)}$ associated with each instant of production; hence, the natural configurations also evolve.

configurations (see Figure 3.2). Also, if we know how the newly produced collagen fiber is laid down, in the stressed state in which it was produced, then the natural configuration for the newly produced collagen fibers can be inferred. Hence, we postulate that the mechanical properties and the "deposition stretch" of the newly synthesized collagen fibers are always the same; that is, constituents are produced at set homeostatic values in each current configuration (c.f. [HUc, GLa]). Let the prestretch of the $k$th new constituent be given by a tensor $\mathbf{G}^k(\tau)$, which is associated with a mapping from the natural (i.e. stress-free) configuration of the newly produced $k$th constituent to the overall loaded configuration $\kappa_\tau$ at time $\tau$ (see Figure 3.2). Moreover, let the aneurysmal wall be subjected to a transmural pressure $P$ at time $\tau \in [0, t]$, where $t$ is the current time. Although any configuration in Figure 3.2 can serve as the fixed configuration, we set a traction-free configuration at time $\tau = 0$ as the configuration $\kappa_R$, which is convenient computationally. At the time $\tau = t$, the aneurysmal wall consists of constituents (i.e. families of collagen fibers) that were produced during the period $\tau = -\infty$ to $t$ and survived until the current time $t$. The deformation gradient for each constituent $k$ at time $t$, relative to its natural configuration, is $\mathbf{F}^k_{n(\tau)}(t)$, which is associated with mappings of points from the natural configuration of the $k$th constituent (produced at time $\tau$) to the current configuration. We assume a constrained mixture, that is, the individual constituents must move with the mixture (lesion). Hence, we have (see Figure 3.2)

$$\mathbf{F}^k_{n(\tau)}(t) = \mathbf{F}(t)\mathbf{F}^{-1}(\tau)\mathbf{G}^k(\tau), \tag{2.3}$$

where $\mathbf{F}(\tau)$ and $\mathbf{F}(t)$ are associated with mappings from the fixed configuration to subsequent configurations of the lesion at times $\tau$ and $t$, respectively.

### 3.2.2    Fibrous Structure

Let $\mathbf{M}^k_{n(\tau)}$ and $\mathbf{m}^k_{n(\tau)}(t)$ be unit vectors in the directions of a family of collagen fibers (i.e. $k$th constituent produced at time $\tau$) in the natural and loaded configurations, respectively. These unit vectors are related via,

$$\mathbf{m}^k_{n(\tau)}(t) = \frac{\mathbf{F}^k_{n(\tau)}(t)\mathbf{M}^k_{n(\tau)}}{|\mathbf{F}^k_{n(\tau)}(t)\mathbf{M}^k_{n(\tau)}|}. \tag{2.4}$$

The unit vector in the fiber direction in $\kappa_R$ is thus $\mathbf{M}^k_R(\tau) = (\mathbf{F}^{-1}(\tau)\mathbf{G}^k \mathbf{M}^k_{n(\tau)})/|\mathbf{F}^{-1}(\tau)\mathbf{G}^k\mathbf{M}^k_{n(\tau)}|$. Let $\{\mathbf{E}_1, \mathbf{E}_2\}$ and $\{\mathbf{e}_1, \mathbf{e}_2\}$ be two orthonormal bases in fixed and current configurations, respectively. Also let $\alpha^k(t)$ be the angle between $\mathbf{m}^k_{n(\tau)}(t)$ and $\mathbf{e}_1$, and $\alpha^k_R(\tau)$ be the angle between $\mathbf{M}^k_R(\tau)$ and $\mathbf{E}_1$ for the $k$th collagen fiber that was produced at time $\tau$. When the principal directions remain principal, the stretch experienced by the $k$th constituent, along the fiber direction, can be computed as

$$\lambda^k(t) = \sqrt{(\lambda_1 \cos \alpha^k_R)^2 + (\lambda_2 \sin \alpha^k_R)^2}, \tag{2.5}$$

where $\lambda_1$ and $\lambda_2$ are two principal stretches at any $t$. Alternatively, the stretch of the $k$th constituent, relative to its individual fiber direction in the natural configuration $\kappa^k_{n(\tau)}$, is given as

$$\lambda^k_{n(\tau)}(t) = \sqrt{\mathbf{M}^k_{n(\tau)} \cdot \mathbf{F}^k_{n(\tau)}(t)^T \mathbf{F}^k_{n(\tau)}(t)\mathbf{M}^k_{n(\tau)}}. \tag{2.6}$$

As noted above, we assume that a newly produced family of collagen fibers is always incorporated within the wall at a homeostatic stretch; that is, the value of the stretch of the constituent is $G_h$ when it is produced ($\mathbf{G}^k\mathbf{M}^k_{n(\tau)} = G_h\mathbf{m}^k_{n(\tau)}(\tau)$). Of course, the fibers are stretched farther during the enlargement of the lesion, this additional stretch being $\lambda^k(t)/\lambda^k(\tau)$, where $\lambda^k(\tau)$ and $\lambda^k(t)$ are stretches calculated from the fixed configuration to the pressurized configurations at time $\tau$ and time $t$, respectively. Thus, the stretches of fibers of the $k$th constituent become

$$\lambda^k_{n(\tau)}(t) = G_h \frac{\lambda^k(t)}{\lambda^k(\tau)}. \qquad (2.7)$$

Let the behavior of the $k$th family of collagen fibers be describable via an exponential-type strain energy function per unit volume in $\kappa_R$

$$W^k(\lambda^k_{n(\tau)}) = c\left\{ \exp\left[ c_1 (\lambda^k_{n(\tau)}{}^2 - 1)^2 \right] - 1 \right\}, \qquad (2.8)$$

where the material parameters $c$ and $c_1$ are assumed to be the same for all families.

### 3.2.3   Kinetics of G&R

The mass of the individual constituents, and thus that of the lesion, changes due to local production and removal of collagen. The total mass per unit area in $\kappa_R$ at time $t$, $M(t)$, can be calculated as

$$M(t) = \sum_k M^k(t) = \sum_k \left[ M^k(0)Q^k(t) + \int_0^t m^k(\tau)q^k(t-\tau)d\tau \right], \qquad (2.9)$$

where $m^k(\tau)$ is the rate of production of the $k$th constituent at time $\tau$ per unit area and $q^k(t-\tau)$ is its survival function; that is, the fraction produced at time $\tau$ that remains at time $t$, $Q^k(t)$ is the fraction of the $k$th constituent that was present at time 0 and still remains at time $t$ (i.e. has not yet been removed). Although we model the aneurysm as a membrane mechanically, we can calculate the thickness in postprocessing. Assuming the overall mass density of the wall remains constant (i.e. $\rho \equiv \rho_o \ \ \forall \tau$; [ROa]), the thickness of the wall is given as

$$h(t) = \frac{M(t)}{J\rho}, \qquad (2.10)$$

where $J = \det(\mathbf{F}(t))$. Although the transient response to loads applied during the cardiac cycle may be isochoric, volume need not be conserved during G&R.

### 3.2.4   Stress-Mediated G&R

The production and removal of each constituent results from biological activity. For example, collagen is produced and organized by fibroblasts and degraded by enzymes such as matrix metalloproteinases (or MMPs).

Recent studies show that many cell types (including fibroblasts that populate the aneurysmal wall) can sense and convert mechanical stimuli into biological signals, and thereby effect growth and remodeling. As an example, we postulate that the production of each constituent is a function of the number of cells $n(t)$ per unit reference area and the stress experienced by the cells via the local collagen matrix, namely

$$m^k(t) = n(t)\Big(f^k(\sigma^k(t) - \sigma_h) + f_h\Big), \tag{2.11}$$

where $\sigma^k$ is a time-averaged (over a cardiac cycle) mean value of a scalar-measure of the stress, and $\sigma_h$ is a homeostatic value of this stress-measure. Here, we assume

$$\sigma^k(t) = \frac{|\mathbf{T}(t)\mathbf{m}^k_{n(\tau)}(t)|}{h(t)}, \tag{2.12}$$

where $\mathbf{T}(t)$ is the overall membrane stress at time $t$. If cells proliferate such that cell density (per unit volume in the current loaded configuration) is constant, then the number of the cells increases proportionately with volume changes; that is, $n(t) = n(0)M(t)/M(0)$. A special case allows a linear dependence on the stress difference (cf. [RAa,TAa]), whereby the production rate of a constituent can be expressed as

$$m^k(t) = \frac{M(t)}{M(0)}\Big(K_g(\sigma^k(t) - \sigma_h) + \tilde{f}_h\Big), \tag{2.13}$$

where $\tilde{f}_h$ is $f_h$ multiplied by the initial cell density $n(0)$ and $K_g$ is a scalar parameter that controls the stress-mediated growth. For illustrative purposes, we assume a simple form for the survival function $q(\tilde{\tau})$ in Eq. (2.9). After its production, let there be no removal of a constituent until time $t_1$ and, then, let the constituent degrade gradually until all of the constituent is removed by time $t_2$. Toward this end, let

$$q(\tilde{\tau}) = \begin{cases} 1 & 0 \le \tilde{\tau} < t_1 \\ \frac{1}{2}\Big\{\cos\Big(\frac{\pi}{t_2-t_1}(\tilde{\tau} - t_1)\Big) + 1\Big\} & t_1 \le \tilde{\tau} \le t_2 \\ 0 & t_2 < \tilde{\tau} \end{cases} . \tag{2.14}$$

We introduce a nondimensional parameter for a stress mediation parameter:

$$\hat{K}_g = \Big(\frac{\sigma_h t_2}{M(0)}\Big)K_g. \tag{2.15}$$

### 3.2.5   Stress and Strain Energy Function

The total Cauchy membrane stress $\mathbf{T}$ (i.e. tension, or force per current length) is:

$$T_{11}(t) = \frac{1}{\lambda_2(t)}\frac{\partial w}{\partial \lambda_1(t)} \qquad T_{22}(t) = \frac{1}{\lambda_1(t)}\frac{\partial w}{\partial \lambda_2(t)}, \tag{2.16}$$

where $w$ is the strain energy per unit area in the fixed configuration $\kappa_R$. We postulate that $w = \sum_k w^k$ and it evolves as

$$w^k(t) = \frac{M^k(0)}{\rho} Q^k(t) W^k(\lambda_{n(0)}^k(t)) + \int_0^t \frac{m^k(\tau)}{\rho} q(t - \tau) W^k(\lambda_{n(\tau)}^k(t)) d\tau,$$
(2.17)

where $W^k$ is the aforementioned strain energy of the $k$th constituent per unit volume and $q^k(\tilde{\tau}) = q(\tilde{\tau})$. Substituting Eq. (2.17) into Eq. (2.16) and using Eq. (2.3), the Cauchy membrane stress is

$$T_{11}(t) = \sum_k \frac{1}{\lambda_2(t)} \left\{ \frac{M^k(0)Q^k(t)G_h}{\rho\lambda^k(0)} \frac{\partial W^k}{\partial\lambda_{n(0)}^k(t)} \frac{\partial\lambda^k(t)}{\partial\lambda_1(t)} \right.$$
$$\left. + \int_0^t \frac{m^k(\tau)q(t-\tau)G_h}{\rho\lambda^k(\tau)} \frac{\partial W^k}{\partial\lambda_{n(\tau)}^k(t)} \frac{\partial\lambda^k(t)}{\partial\lambda_1(t)} d\tau \right\},$$
(2.18)

$$T_{22}(t) = \sum_k \frac{1}{\lambda_1(t)} \left\{ \frac{M^k(0)Q^k(t)G_h}{\rho\lambda^k(0)} \frac{\partial W^k}{\partial\lambda_{n(0)}^k(t)} \frac{\partial\lambda^k(t)}{\partial\lambda_2(t)} \right.$$
$$\left. + \int_0^t \frac{m^k(\tau)q(t-\tau)G_h}{\rho\lambda^k(\tau)} \frac{\partial W^k}{\partial\lambda_{n(\tau)}^k(t)} \frac{\partial\lambda^k(t)}{\partial\lambda_2(t)} d\tau \right\}.$$
(2.19)

## 3.3 Simulations for Saccular Aneurysms

### 3.3.1 Method

We assume axisymmetric ellipsoidal membrane geometries for a saccular aneurysm in both the fixed and current configuration. Furthermore, parametric relations in the fixed configuration can be described by continuous functions $Z = Z(\Phi)$ and $R = R(\Phi)$ (Figure 3.3). We also use two sets of two-dimensional curvilinear coordinates, $\Xi = \{\Phi, \Theta\}$ for the fixed configuration and $\xi = \{\phi, \theta\}$ for the current configuration. The associated bases are given by

$$\mathbf{G}_i = \frac{\partial\mathbf{X}}{\partial\Xi_i}, \qquad \mathbf{g}_i = \frac{\partial\mathbf{x}}{\partial\xi_i},$$
(3.1)

where $i = 1, 2$. Locally orthonormal bases are obtained by

$$\mathbf{E}_i = \frac{\mathbf{G}_i}{|\mathbf{G}_i|}, \qquad \mathbf{e}_i = \frac{\mathbf{g}_i}{|\mathbf{g}_i|}$$
(3.2)

and the outward unit normal directions are

$$\mathbf{N} = \frac{\mathbf{G}_1 \times \mathbf{G}_2}{|\mathbf{G}_1 \times \mathbf{G}_2|}, \qquad \mathbf{n} = \frac{\mathbf{g}_1 \times \mathbf{g}_2}{|\mathbf{g}_1 \times \mathbf{g}_2|}.$$
(3.3)

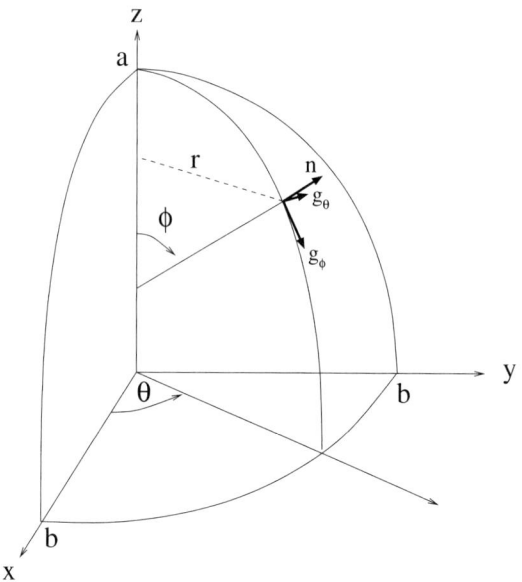

Figure 3.3. Coordinate system for an axisymmetric saccular lesion in a current configuration. Similar values exist for the reference configuration: $X$, $Y$, $Z$, $A$, $B$, $\Theta$, $\Phi$, and $R$.

When the motion due to growth is assumed as

$$\phi = \phi(\Phi) \qquad \theta = \Theta, \tag{3.4}$$

then the components of the 2-D deformation gradient $\mathbf{F} = F_{iJ}\mathbf{e}_i \otimes \mathbf{E}_J$ can be calculated by

$$F_{iJ} = \begin{bmatrix} \frac{h_1}{H_1}\phi' & 0 \\ 0 & \frac{h_2}{H_2} \end{bmatrix} = \begin{bmatrix} \lambda_1 & 0 \\ 0 & \lambda_2 \end{bmatrix}, \tag{3.5}$$

where $(\cdot)' = \partial(\cdot)/\partial\Phi$ and

$$H_1 = \frac{A^2 B^2}{(A^2 \cos^2 \Phi + B^2 \sin^2 \Phi)^{3/2}}, \qquad H_2 = R \tag{3.6}$$

$$h_1 = \frac{a^2 b^2}{(a^2 \cos^2 \phi + b^2 \sin^2 \phi)^{3/2}}, \qquad h_2 = r. \tag{3.7}$$

For numerical simulations, we consider an aneurysm approximated as an initially ellipsoidal membrane with two primary axes: dimensions $2A$ for the height and $2B$ for the diameter of the equator. Moreover, let the wall initially consist of two families of fibers (i.e. two constituents) and thus two preferred directions, at $0$ and $\pi/2$, and initially the same mass fraction for both constituents. The strain energy (2.17) is not only a function of the deformation (and thus position) at any time, but also of the past history of the deformations and the rate of mass production. In general, therefore,

$$w(t) = w(\lambda_i(\Phi, t); \lambda_i(\Phi, \tau), m^k(\Phi, \tau)), \qquad (i = 1, 2, \ 0 \le \tau < t) \qquad (3.8)$$
$$= \hat{w}(\phi(\Phi, t), a(t), b(t); \lambda_i(\Phi, \tau), m^k(\Phi, \tau)), \qquad (3.9)$$

where $a$ and $b$ are dimensions in the deformed primary axes. Such inflation problems can be solved using the principle of virtual work, the governing equation for which is

$$\delta I = \int_S \delta w \, dA - \int_s \mathbf{Pn} \cdot \delta \mathbf{x} da = 0, \qquad (3.10)$$

where $\delta \mathbf{x}$ represents virtual changes in position. The surfaces $S$ and $s$ correspond to the surface area of the fixed and current configuration. Next, let the function $\phi(\Phi)$ be approximated via

$$\phi = \sum_{j=1}^{n} \phi_j \psi_j, \qquad (3.11)$$

where $\phi_j$ is the $j$th nodal value of $\phi(\Phi)$ and $\psi_j$ is a quadratic interpolation function; a variational procedure for (3.10) with respect to $\phi$ yields a nonlinear algebraic equation. Using similar approximations for $a$ and $b$, each yielding associated algebraic equations, we thus formulate the weak form (3.10) as

$$\mathcal{F} \equiv \left\{ \begin{array}{c} 2\pi \displaystyle\int_0^{\Phi_o} \left\{ \dfrac{\partial w}{\partial \phi} \psi_i + \dfrac{\partial w}{\partial \phi'} \psi_i' \right\} RH_1 d\Phi \\[2ex] 2\pi \displaystyle\int_0^{\Phi_o} \left\{ \dfrac{\partial w}{\partial a} RH_1 - P(r_{,\phi} z_{,a} - z_{,\phi} r_{,a}) \phi' r \right\} d\Phi \\[2ex] 2\pi \displaystyle\int_0^{\Phi_o} \left\{ \dfrac{\partial w}{\partial b} RH_1 - P(r_{,\phi} z_{,b} - z_{,\phi} r_{,b}) \phi' r \right\} d\Phi \end{array} \right\} = \left\{ \begin{array}{c} 0 \\ 0 \\ 0 \end{array} \right\}. \qquad (3.12)$$

The nonlinear finite element equations (3.12) are solved using a Newton–Raphson procedure.

### 3.3.2   Results

To simulate an initial insult, we prescribe a reduction in mass from the total mass of the stable lesion as an initial condition. Such a mass reduction could be caused by a proteolytic weakening of the wall, with an associated loss of elastin and then smooth muscle. Mass reduction induces larger values of stretch in the wall than the homeostatic value and increases wall stress. This initial perturbation initiates enlargement due to G&R in the wall. The enlargement of an ellipsoidal aneurysm is plotted (Figure 3.4) for different

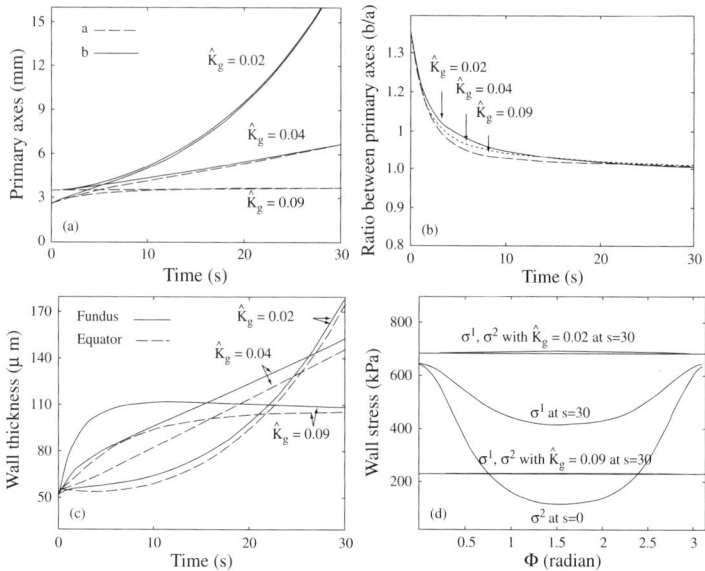

Figure 3.4. G&R of an initially ellipsoidal aneurysm for different values of the parameter $K_g$ for stress mediation. Note that $a$ and $b$ tend to become equal early on (panels a and b), thus yielding a spherical saccular lesion with more uniform wall stress (panel d).

stress mediation parameters $\hat{K}_g$ for a 20% mass reduction. The time scale is normalized by a collagen life span $t_2$; that is, $s = t/t_2$. The rate of enlargement was higher for a 20% mass reduction than a 5% reduction for the same value of $\hat{K}_g$ (not shown), which suggests that a more severe initial perturbation may cause a faster enlargement of an aneurysm. The rate of enlargement decreases with larger values of $\hat{K}_g$ (Figure 3.4a). When $\hat{K}_g$ is 0.09, the aneurysm quickly reaches its (biologically) stable state and there is no more enlargement. In contrast, when $\hat{K}_g$ is smaller than 0.04, the aneurysm grows in a unbounded manner. Stabilization depends on changes in the ratio of the primary axis ($a$ or $b$) to the wall thickness $h$ during G&R (recall that the stress in a spherical membrane is $Pa/2h$). For a larger value of $\hat{K}_g$, thickness increases faster than the rate of enlargement, hence wall stress can reach its homeostatic value and the aneurysm can become stable (Figure 3.4c). For a value of $\hat{K}_g$ smaller than 0.04, the ratio between the radius and the thickness keeps increasing, hence the stress increases similarly. When $\hat{K}_g = 0.04$, both radius and thickness increase linearly with respect to time, but the ratio of the radius to the thickness remains nearly constant. Thus the stress also remains the same despite continued enlargement. The tendency toward a spherical shape is similar for all three values of $\hat{K}_g$ (Figure 3.4b). Moreover, this tendency is strong early on, with lesions becoming almost spherical by $s = 10$. Aneurysmal wall stress thus becomes more uniform over time (Figure 3.4d).

## 3.4 Simulations for Fusiform Aneurysms

### 3.4.1 Method

We assume that initial geometries for fusiform aneurysms are axisymmetric cylindrical membranes. The parametric relation in the fixed configuration can be described by continuous functions $R = R(Z)$ (Figure 3.5). The positions of a point in the fixed and current configurations of the lesion, $\mathbf{X}$ and $\mathbf{x}$, can thus be expressed by two sets of cylindrical polar coordinates $(Z, \theta, R)$ and $(z, \theta, r)$, respectively:

$$\mathbf{X} = R \cos \Theta \mathbf{i} + R \sin \Theta \mathbf{j} + Z \mathbf{k}, \tag{4.1}$$

$$\mathbf{x} = r \cos \theta \mathbf{i} + r \sin \theta \mathbf{j} + z \mathbf{k}. \tag{4.2}$$

We also use a set of two-dimensional curvilinear coordinates, $\Xi_i = \{Z, \Theta\}$ whereby the current position can be expressed by $\mathbf{x} = \mathbf{x}(z(Z, \Theta),$

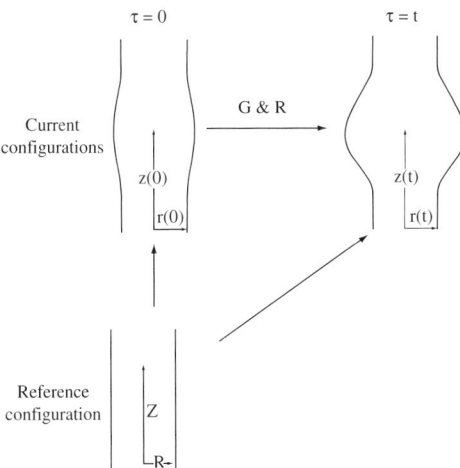

Figure 3.5. Axisymmetric geometries of a fusiform lesion in current (i.e. pressurized) configurations at time $\tau = 0$ and $t$ as well as a convenient unpressurized reference configuration (which need not be stress-free if different families of collagen–proteoglycans are in tension/compression and self-equilibrate).

$\theta(Z, \Theta)$, $r(Z, \Theta)) = \mathbf{x}(Z, \Theta)$. The associated bases and orthonormal bases are given by

$$\mathbf{G}_i = \frac{\partial \mathbf{X}}{\partial \Xi_i}, \qquad \mathbf{g}_i = \frac{\partial \mathbf{x}}{\partial \Xi_i}, \qquad \mathbf{E}_i = \frac{\mathbf{G}_i}{|\mathbf{G}_i|}, \qquad \mathbf{e}_i = \frac{\mathbf{g}_i}{|\mathbf{g}_i|}, \qquad (4.3)$$

where $i = 1, 2$. When the deformation due to growth is assumed as

$$z = z(Z), \qquad \theta = \Theta, \qquad r = r(Z), \qquad (4.4)$$

then the components of the 2-D deformation gradient $\mathbf{F}(t) = F_{iJ} \mathbf{e}_i \otimes \mathbf{E}_J$, where

$$F_{iJ} = \begin{bmatrix} \lambda_1 & 0 \\ 0 & \lambda_2 \end{bmatrix}, \qquad \lambda_1 = \frac{\sqrt{(z')^2 + (r')^2}}{\sqrt{1 + (R')^2}}, \qquad \lambda_2 = \frac{r}{R}, \qquad (4.5)$$

where $(\cdot)' = \partial(\cdot)/\partial Z$. For computations of the enlargement of fusiform aneurysms, we allow the orientation of new collagen to change during G&R. It is not known how the alignment of newly produced collagen is decided, however, thus we consider multiple hypotheses and compare their consequences. Let us define a unit vector for the preferred alignment $\mathbf{e}^p$ and assume that new collagen is deposited with this preferred alignment (similar to Driessen et al. [DRa]). Let the vector $\mathbf{e}^p = f_1 \mathbf{e}_1 + f_2 \mathbf{e}_2$, where $\mathbf{e}_1$ and $\mathbf{e}_2$ are unit vectors in the axial and circumferential directions, respectively. When $f_1$ and $f_2$ are functions of principal stresses, the preferred alignment will be dictated by the mixture stress at each time $t$. Conversely, when $f_1$ and $f_2$ are functions of principal stretches, the preferred alignment will be dictated by the mixture stretch. We compare several cases:

- Case 1: The preferred alignment is dictated by principal stresses, and the angle between $\mathbf{e}^p$ and a principal axis decreases when the principal stress along that axis becomes larger, specifically, $f_1 = \sigma_1/\sqrt{\sigma_1^2 + \sigma_2^2}$ and $f_2 = \sigma_2/\sqrt{\sigma_1^2 + \sigma_2^2}$, where $\sigma_1$ and $\sigma_2$ are principal stresses in $\mathbf{e}_1$ and $\mathbf{e}_2$ directions, respectively.

- Case 2: The preferred alignment is dictated by the lesser principal stress, thus in contrast to Case 1, let $f_1 = \sigma_2/\sqrt{\sigma_1^2 + \sigma_2^2}$ and $f_2 = \sigma_1/\sqrt{\sigma_1^2 + \sigma_2^2}$.

- Case 3: The preferred alignment is dictated by principal stretches and the angle between $\mathbf{e}^p$ and a principal axis decreases when that principal stretch is larger: $f_1 = \lambda_1/\sqrt{\lambda_1^2 + \lambda_2^2}$ and $f_2 = \lambda_2/\sqrt{\lambda_1^2 + \lambda_2^2}$.

Similar to the simulation for saccular aneurysms, we use the principle of virtual work (3.10). Let functions $r(Z)$ and $z(Z)$ be approximated via

$$r = \sum_{j=1}^{n} r_j \psi_j, \qquad z = \sum_{j=1}^{n} z_j \psi_j, \tag{4.6}$$

where $r_j$ and $z_j$ are the $j$th nodal values of $r(Z)$ and $z(Z)$, respectively, and $n$ is the number of nodes. The function $\psi_j$ is a global quadratic interpolation function corresponding to the $j$th node. A variation of Eq. (3.10) with respect to $r$ and $z$ yields two sets of nonlinear algebraic equations:

$$\mathcal{F} \equiv \begin{Bmatrix} 2\pi \int_0^{Z_o} (\frac{\partial w}{\partial r}\psi_i + \frac{\partial w}{\partial r'}\psi_i')R\sqrt{1+(R')^2} - Pz'r\psi_i \; dZ \\ 2\pi \int_0^{Z_o} (\frac{\partial w}{\partial z}\psi_i + \frac{\partial w}{\partial z'}\psi_i')R\sqrt{1+(R')^2} + Pr'r\psi_i \; dZ \end{Bmatrix} = \begin{Bmatrix} 0 \\ 0 \end{Bmatrix}. \tag{4.7}$$

The associated algebraic equations (4.7) are solved using a Newton–Raphson procedure.

### 3.4.2 Results

An initial mass reduction allows a slight bulge under the constant transmural pressure as well as thinning of the wall in the middle of the vessel at $s = 0$. These changes alter stresses from homeostatic values and the lesion starts to enlarge (e.g. see Figure 3.6). Similar to results for saccular aneurysms, a larger value of $\hat{K}_g$ decreases the rate of enlargement of the lesions (Figure 3.7). Here, however, the alignment of new collagen is allowed

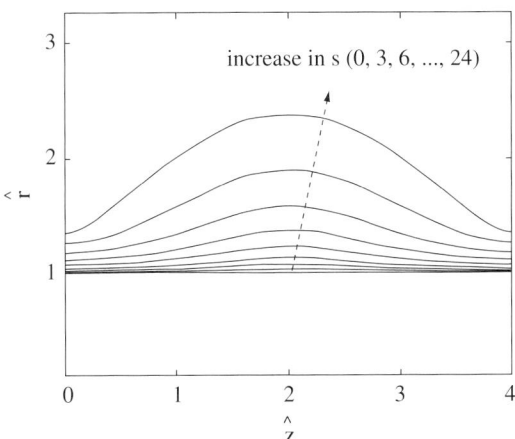

Figure 3.6. Simulation of the enlargement of an axisymmetric fusiform lesion, as a function of nondimensional time $s = t/t_2$, for the case 3 preferred deposition (i.e. preferred direction of new collagen dictated by larger principal stretch) and $\hat{K}_g = 0$.

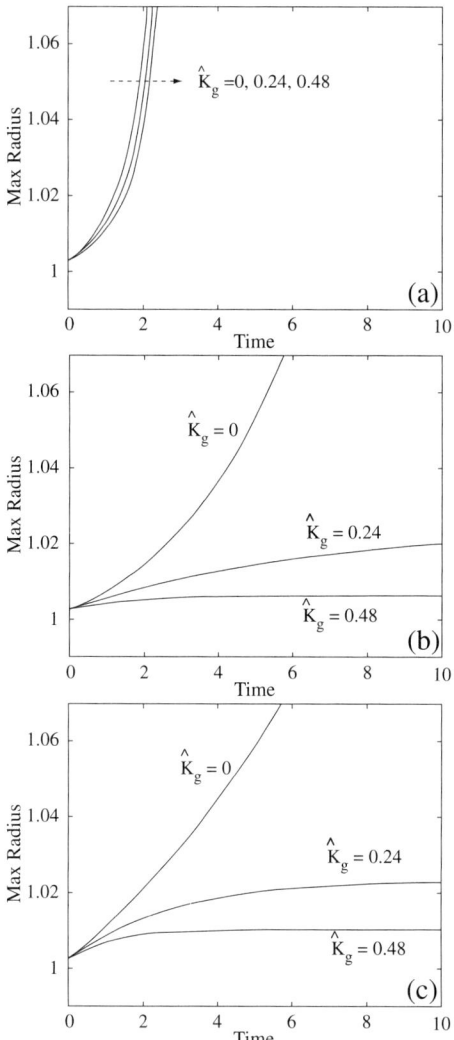

Figure 3.7. Effect of the stress mediation parameter $\hat{K}_g$ on the enlargement of a fusiform lesion due to a 20% initial mass reduction within the central region of the wall and growth and remodeling for different hypotheses on the alignment of newly deposited collagen fibers: (a) case 1; (b) case 2; (c) case 3. Recall that radius and time are nondimensionalized via $r/r_h$ and $t/t_2$.

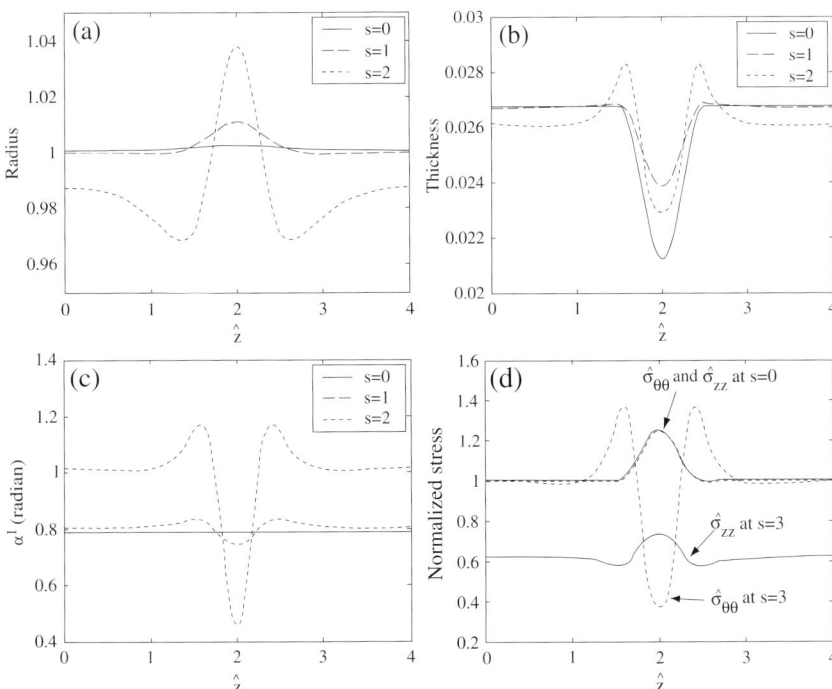

Figure 3.8. Evolution of a fusiform lesion ($\hat{K}_g = 0.48$) with a 20% initial mass reduction within the central region of the wall and case 1 preferred deposition: (a) radius; (b) thickness; (c) fiber orientation of new collagen; (d) principal stresses. The simulation shows that fiber reorientation by case 1 causes an unstable enlargement. Results are similar for $\hat{K}_g = 0$ and 0.24. Recall that radius, thickness, and stress are nondimensionalized via $r/r_h$, $h/r_h$, and $\sigma_{ii}/\sigma_h$ ($i = \theta, z$).

to change corresponding to the stress or strain. When the fiber alignment is assumed according to case 2 or 3, the lesion is stable for $\hat{K}_g = 0.24$ and 0.48 but is unstable for $\hat{K}_g = 0$ (e.g. G&R transitions from unstable to stable at $\hat{K}_g \approx 0.06$ for case 3). For case 1, however, the lesion shows unstable growth for $\hat{K}_g = 0$, 0.24, and 0.48. In this case, the angle between the preferred alignment and z-axis, $\alpha^1(s)$ (note that $\alpha^2(s) = -\alpha^1(s)$), decreases because the principal stress $\sigma_{zz}$ is slightly higher than $\sigma_{\theta\theta}$ at $s = 0$ (see Figures 3.8(c) and 3.8(d)). Further remodeling causes $\alpha^1(s)$ to decrease in the middle but to increase in other regions ($s = 2$ in Figure 3.8c). Because the principal stress $\sigma_{zz}$ remains higher than $\sigma_{\theta\theta}$, however, the fiber orientation changes further, which makes the enlargement of the lesion unstable even for higher values of $\hat{K}_g$. In contrast, for the case 2 and 3 preferred depositions, the thickness increases in the center and the stresses

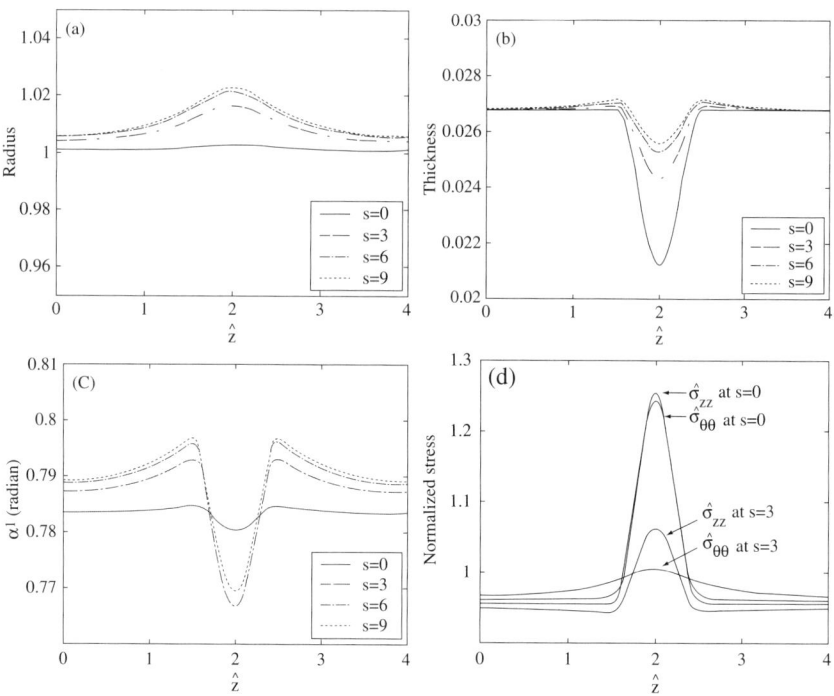

Figure 3.9. Evolution of a fusiform lesion ($\hat{K}_g = 0.24$) with a 20% initial mass reduction and case 3 preferred deposition: (a) radius; (b) thickness; (c) fiber orientation of new collagen; (d) principal stresses. Recall from Figure 3.7 that this is a stable enlargement.

tend to return to the homeostatic value through G&R with $\hat{K}_g = 0.24$ and 0.48 (Figure 3.9). Thus, the postulated alignment of newly deposited collagen fibers had a significant influence on the potential stability of the enlargement.

## 3.5   Fluid–Solid Interaction

Although conventional analyses of arterial wall mechanics require the deformation to be computed relative to a suitable reference configuration (e.g. a stress-free sector obtained by introducing multiple cuts in an excised segment [HUb,FUa]), the focus of most computational biofluid mechanical analyses is on changes from diastole to systole. Because of the highly nonlinear material behavior of arteries, the deformation from a suitable reference configuration to an intact diastolic configuration is "large" whereas that from diastolic to systolic is typically "small." Hence, if we know the state of stress in the artery at any configuration between diastole and systole, then the primary need in computational biofluid mechanics is to determine changes in stiffness during the cardiac cycle. We suggest here that the theory of small deformations superimposed on large can be exploited when solving coupled fluid–solid interaction problems during the cardiac cycle. In particular, this approach allows one to include the effects of residual stress, nonlinear material behavior, anisotropy, smooth muscle contractility, and finite deformations of the arterial wall while recovering equations relevant throughout the cardiac cycle that can be solved using methods common to linearized elasticity.

Let the body occupy a configuration $\kappa_{t_o}(\mathcal{B})$ at an intermediate time $t_o$ characterized by a large strain measured from a reference configuration $\kappa_R(\mathcal{B})$. Then, let the position in the intermediate (stressed) configuration be denoted by $\mathbf{x}_o = \chi_{\kappa_R}(\mathbf{X}, t_o)$. Hence, we can consider that a small displacement $\mathbf{u} = \mathbf{u}(\mathbf{x}_o, t)$, superimposed upon the large deformation, yields the "current" position $\mathbf{x}$ at time $t$, namely

$$\mathbf{x} = \mathbf{x}_o + \mathbf{u}(\mathbf{x}_o, t). \tag{5.1}$$

Deformation gradients associated with mappings from the reference to the intermediate and current configurations are thus given by

$$\mathbf{F}^o = \frac{\partial \chi_{\kappa_R}(\mathbf{X}, t_o)}{\partial \mathbf{X}}, \qquad \mathbf{F} = \frac{\partial \chi_{\kappa_R}(\mathbf{X}, t)}{\partial \mathbf{X}}. \tag{5.2}$$

The deformation gradient representing a mapping from the intermediate configuration to current configurations is similarly,

$$\mathbf{F}^* = \frac{\partial \mathbf{x}}{\partial \mathbf{x}_o} = \mathbf{I} + \mathbf{H}, \qquad \text{where} \quad \mathbf{H} = \frac{\partial \mathbf{u}}{\partial \mathbf{x}_o}. \tag{5.3}$$

The displacement gradient $\mathbf{H}$ can be divided into a symmetric part $\epsilon = \frac{1}{2}(\mathbf{H} + \mathbf{H}^T)$ and a skew-symmetric part $\Omega = \frac{1}{2}(\mathbf{H} - \mathbf{H}^T)$. If $\mathbf{H}$ is small, $\epsilon$ and $\Omega$ are identified as the infinitesimal strain and infinitesimal rotation. Regardless, for the successive motions

$$\mathbf{F} = \mathbf{F}^* \mathbf{F}^o. \tag{5.4}$$

For an isochoric motion, the material is subject to a kinematic constraint: $\det \mathbf{F} = 1$ in general, which reduces to $\mathrm{tr}(\epsilon) = 0$ for an infinitesimal strain. The Cauchy stress $\mathbf{t}$ for an incompressible Green (hyper)elastic material can be written as

$$\mathbf{t} = -p\mathbf{I} + \hat{\mathbf{t}}, \qquad \hat{\mathbf{t}} = \mathbf{F}\hat{\mathbf{S}}\mathbf{F}^T, \qquad \hat{\mathbf{S}} = 2\frac{\partial \hat{W}}{\partial \mathbf{C}}, \tag{5.5}$$

where $p$ is a Lagrange multiplier that enforces the constraint that the motion is isochoric, $\mathbf{C} = \mathbf{F}^T\mathbf{F}$ is the total right Cauchy–Green tensor, and $\hat{\mathbf{t}}$ is the deformation-dependent (or extra) part of the Cauchy stress. For purposes herein, it is convenient to relate $\hat{\mathbf{t}}$ to the extra part of the second Piola–Kirchhoff stress $\hat{\mathbf{S}}$, which in turn is computed directly from a stored energy function $W = \tilde{W}(\mathbf{F})$, or by material frame indifference, $W = \hat{W}(\mathbf{C})$. Now, let the deformation gradient and Cauchy stress of arterial wall in any convenient intermediate configuration during the cardiac cycle be represented by $\mathbf{F}^o$ and $\mathbf{t}^o$ whereas that in any "current" configuration between diastolic and systolic be denoted as $\mathbf{F}$ and $\mathbf{t}$. Using (5.3) and (5.4), (5.5) can be written as

$$\mathbf{t} = -(p^o + p^*)\mathbf{I} + (\mathbf{F}^o + \mathbf{H}\mathbf{F}^o)(\hat{\mathbf{S}}^o + \mathbf{S}^*)(\mathbf{F}^{oT} + \mathbf{F}^{oT}\mathbf{H}^T), \qquad (5.6)$$

where

$$\mathbf{S}^* = \frac{\partial\hat{\mathbf{S}}}{\partial\mathbf{C}}\Big|_{\mathbf{C}^o}\mathbf{C}^* \quad \text{with} \quad \mathbf{C}^* = 2\mathbf{F}^{oT}\boldsymbol{\epsilon}\mathbf{F}^o. \qquad (5.7)$$

Hence, neglecting higher-order terms in the "small displacement gradient" $\mathbf{H}$, (5.6) can be written in terms of physical components of the tensors as (per the usual summation convention)

$$t_{ij} = t_{ij}^o + H_{ik}\hat{t}_{kj}^o + \hat{t}_{ik}^o H_{jk} - p^*\delta_{ij}$$

$$+ 4F_{iA}^o F_{jB}^o F_{kP}^o F_{lQ}^o \frac{\partial\hat{W}}{\partial C_{AB}\partial C_{PQ}}\Big|_{\mathbf{C}^o}\epsilon_{kl}. \qquad (5.8)$$

Moreover, recalling that $\mathbf{H} = \boldsymbol{\epsilon} + \boldsymbol{\Omega}$, (5.8) can be written as

$$t_{ij} = t_{ij}^o - p^*\delta_{ij} + \mathcal{C}_{ijkl}\,\epsilon_{kl} + \mathcal{D}_{ijkl}\,\Omega_{kl}, \qquad (5.9)$$

where

$$\mathcal{C}_{ijkl} = \delta_{ik}\hat{t}_{lj}^o + \hat{t}_{il}^o\delta_{jk} + 4F_{iA}^o F_{jB}^o F_{kP}^o F_{lQ}^o \frac{\partial\hat{W}}{\partial C_{AB}\partial C_{PQ}}\Big|_{\mathbf{C}^o} \qquad (5.10)$$

$$\mathcal{D}_{ijkl} = \delta_{ik}\hat{t}_{lj}^o + \hat{t}_{il}^o\delta_{jk}. \qquad (5.11)$$

Because $\epsilon_{ij} = \epsilon_{ji}$ and $\Omega_{ij} = -\Omega_{ji}$ $(i \neq j)$, $\mathcal{C}_{\alpha\beta ij}\epsilon_{ij} + \mathcal{C}_{\alpha\beta ji}\epsilon_{ji} = (\mathcal{C}_{\alpha\beta ij} + \mathcal{C}_{\alpha\beta ji})\epsilon_{ij}$ and similarly $\mathcal{D}_{\alpha\beta ij}\Omega_{ij} + \mathcal{D}_{\alpha\beta ji}\Omega_{ji} = (\mathcal{D}_{\alpha\beta ij} - \mathcal{D}_{\alpha\beta ji})\Omega_{ij}$. Finally, let new quantities $\tilde{\mathcal{C}}_{\alpha\beta ij} = \frac{1}{2}(\mathcal{C}_{\alpha\beta ij} + \mathcal{C}_{\alpha\beta ji})$ and $\tilde{\mathcal{D}}_{\alpha\beta ij} = \mathcal{D}_{\alpha\beta ij} - \mathcal{D}_{\alpha\beta ji}$. Now, we see, from (5.9) and (5.11), that the stress response of a nonlinearly elastic material, from a finitely deformed intermediate configuration such as that at the mean arterial pressure in a large artery, depends strongly on the prestress $\mathbf{t}^o$, initial finite deformation $\mathbf{F}^o$, and possibly small rotations $\boldsymbol{\Omega}$. As an example of the linearization, Figure 3.10 shows the pressure and axial forces for both a nonlinear constitutive relation and its linearized elastic response to the radius changes. The radius changes 3.7% during a cardiac cycle, and the linearized elastic response is close to that determined using the finite nonlinearly elastic response within the cardiac cycle.

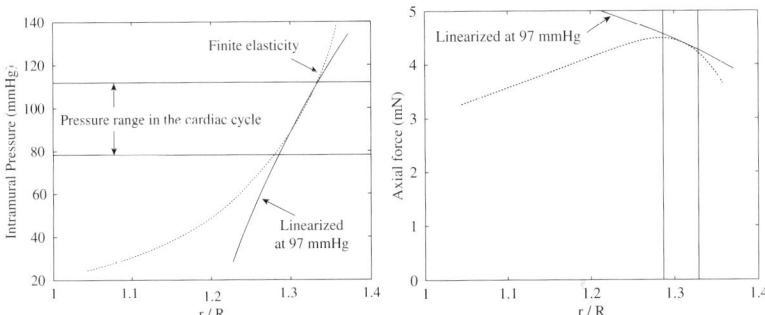

Figure 3.10.   Pressure-circumferential stretch $(r/R)$ and axial force-circumferential stretch during inflation at a fixed length. The values are calculated for both the nonlinear and linearized elastic response with respect to the radius changes. The material parameters for the nonlinear response is obtained by a best-fitting to an experimental result of a rabbit basilar artery and the linearized elastic parameters are calculated at 97 mmHg and $\Lambda = 1.3$.

## 3.6   Discussion

Over the last few decades, many studies have revealed a great deal about the biochemistry related to proteins, DNA, and cell behaviors, and more information is accumulating every day. However, much less is known about how these biochemical and cellular mechanisms result in changes at the organ level. As Hunter and Borg [HUd] emphasized, there is a need to integrate information from proteins to organs, and particularly to develop a framework for computational methods that incorporates biochemical, biophysical, and anatomical information on cells, tissues, and organs. Such frameworks not only integrate information, they also help identify missing data, test hypotheses, and suggest new theoretical and experimental studies. For this purpose, we chose a constrained mixture theory that can model G&R of soft tissue that results from mechanosensitive reactions of cells, including their control of the degradation and deposition of collagen fibers. We specialized the model to a 2-D formulation for the enlargement of intracranial aneurysms with numerical simulations for two idealized geometries: an ellipsoidal shape for saccular aneurysms and a cylindrical shape for fusiform aneurysms. We hypothesized that enlargement results primarily from the coordinated degradation and synthesis of collagen by fibroblasts; moreover, we hypothesized that newly synthesized collagen is incorporated within the wall at a preferred, or homeostatic, deposition stretch. Given these basic hypotheses, our model predicts that stress-mediated enlargement proceeds via a competition between a local thickening and radial expansion for both saccular and fusiform aneurysms. For a saccular aneurysm, an initially ellipsoidal lesion tends to enlarge toward a spherical shape (Figure 3.4b), which agrees with a statistical study by Parlea et al. [PAa]. For a fusiform aneurysm, we postulated three different hypotheses for the alignment of newly deposited collagen fibers and found that the alignment had a significant influence on the potential stability of the enlargement.

As a first step, it is prudent to use simplified models and idealized geometries to capture salient features of stress-mediated G&R of the aneurysms. In the future, however, we hope to expand the model to incorporate more biochemical, biophysical, and cellular information for G&R of arteries with real anatomical geometries. Also there is no doubt that we have to include fluid–solid interactions for a better understanding. Hence, we suggested that the theory of small on large deformations will be a useful tool for coupled fluid–solid computation within complex geometries.

**Acknowledgments** This research was supported, in part, by grants HL-64372 and HL-80415 from the NIH.

## 3.7    References

[CAa] Canham, P.B., Finlay, H.M., Kiernan, J.A., and Ferguson, G.G., Layered structure of saccular aneurysms assessed by collagen birefringence, *Neurol. Res.*, **21** (1999), 618–626.

[CAb] Canham, P.B., Finlay, H.M., and Tong, S.Y., Stereological analysis of the layered collagen of human intracranial aneurysms, *J. Microscopy*, **183** (1996), 170–180.

[DRa] Driessen, N.J.B., Wilson, W., Bouten, C.V.C., and Baaijens, F.P.T., A computational model for collagen fibre remodelling in the arterial wall, *J. Theor. Biol.*, **226** (2004), 53–64.

[FUa] Fung, Y.C., **Biomechanics: Motion, Flow, Stress, and Growth**, Springer-Verlag, New York (1990).

[HOa] Holzapfel, G.A., Gasser, T.G., and Ogden, R.W., A new constitutive framework for arterial wall mechanics and a comparative study of material models, *J. Elasticity*, **61** (2000), 1–48.

[HUa] Humphrey, J.D. and Canham, P.B., Structure, mechanical properties, and mechanics of intracranial saccular aneurysms, *J. Elasticity*, **61** (2000), 49–81.

[HUb] Humphrey, J.D., **Cardiovascular Solid Mechanics: Cells, Tissues, and Organs**, Springer-Verlag, New York (2002).

[HUc] Humphrey, J.D., and Rajagopal, K.R., A constrained mixture model for growth and remodeling of soft tissue, *Math. Models Meth. Appl. Sci.*, **12** (2002), 407–430.

[HUd] Hunter, P.J. and Borg, T.K., Integration from proteins to organs: The Physiome Project, *Nature Rev: Molec. Cell Biol.*, **4** (2003), 237–243.

[KYa] Kyriacou, S.K. and Humphrey, J.D., Influence of size, shape and properties on the mechanics of axisymmetric saccular aneurysms, *J. Biomech.*, **29** (1996), 1015–1022. Erratum, *J. Biomech.*, **30**, 761.

[PAa] Parlea, L., Fahrig, R., Holdsworth, D.W., and Lownie, S.P., An analysis of the geometry of saccular intracranial aneurysms, *Amer. J. Neuroradiol.*, **20** (1999), 1079–1089.

[RAa] Rachev, A., Manoach, E., Berry, J., and Moore, Jr., J.E., A model of stress-induced geometrical remodeling of vessel segments adjacent to stents and artery/graft anastomoses, *J. Theore. Biol.*, **206** (2000), 429–443.

[ROa] Rodriguez, E.K., Hoger, A., and McCulloch, A.D., Stress-dependent finite growth in soft elastic tissues, *J. Biomech.*, **27** (1994), 455–467.

[SAa] Saravanan, U. and Rajagopal, K.R., A comparison of the response of isotropic inhomogeneous elastic cylindrical and spherical shells and their homogenized counterparts, *J. Elasticity*, **71** (2003), 205–233.

[SKa] Skalak, R., Growth as a finite displacement field, in *Proc. IUTAM Symp. on Finite elasticity*, Carlson, D.E. and Shield, R.T., Eds. Martinus Nijhoff (1981).

[TAa] Taber, L.A., A model for aortic growth based on fluid shear stresses and fiber stresses, *ASME J. Biomed. Eng.*, **120** (1998), 348–354.

# 4

## Theoretical Modeling of Cyclically Loaded, Biodegradable Cylinders

J. S. SOARES,

*Department of Mechanical Engineering*
*Texas A&M University*
*College Station, TX 77843, USA*

J. E. MOORE, JR.,

*Department of Biomedical Engineering*
*Texas A&M University*
*College Station, TX 77843, USA*

AND

K. R. RAJAGOPAL

*Department of Mechanical Engineering and*
*Department of Biomedical Engineering*
*Texas A&M University*
*College Station, TX 77843, USA*

ABSTRACT.  The adaptation of fully biodegradable stents, thought to be the next revolution in minimally invasive cardiovascular interventions, is supported by recent findings in cardiovascular medicine concerning human coronaries and the likelihood of their deployment has been made possible by advances in polymer engineering. The main potential advantages of biodegradable polymeric stents are: (1) the stent can degrade and transfer the load to the healing artery wall which allows favorable remodeling, and (2) the size of the drug reservoir is dramatically increased. The in-stent restenotic response usually happens within the first six months, thus a fully biodegradable stent can fulfill the mission of restoring flow while mitigating the probability of long-term complications. However, it is a key concern that the stent not degrade away too soon, or develop structural instabilities due to faster degradation in key portions of the stent. We present here a preliminary model of the mechanics of a loaded, biodegradable cylindrical structure. The eventual goal of this research is to provide a means of predicting the structural stability of biodegradable stents.

As a first step towards a fully nonlinear model, biodegradable polymers are modeled as a class of linearized materials. An inhomogeneous field that reflects the degradation, which we henceforth refer to as degradation, and a partial differential equation governing the degradation are defined. They express the local degradation of the material and its relationship to the strain field. The impact of degradation on the material is accomplished by introducing a time-dependent Young's modulus function that is influenced by the degradation field. In the absence of degradation, one recovers the classical linearized elastic model. The rate of increase of degradation was assumed to be dependent on time and linearized strain with the following characteristics: (1) a material degrades faster when it is exposed to higher strains, and (2) a material that is strained for a longer period of time degrades more rapidly than a material that has been strained by the same amount for a shorter period of time.

The initial boundary value problem considered is that of an infinitely long, isotropic, nearly incompressible, homogeneous, and strain-degradable cylindrical annulus subjected to radial stresses at its boundaries. A semi-inverse method assuming a specific form of the displacement field was employed and the problem reduced to two coupled nonlinear partial differential equations for a single spatial coordinate and time. These equations were solved simultaneously for the displacement and degradation fields using a time marching finite element formulation with a set of nonlinear iterations for each time step.

The main features that were observed were: (1) strain-induced degradation showed acceptable phenomenological characteristics (i.e. progressive failure of the material and parametric coherence with the defined constants); (2) an inhomogeneous deformation leads to inhomogeneous degradation and therefore in an initially homogeneous body the properties vary with the current location of the particles; and (3) the linearized model, in virtue of degradation, exhibits creep, stress relaxation, and hysteresis, but this is markedly different from the similar phenomena exhibited by viscoelastic materials.

## 4.1 Cardiovascular Stents

Since the introduction of angioplasty by Dotter in the 1960s [DOa], catheter-based technologies have improved health care for atherosclerosis. Well over a million balloon dilatations were performed by the early 1990s and as of today more than 600,000 a year are carried out in coronary arteries alone [LIf]. Yet, after more than 20 years of clinical experience and many catheter designs, angioplasty was far from being perfect and the incidence of restenosis remained unchanged. Many studies reported acute complications in 3% to 5% of the patients and restenosis rates at 3 to 6 months between 25 and 50% [GLb, MUa]. Restenosis seemed to be largely independent of the technique, device, or the clinician's skill [ROc, DOb, SAb]. The pathophysiology of restenosis is complex and incompletely understood. Early events in restenosis are thought to consist of immediate elastic recoil, platelet deposition, and thrombus formation, followed by smooth muscle cell proliferation and matrix formation [LIc, CUa, MIc].

Percutaneous implantation of metallic stents in the coronary vessels was first performed in humans in 1987 by Sigwart et al. [SIb]. During the late 1990s, stents revolutionized the field of interventional cardiology and stent implantation has become the new standard angioplasty procedure [ALa]. The major design concept behind cardiovascular stents was to prevent post-traumatic vasospasm [SIb]. Besides keeping the artery patent immediately after intervention, a stent also tackles injured flaps of the lumen preventing downstream embolic complications [PAa, SCc]. Although the concept seemed to be flawless and a significant reduction of the incidence of restenosis was promptly reported [SEa, FIb], all cardiovascular stents have two distinct and significant modes of chronic failure. Immediately after deployment acute thrombosis can occur due to the thrombogenic aspect of the stent promoting a foreign body response, but it can be promptly treated with anticoagulant drug therapy [SCb]. Also the most critical failure mode is in-stent restenosis which still occurs at intolerable rates. Despite the success and growth of stent implantation procedures, there are patients in whom in-stent restenosis is a chronic and recurrent problem [MIb]. The mechanism of in-stent restenosis can be obviously related with restenosis after angioplasty as well as with atherosclerosis and has been shown to be neointimal proliferation in response to injury [SCf] and not chronic stent recoil [SCg].

The reaction of the artery to a stent is a multistage process [EDa]. First, the exposure of the subendothelium and the stent material to the blood stream activates platelets and leads to thrombus formation. This process is initiated immediately after deployment and the extent to which the thrombus deposition occurs is highly correlated not only with the surface characteristics of the stent but also with its design. Areas of flow stagnation, which depend heavily on strut design, influence the degree of platelet adhesion [ROa]. The second stage is inflammation. Stenting overstretches and may even rupture the internal elastic membrane inducing leukocyte adhesion and consequent inflammatory reaction [ROb]. The peak of this process occurs approximately one week after deployment. Deposits of surface adherent and tissue infiltrating monocytes can be seen around stent struts, demonstrating the degree to which the struts are injuring the wall. These monocytes release cytokines, mitogens, and tissue growth factors that further increase neointimal formation [FAa]. The third stage is the proliferation of vascular smooth muscle cells in the media and neointima. This process can be thought as the short-term response to the change in hemodynamics and the wall's response after stent placement [WEc, WEd]. Cellular proliferation provides additional tissue to shore up stress concentrations due to stent deployment [MOa]. The final stage of arterial adaptation is remodeling. One can think of this phase as the artery's attempt to reach a new homeostatic state in the presence of persistent injury and change in the normal environment caused by the stent [GLa].

Systemically administered pharmaceutical agents, besides pre- and post interventional anticoagulant therapies, fail to prevent restenosis because the tolerated dose for such agents is too low to achieve a sufficient drug concentration at the targeted site [LIe]. The problem of in-stent restenosis is currently being addressed by coating stents with polymers in which drugs can be impregnated and locally delivered. Polymers provide a stable medium into which drugs can be either uniformly distributed or surface layered and then locally released over a specific and controlled period of time, usually between weeks to months [WHa]. The first reports of local drug delivery in the cardiovascular system date back only to the mid 1990s with forskolin [LAb] and heparin [AHa]. Success with anti-inflammatory dexamethasome was reported by Lincoff et al. in 1997 [LId]. Suppression of restenotic proliferative stimuli was achieved by Yamawaki et al. in 1998 [YAa]. Successful gene transfer and expression following implantation of polymer stents impregnated with a recombinant adenovirus gene was demonstrated by Ye et al. in 1998 [YEa]. The objective of the pharmacological agents used in drug eluting stents is to address a particular stage of the restenotic cascade: heparin is loaded into stents in order to inhibit the thrombus formation [AHa] and inflammation is prevented with dexamethasone [LId]. The most effective drugs are antimitotic agents that prevent the proliferation stage. As of now, the two most effective and well-studied pharmacological agents for this outcome are Paclitaxel and Sirolimus. Paclitaxel inhibits microtubule depolymerization, and thereby has potent effects in cell division and migration [AXa]. Sirolimus is a macrolide antibiotic with potent antiproliferative effects on vascular smooth muscle cells preventing the initiation of DNA synthesis [MAa, GAa].

The use of drug eluting stents generally improves the success of coronary interventions. In fact, drug eluting stents are now considered general practice [SAa]. Several randomized studies have been carried out and are still ongoing with the objective to evaluate the efficacy of the drugs with regard to their release kinetics, effective dosage, and the benefit of such a particular pharmacological approach [GRc]. Two major randomized trials have been carried out: the RAVEL randomized trial with several followup SIRIUS studies, with Sirolimus eluting stents, showed promising zero restenosis at six months in 238 patients [MOb, MOc], and good results all across the most common subsets of patients and lesion types [MOc, SCe, ARa]; and the TAXUS series of randomized trials, a Paclitaxel eluting stent, also exhibited good restenosis results [GRd, TAe, STb]. As of today, only two polymer-coated drug eluting stents (Cypher$^{TM}$ Sirolimus eluting stent from Cordis, Johnson & Johnson, Miami Lake, FL, USA, and the Taxus$^{TM}$ Paclitaxel eluting stent, Boston Scientific, Natick, MA, USA, introduced in February 2003) are commercially available. Several registries are in effect in several European countries [LEa, ZAa, ONa] (less data on the Paclitaxel eluting stent are available: Cypher$^{TM}$ was introduced in 2002, Taxus$^{TM}$ in February 2003) and show similar results with regard to the two stents and good "real world" success rates ($\approx 1\%$ stent thrombosis, $\approx 10\%$ angiographic restenosis, $\approx 7\%$ target lesion revascularization) [KAa].

Obviously, some problems have been reported for drug eluting stents. Delayed stent thrombosis due to incomplete endothelialization of the stent struts is still a problem with stents, drug eluting or not [JEa]. The "catch-up" effect after the complete elution of the drug raised some concerns [VIa], and the lack of long-term followup studies still haunts this technology. Another limitation is the emphasis given to drug eluting stent implantations on coronaries; only a limited amount of data exists on their application in peripheral arteries. In the SCIROCCO trials, Sirolimus eluting stents deployed in long lesions in peripheral arteries, showed promising short-term results (6% restenosis at 6 months) but no difference relative to bare metal stents after 18 months [DUa, DUb].

## 4.2 Biodegradable Stents

There are some theoretical concerns with metallic stents: (1) most metals are electropositively charged, resulting in high thombogenicity [DEc],

(2) in addition, metal stents remain in the body indefinitely and may interfere with future clinical procedures [AGb]; and (3) due to their microstructural properties, metals are not feasible materials to act as loadable drug carriers. All these problems have encouraged significant efforts in the development of new stent materials, either used in coatings [VAe] or in stents completely made of polymeric materials [MUc]. Polymers can also act as optimal carriers for the controlled release of drugs [PEa].

One possible objective in coating a metal stent is to diminish its thrombogenic properties [DEb]. Experience with Nylon$^{TM}$, silicone, polyurethane, and other materials have been reported in the literature since the beginning of the 1990s [BEc]. Either naturally occurring polymers (fibrin [HOa]) or pharmacological agents (heparin [SEb], dexamethasone [LId], and others [BAa]) relevant to the local biochemistry of the lesion were tested in vivo as coatings. The use of polymers in stent coatings requires less mechanical requisites from the polymer by itself and shifts the attention mostly to biocompatibility and to manufacturability. Still, poor adherence of the coating to the metal, possible delamination with strain, or damage during implantation are problems that may occur [SIc].

Interest in polymeric stents started in the 1990s. Significant progress has been achieved in increasing the level of biocompatibility of polymers tailoring surface characteristics and mechanical strength through advancements in polymerization procedures and processing techniques [PEa]. In 1992, Murphy et al. demonstrated the technical feasibility of polyethylene terephthalate stents but obtained poor results in porcine coronaries, particularly an intense proliferative neointimal response that resulted in complete vessel occlusion [MUb]. On the other hand, around the same time, van der Giessen et al. showed acceptable results with stents made of the same material deployed in the same animal model [VAd]. The extent of neointimal proliferation was similar to that observed after the placement of metal stents (obviously, compared with the standards of that time), despite the presence of a more pronounced inflammatory reaction [VAc]. Later in 1996, van der Giessen et al. investigated the biocompatibility of an array of both biodegradable and nonbiodegradable polymers (polyglycolic acid/polylactic acid, polycaprolactone, polyhydroxybutyrate valerate, polyorthoester, and polyethyleneoxide/polybutylene terephthalate) for stent coating and found a marked inflammatory reaction with subsequent neointimal thickening in all of them [VAb]. The experimental procedure used was completely inappropriate, in that the stents were not sterilized before implantation [FIa]. The biocompatibility of these polymers has been proven in other in vitro and in vivo tests [DEb, DEa].

Because of this general disagreement on the biocompatibility of polymers, the idea of either biodegradable or biostable polymers, which had considerable appeal during the early 1990s, was set aside. The interest peaked in

1994 with Zidar's chapter included in the second edition of the *Textbook of Interventional Cardiology* dedicated in full to the topic of biodegradable stents [ZIa]. Later in 2000, Tamai et al. should be credited with rekindling the resurgence of employing fully biodegradable stents. They provided the first report on the immediate and six-month results after implantation of biodegradable poly-L-lactic acid stents in humans. With their good initial results [TAb] (obviously, compared with the standards before the drug eluting stents era), the motivation for fully biodegradable stents was flourishing once again [FAb].

The rationale behind biodegradable stents can be simply explained in the wonderful allegory of Lucius Quintius Cincinnatus by Colombo and Karvouni in their 2000 *Circulation* editorial: a "biodegradable stent fulfills the mission and steps away" [COa]. Because development of restenosis usually happens during the first six months after deployment [KIa, KAb], a permanent prosthesis that is in place beyond this initial period has no clear function. However, it is worth recognizing that besides leading to unpredictable complications (e.g. stent failure due to fatigue, obstacles for other treatments, and infection due to the presence of a foreign body inside the lumen), there are no demonstrable clinical complications with a permanent intracoronary stent. Thus, the question should be turned around and one should ask what the advantages of a temporary stent are [COa]. The answer is manifold: (1) if a stent degrades and is absorbed by the body it will not be an obstacle for future treatments; (2) if a stent degrades in a controlled manner, its desired failure can be predicted and prescribed; (3) also, the gradual softening of the stent would permit a smooth transfer of the load from the stent to the healing wall; (4) because of the viscoelastic behavior of most polymers, a nonchronic deployment could be designed, preventing arterial injury inherent to balloon inflation; and (5) a biodegradable stent may act as an optimal vehicle for specific therapy with drugs or genes.

Also, there are some drawbacks with regard to permanent metal stents that biodegradable stents would not have. Metal stents remain inside the body indefinitely, becoming a potential nidus for infection [THa], and can be an adverse obstacle for subsequent treatments making bypass surgery almost the only hope for treatment of in-stent restenosis [AGb]. A significant challenge in the development of a novel biodegradable stent is the lack of precise engineering modeling tools [BLa].

Essentially, three different steps are usually taken in the design of such devices. First, only a limited number of biodegradable materials have been tried, and in many cases the materials are picked based on the designer's past experience [STa, HAb]. Scant data are available on the mechanical behavior of polymers during degradation. In most studies, emphasis is given to chemical quantities or phenomenological measurements. Examples are the temporal evolution of molecular weight distribution (MWD) and quantification of mass loss over time. This results in a considerable amount of uncertainty with regard to the design of a biodegradable stent. Secondly, the usual procedure is to pick pre-existing stent designs and manufacture them with biodegradable materials. There are some concerns with manufacturing and sterilization of these polymeric devices when compared with stainless steel counterparts because polymers cannot usually be processed using metal stent techniques [GRb]. Also, the usual forms of solid polymers are fibers, films, or matrices. From these building blocks, the stent must be woven or assembled. The sophistication of the existing designs is variable, ranging from the simplest single fiber helicoidal stents [VAa] to the more complex interwoven stents [GRb, UUa, SUa]. The last step is then to conduct experiments, either in vivo [HIa, UNa], or in vitro [AGb, ZIb], analyze the results, and draw conclusions. Computational simulations with biodegradable stents are either nonexistent or simplistic in virtue of the inability to account for the complexity of the constitutive modeling. Grabow et al. used a finite element analysis to investigate the mechanical properties of a balloon-expandable PLLA stent under various load conditions, whereas Nuutinen et al. used an analytical method for calculating the mechanical properties of braided stents [GRa, NUa]. Both models consider the polymer as being a linearized elastic material with no effect due to the degradation being taken into account.

Because of this inductive way of dealing with the problem, the number of materials and designs used in biodegradable stents is as large as the number of people working in the field. Certain main trends can be identified: (1) biodegradable stents made of bioerodible metals, for example, magnesium [HEa, DIa], are currently in use (a choice that evolved from corrodable iron [PEb]), (2) natural polymers such as type I collagen were used to make tubes [BIa], and lastly, (3) a somewhat large number of biodegradable polymers were tried, more commonly aliphatic polyesters (e.g. polyglycolic acid and polylactic acid) [PEa, VEc]. Poly-L-lactic acid is probably the most commonly used of all these polymers. It was used for the Duke stent [AGb, AGa, LAa], and is being used in the Igaki–Tamai stent [TAb, TAc, TSa, TSb], by Eberhart et al. in their biodegradable stent [YEa, SUa, ZIb, ZIc], and in the Tampere stent for urethral applications [UUa, TAd, ISa]. Unfortunately, almost all of these previous studies focused on the chemical aspects of degradation and not in the mechanical changes occurring during degradation. Regardless of the material that the stent is made of, the issue of structural integrity is the most important for its performance. Structural collapse can take place if weakness occurs in particular regions, so understanding the impact of degradation on local mechanical properties should be the ultimate goal of biodegradable stent design. Obviously, this question does not have an easy answer. Lastly, drug delivery modeled with diffusion kinetics is another important aspect that needs to be addressed and is closely related with degradation, erosion, and mechanical response.

## 4.3  Degradation, Erosion, and Elimination

The number, availability, and utilization of synthetic biodegradable polymers has increased dramatically over the last 50 years, with applications ranging from the field of agriculture to biomedical devices. The first reported biomedical application of biodegradable polymers was absorbable sutures in the 1970s [LAd], and this remains today to be the most widespread use of this family of materials. After 30 years of growth and development, many other devices have become available to the practicing surgeon. Absorbable internal fixation devices for orthopedic surgery, such as pins, screws, suture anchors, and osteosynsthesis plates [PIb]. Biodegradable polymers have been chosen to be the optimal carriers for local drug delivery [LAc] and are widely used in tissue engineering applications [LEb]. The interest in these applications continues to increase as the number of biodegradable polymers evaluated with respect to the concept of biomaterials increases [VEb]. However, the number of compounds having reached the stage of clinical and commercial applications is still small [KHa].

Basically, one can distinguish between the two major applications for biodegradable polymers in the medical field. When used for prosthetic purposes, the contribution of the polymer is required for a limited period of time, especially the healing time, and the polymer can be engineered to degrade at a rate that will transfer load to the healing bone [ATa]. Also, there is no need for a second surgical event for removal [MIa]. To accomplish all of these requirements, the main concern behind the design of the device is its load-bearing capabilities as well as its evolution during degradation over time. On the other hand, for drug delivery implants, the attention is shifted to delivery kinetics and their changes during degradation. The case of a biodegradable drug eluting stent is a bridge connecting the two approaches. The stent must perform mechanically, maintaining the artery patent after deployment and during degradation, and be capable of effective drug delivery.

It is important to make distinctions between the terminologies often encountered in the literature. Biodegradable polymers are polymers that are decompose in the living body but whose degradation products remain in the tissues long-term. On the other hand, bioresorbable polymers can be defined as polymers that degrade after implantation into nontoxic products which are then eliminated from the body or metabolized therein. Although this last term is more precise, it is often used interchangeably with other terms, including absorbable, resorbable, bioabsorbable, and biodegradable [HAb].

In their book, *Biodegradable Polymers and Plastics*, Ottenbrite et al. present a discussion aimed at settling the terminology for such polymers. The conclusion of the discussion board was a set of working definitions. Polymer degradation is a deleterious change in the properties of a polymer due to a change in its chemical structure. A biodegradable polymer is a polymer in which the degradation is mediated at least partially by a biological system. Also, a distinction between degradation and erosion was made. Degradation, defined as the change in chemical structure, is a process different from erosion, defined to be the process of dissolution or wearing away of a polymer [OTa]. Thus, a bioabsorbable polymer automatically implies degradation mediated by a biological system as well as its erosion into nontoxic byproducts that will be then absorbed by the body.

More precisely, polymer degradation is the chain scission process that breaks polymer chains down to oligomers and finally monomers. Degradation leads to erosion, which is the process of material loss from the polymer bulk. Such materials can be monomers, oligomers, parts of the polymer backbone, or even parts of the polymer bulk. Thus, degradation and erosion are distinct but related processes [PIa]. It is worth noting that all polymers undergo backbone chain scission; that is, all polymers "degrade." Only the time they require for degradation is different, and it can range from hours in the case of the hydrolytic degradation of poly-anhydrides, to many years for poly-amines [GOc]. The relationship between the actual life of the polymer and the intended life to perform its function will ultimately dictate the distinction between a polymer being degradable or nondegradable.

Polymers degrade by several different mechanisms, depending on their inherent chemical structure and on the environmental conditions to which they are subjected. Degradation results from an irreversible change of the material which eventually leads to its breakdown or failure. There are five major mechanisms of polymer degradation: thermal, radiation induced, mechanical, enzymatic, and chemical [GOb].

Covalent bonds of the backbone chain of the polymer have a limited strength. In thermal degradation, scission is due to the highly excited vibrational state of bonds attained with increases of temperature. When the energy associated with the vibrational state overcomes the bond dissociation energy, scission and consequently degradation occur. Although these processes cause rapid decomposition of polymers only at highly elevated temperatures (around $500°C$), the pronounced temperature dependence of the rates of chemical reactions can cause a significant and rather rapid degradation already under milder conditions [WEb]. Radiation induced degradation occurs when polymers undergo chemical reactions upon irradiation with ultraviolet light or gamma radiation [SCd]. In general isothermal biomedical applications, such as the use of endovascular biodegradable stents, thermal degradation and radiation-induced degradation are not assumed to be relevant.

Mechanical degradation of polymers comprises a large number of different phenomena, ranging from fracture to chemical changes induced by the mechanical environment. Mechanical energy transferred to a polymeric system can be dissipated via two main relaxation processes: enthalpy relaxation, defined to be the slippage of chains relative to surrounding molecules, and entropy relaxation, changes of chain conformation. These relaxations are harmless to the polymer because they do not induce chemical changes. In competition with these relaxation processes, the scission of chemical bonds can occur. Obviously, the probability for bond scission should increase as relaxation is impeded. A single, generally applicable mechanism of stress induced chemical reactions does not appear to exist. It seems that different bond scission mechanisms are operative depending on the state of the polymer (solid, rubbery, or molten) and the mode of imposition of stress. In solid polymers, fracture planes and voids might give rise to the rupture of chemical bonds. In the rubbery state or molten in solution, inter- and intrachain entanglements might cause stretching of parts of the macromolecules, resulting eventually in bond scission. Strain is a prerequisite for bond rupture in polymer chains regardless of the state of the material; that is, bond rupture occurs when sufficient energy is concentrated in a certain segment of a macromolecule as a consequence of the nonuniform distribution of internal stresses [SCd].

Enzymatic degradation is mainly relevant for natural polymers such as proteins, polysaccharides, or poly $\beta$-hydroxy esters, for which specific enzymes exist [GOc]. Chemical degradation is a general classification of molecular weight reduction due to chemical reactions that start spontaneously when certain low molecular weight compounds are brought in contact with the polymer [SCd]. Hydrolysis and oxidation are classic examples of chemical degradation.

The prevailing mechanism of biological degradation for synthetic biodegradable polymers is scission of the hydrolytically unstable backbone chain by passive hydrolysis, because for most of them, no specific enzymes exist [WEa]. By tailoring the polymer backbone with hydrolyzable functional groups, the polymer chains become labile to an aqueous environment and their ester linkages are cleaved by absorbed water [HAa]. There are several factors that influence the rate of this reaction: the type of chemical bond, pH, co-polymer composition, and water uptake are the most important [GOb]. Other factors can also be relevant: residual monomer concentration [HYa], autocatalysis [SId], temperature [WEb], chemical environment [ZHa], and initial molecular weight [IVa], just to name a few. On the other hand, one must realize that although the number of factors that influence the degradation of polymers might be infinite, under the conditions of interest only some might be relevant. Moreover, inherent chemical and physical changes to the polymer and to the surrounding environment might have a substantial feedback on the degradation rate [GOc].

For semi-crystalline polymers, hydrolysis occurs in two distinct stages: initially, water penetrates the polymer, preferentially attacking the more accessible chemical bonds in the amorphous phase and converting long polymer chains into shorter, ultimately water-soluble fragments [GOb]. Because the amorphous phase is degraded in the first place, there is a reduction in molecular weight without a loss of apparent physical properties as the polymer matrix is still held together by the crystalline regions. The reduction in molecular weight is soon followed by a reduction in physical properties as water begins to fragment the polymer bulk [ALb, ALc].

The diffusion of water into the polymer bulk and polymer degradation compete against each other in the process of polymer erosion. If degradation is fast, the diffusing water is absorbed quickly by hydrolysis and hindered from penetrating deep into the polymer bulk. In this case, degradation and consequently erosion are restricted to the surface of the polymer, a phenomenon referred to as heterogeneous or surface erosion [TAa]. This type of erosion changes if degradation is slower than the rate of diffusion of the water through the polymer. In this case water cannot be absorbed quickly enough to be hindered from reaching deep into the polymer and the polymer degrades and erodes through its cross-section, a behavior which has been termed homogeneous or bulk erosion [TAa]. It must be stressed, however, that surface and bulk erosion are two extremes and the erosion mechanism in a degradable polymer usually shows characteristics of both.

In addition to water diffusion and bond stability, other factors such as water uptake which depends on the hydrophilicity of the polymer affect the hydrolysis rate and the erosion behavior of polymers substantially [BUa]. As should be expected, many different types of morphological changes occur upon erosion. An increase in surface roughness and the formation of cracks, macropores, and micropores are common phenomena observed in degrading polymers. Erosion fronts, which separate eroded from noneroded polymer, have been reported for surface eroding polymers such as poly(anhydrides) [KAc]. In contrast, inversely moving erosion fronts have been observed in poly(DL-lactide) [LIa], where polymer degradation is increased inside eroding polymer matrices due to the autocatalytic activity of monomers that have been created [SId]. Due to the preference for the amorphous phase, the degree of crystallinity of degradable polymers can change tremendously during erosion [PIc]. Additional changes in crystallinity are a consequence of the recrystallization of oligomers and monomers [LIb].

Elimination is the concluding stage of the complete function of a biodegradable implant. The obvious requirement is to have a polymer that is biocompatible during the whole time of permanence inside the body as well as its breakdown products being eliminated through metabolism in a nontoxic manner [ZIa]. The biocompatibility of aliphatic polyesters, especially polyglycolic and polylactic acid, is well established in the literature [SCa, PId, NGa]. On the other hand, the elimination of the byproducts of degradation and erosion appears to follow different mechanisms for different polymers and is controversial. Ultimately, the elimination involves the solubilization of the degradation products which are then carried away from the implantation site and eliminated [KAc]. The surrounding tissue (in the case of a biodegradable stent, the artery wall) must be capable of absorption, digestion, and elimination of the resulting oligomers and monomers [GUa]. The last step is the removal of these waste products from the blood. Lysosomal degradation is the major pathway for the elimination of polymers that cannot be excreted directly via the kidney [WIa].

## 4.4 Models of Degradation and Erosion

Theoretical models to predict polymer degradation and erosion would seem to be important tools for a number of different applications. If drug elution is to be part of the therapy, drug delivery profiles should be programmable at the design stage. For orthopedic applications, load-bearing capabilities as well as their evolution with time must be determined. A drug eluting biodegradable stent should ideally be designed accounting for all of these criteria.

Hydrolysis degradation is the breakage of backbone bonds caused by incoming water and is a phenomenon that occurs at the molecular level [VEa]. It is a very intricate process, as a variety of different degradation mechanisms can occur simultaneously and concurrently. Also, the reactivity of each bond might be equal when considered individually, but the large number of repeating units and their inherent steric environment, weak links, and branches, may influence the local rate of reaction [NGb]. The probability of hydrolysis and consequent scission of a particular bond is expressed as a distribution function (commonly random scission, central Gaussian and parabolic). Monte Carlo or other more complex techniques are applied to populations of simple polymers to predict the theoretical evolution of MWD [NGb]. Experiments with gel permeation chromatography provide data to model the mechanism of degradation [NGc]. Other degradation models based on kinetics have been reported [BRa, BEb]. Lastly, complex degradation schemes depicting possible mechanisms can be developed and computationally solved to obtain realistic MWD evolutions [BOa, YOa].

Erosion is the dissolution of oligomers and monomers resulting from degradation. Joshi and Himmelstein [JOa] proposed a reaction-diffusion model for degradation and drug release, consisting of Fick's law of diffusion coupled with a reaction equation describing the kinetics of the degradation mechanism. Theoretical results for drug release, water penetration, and erosion were obtained as a consequence of degradation and were corroborated with experimental results [JOa, THb]. One drawback of this model is that it does not take into account changes in the microstructure caused by the preferential erosion of the amorphous phase compared with crystalline phase.

Gopferich and Langer developed different models for erosion [GOa]. They describe erosion as being a probabilistic event and the polymer matrix as a grid of pixels. Monte Carlo simulations coupled with a reaction equation describing random scission were performed. Different properties can be assigned to each pixel, so a distinction between the crystalline and amorphous phase was considered. By removing eroded pixels continuously from the grid, temporal evolutions of a degradable polymer matrix can be determined stochastically. From such simulations, many experimentally measurable parameters can be calculated, such as porosity or weight loss. Erosion fronts and erosion modes can also be inferred from the results of the simulation. The fit of experimental data allows the determination of the erosion rate constants and demonstrates that the stochastic model is quite well able to adjust to the experimental data. Later models by the same group included diffusion equations to obtain theoretical results on the release of drugs through the pores [SIa].

Up until now, most of the research effort on biodegradable polymers were directed experiments and product development. A fair amount of experimental data concerning biodegradable polymers exists, ranging from MWD evolutions, mass loss, and amount of drug eluted. Because of the complexity of these materials and the variety of processes to which they are subjected, the modeling effort has been very limited. The existing models are based on widely different approaches, certainly driven by the field of application. Drug delivery and erosion for drug delivery implants are far better understood when compared to the impact of degradation on the load-bearing performance in orthopedic applications, usually based on phenomenological models with data from in vitro and in vivo experiments. We were not able to identify any previous study of the impact of degradation and erosion on the mechanical response of polymers.

As can be expected, the biodegradation of the polymers that constitute a stent depend on two classes of factors: the mechanical environment and the biochemical environment. One could easily imagine an astonishingly large number of parameters in these broad categories that potentially could influence the degradation, ranging from stress or strain on a strut to the concentration of a particular compound present in the blood. Besides degradation modeling (how the polymer chains are broken) and erosion modeling (how monomeric and oligomeric products are washed out), the modeling of the mechanical response is equally relevant. To know how the degradation influences the mechanical response requires significant effort. The next sections outline our initial modeling efforts, which are aimed at developing a tool for biodegradable stent design.

## 4.5   Model Description

As mentioned earlier we introduce a measure of the degradation through a field $d(\mathbf{x}, t)$, which we refer to as the degradation field. It is a scalar field defined over the body and is assumed to be positive. It reflects the bond scission of the polymer backbone chains and results solely in molecular weight reduction. One can think of degradation as a measure of the density of broken bonds. Another important assumption is that degradation should be a consequence of bond scission and the factors on which it depends. Degradation is assumed to depend on the strain field and time and only mechanical degradation is described by our model. We define the evolution of the degradation through

$$\frac{\partial d(\mathbf{x}, t)}{\partial t} = \hat{D}(\epsilon, t), \tag{5.1}$$

where $\epsilon$ is the linearized strain. In a problem involving large strains, we use a nonlinear measure of strain such as the Almansi–Hamel strain. This relation reflects the mechanism of scission caused by strain and describes mechanical degradation.

The rationale for the choice of (5.1) for the evolution of the degradation field is the following: (1) a material degrades faster when exposed to higher strains, and (2) a material that is strained for a longer period of time degrades more rapidly than a material that has been strained by the same amount for a shorter period of time. In other words, materials subjected to larger strains, other things being held equal, degrade faster; and materials subjected to the same strain for a longer time, other things being held equal, degrade faster [RAa]. Obviously, this behavior depends on the

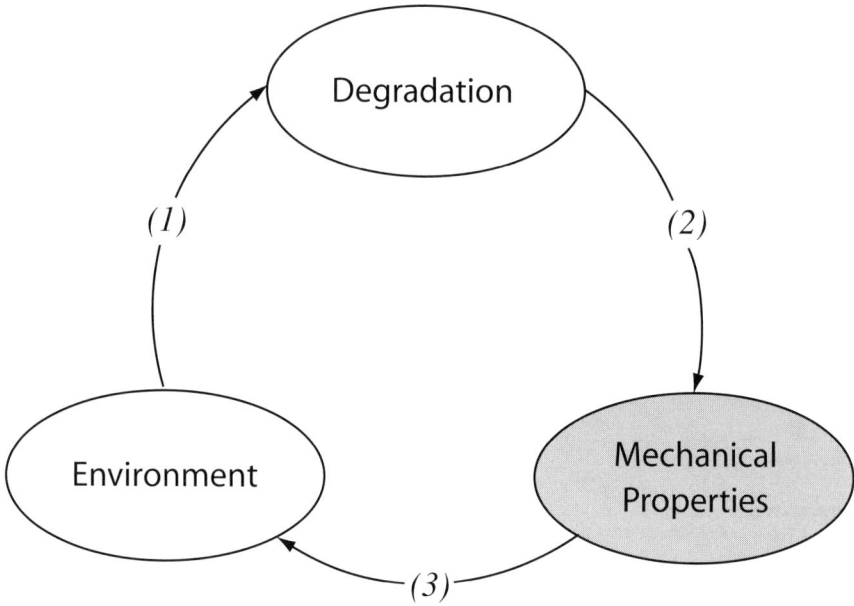

Figure 4.1. General pathway of strain induced degradation. A material degrades depending on the strain to which it is subjected (1). As degradation proceeds its mechanical properties decrease (2), leading to a new equilibrium position (3) that will be responsible for more degradation (1).

choice of $\hat{D}(\epsilon, t)$. Examples of mechanical degradation are common. Aging processes, ultrasonic degradation, stress induced chemical alterations of polymers, and mastication of rubber can be described through these mechanisms. The closed loop cause–effect mechanism of degradation is shown in Figure 4.1.

   The body was considered to be a linearized elastic body when degradation was absent. Although this simple model does not describe polymeric materials undergoing large deformations, the choice of this model was made based in virtue of the simplicity of the governing equations that it yields. The methodology can certainly be extended to the finite deformation case, as will be necessary for a fully realistic model of a stent. For the same reasons, a simple geometry was chosen in order to obtain an easy mathematical problem. The classical linearized elasticity solution of a cylindrical pressure vessel was chosen in order to obtain a problem that involves only one spatial variable. The cylindrical model is representative of stent geometries, and provides a means to obtain results in a tractable framework. Incorporation of this model into a finite element code would allow for more complex geometries to be modeled.

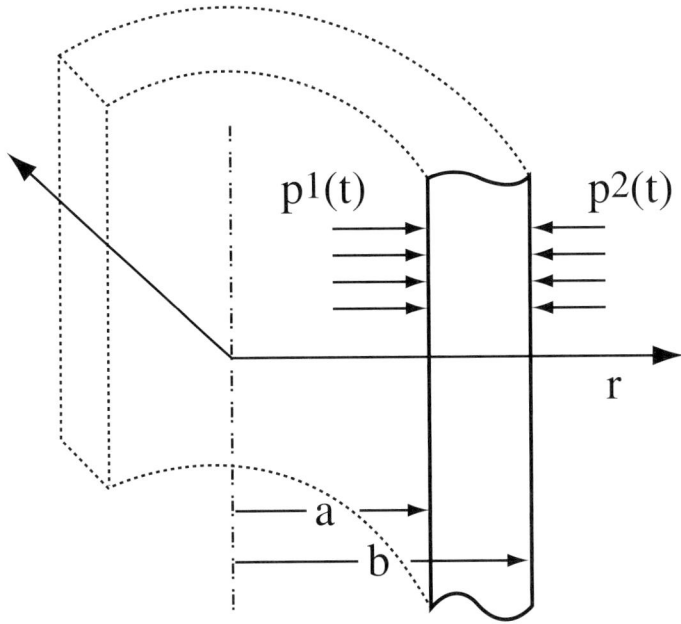

Figure 4.2. Geometry of the body considered.

Consider an infinitely long, homogeneous, isotropic, elastic cylindrical annulus described by the classical linearized theory with inner and outer radii $a$ and $b$, respectively, under radial pressure at both the inner and outer surfaces as shown in Figure 4.2. A semi-inverse method can be used to solve this simple classical linearized elasticity problem. We assume that

$$\mathbf{u} = u(r, t)\mathbf{e_r}. \tag{5.2}$$

The linearized strain $\epsilon$ is defined by

$$\epsilon = \frac{1}{2}\left(\nabla \mathbf{u} + \nabla \mathbf{u}^{\mathsf{T}}\right).$$

In this problem, there are only two nonzero components of the strain tensor (expressed in cylindrical polar coordinates defined along the cylindrical annulus).

$$\epsilon_{rr} = \frac{\partial u}{\partial r}, \tag{5.3}$$

$$\epsilon_{\theta\theta} = \frac{u}{r}. \tag{5.4}$$

The stress field $\sigma$ can be computed in terms of two material parameters. In this case, the Young's modulus $E$ and the Poisson ratio $\nu$ are used and the constitutive relation takes the form

$$\boldsymbol{\sigma} = \frac{E}{1+\nu}\boldsymbol{\epsilon} + \frac{E\nu}{(1+\nu)(1-2\nu)}(\text{tr }\boldsymbol{\epsilon})\mathbf{I}. \tag{5.5}$$

Only three components of the stress tensor are different from zero, $\sigma_{rr}$, $\sigma_{\theta\theta}$, and $\sigma_{zz}$. Substituting (5.3) and (5.4) into (5.5), the stresses are given by

$$\sigma_{rr} = \frac{E\nu}{(1+\nu)(1-2\nu)}\left[(1-\nu)\frac{\partial u}{\partial r} + \nu\frac{u}{r}\right], \tag{5.6}$$

$$\sigma_{\theta\theta} = \frac{E\nu}{(1+\nu)(1-2\nu)}\left[(1-\nu)\frac{u}{r} + \nu\frac{\partial u}{\partial r}\right], \tag{5.7}$$

$$\sigma_{zz} = \frac{E\nu}{(1+\nu)(1-2\nu)}\left(\frac{\partial u}{\partial r} + \frac{u}{r}\right). \tag{5.8}$$

The balance of linear momentum yields the equation

$$\text{div }\boldsymbol{\sigma}^T + \rho\mathbf{b} = \rho\frac{\partial^2 \mathbf{u}}{\partial t^2}. \tag{5.9}$$

The density $\rho$ was assumed not to change during degradation. This assumption can be supported by experimental data on bulk erosion. Degradation takes place over the entire body and only after some time erosion mechanisms are triggered [PIa]. Before that happens, backbone chain scission can be assumed to happen without mass loss. This once again stresses the difference between degradation and erosion. Using the predetermined stress field (5.6)–(5.8) in the local form of the momentum balance (5.9) and assuming no body forces, the only remaining nontrivial component is the radial component, which simplifies to

$$\frac{\partial \sigma_{rr}}{\partial r} + \frac{\sigma_{rr} - \sigma_{\theta\theta}}{r} = \rho\frac{\partial^2 u}{\partial t^2}. \tag{5.10}$$

The degradation field $d(\mathbf{x}, t)$ is assumed to depend only on the radial position and time, $d(r, t)$. This field quantifies the progress of degradation of the material at each current location. It can only assume positive values without any upper bound and if it is zero it means no degradation has taken place.

To model a biodegradable polymer, the parameters describing the mechanical properties are allowed to decrease as degradation progresses. An equation relating the decrease in the mechanical properties with the increase of degradation is assumed. In this model, it was assumed that only the Young's modulus of the material decreases during degradation in the following manner,

$$E(r, t, d) = E_0[1 - \beta d(r, t)], \tag{5.11}$$

where $E_0$ is the Young's modulus associated with the virgin specimen and $\beta$ is a constant that weights how the degradation field leads to a reduction of the mechanical properties. The Poisson ratio was assumed to be constant with a value close to that for an incompressible body ($\nu = 0.49$), so as to reflect the near constant density of the polymeric materials under consideration. Then, the constitutive equation (5.5) assumes the form

$$\boldsymbol{\sigma} = \frac{E(r,t,d)}{1+\nu}\boldsymbol{\epsilon} + \frac{E(r,t,d)\nu}{(1+\nu)(1-2\nu)}(\operatorname{tr}\boldsymbol{\epsilon})\mathbf{I}. \tag{5.12}$$

To relate the increase of degradation to the mechanical stimuli to which the body is subject, the equation that governs the degradation (5.1) is assumed to be of the following form,

$$\frac{\partial d(r,t)}{\partial t} = D(t)\epsilon_{\theta\theta}, \tag{5.13}$$

where $D(t)$ is a strain degradation parameter. Generally speaking, the rate of increase of degradation is assumed to be dependent solely on the linearized strain. Due to incompressibility, the dilation (the trace of the linearized strain tensor) is approximately zero. Thus, because of particular characteristics of the strain field of a pressurized cylindrical annulus (i.e. plane strain), $\epsilon_{zz}$ is zero and $\epsilon_{\theta\theta}$ and $\epsilon_{rr}$ add up to zero. For this problem, the degradation was simplified to be dependent on just one component of the strain field, more precisely the hoop strain $\epsilon_{\theta\theta}$. Also, it is assumed that the rate of increase of degradation would have the separable representation shown in (5.13).

Introducing the stress field (5.6)–(5.8) together with (5.12) into the balance of linear momentum (5.10) yields a second-order partial differential equation

$$A_1 u_{rr} + A_2 u_r - A_3 u = \rho u_{tt} \qquad a < r < b, t > 0, \tag{5.14}$$

where each letter subscript represents a partial derivative with respect to that variable. The coefficients $A_1$ through $A_3$ depend on the degradation field through the following expressions.

$$A_1 = \frac{E_0(1-\nu)}{(1+\nu)(1-2\nu)}[1-\beta d], \tag{5.15}$$

$$A_2 = \frac{E_0(1-\nu)}{(1+\nu)(1-2\nu)}\left[\frac{1-\beta d}{r} - \beta d_r\right], \tag{5.16}$$

$$A_3 = \frac{E_0(1-\nu)}{(1+\nu)(1-2\nu)}\left[\frac{\nu}{1-\nu}\frac{\beta d_r}{r} - \frac{1-\beta d}{r^2}\right]. \tag{5.17}$$

Substituting the hoop strain (5.4) into the equation governing the degradation (5.13) yields

$$d_t = \frac{D(t)}{r}u. \tag{5.18}$$

Equation (5.14) is a nonlinear hyperbolic PDE. The nonlinearity arises from the dependence of the Young's modulus on the degradation field. Equation (5.18) is parabolic. Both must be solved for $d(r,t)$ and $u(r,t)$ simultaneously. Traction boundary conditions are imposed on both surfaces of the annulus:

$$\begin{cases} \sigma_{rr}|_{r=a} = -p_1(t) \\ \sigma_{rr}|_{r=b} = -p_2(t) \end{cases} \qquad t > 0. \tag{5.19}$$

Initial conditions on the displacement and degradation fields must be specified. They are:

$$\begin{cases} u|_{t=0} = 0 \\ \dot{u}|_{t=0} = 0 \qquad a < r < b \\ d|_{t=0} = 0 \end{cases} \tag{5.20}$$

with the understanding that the annulus is at rest and nondegraded initially.

## 4.6   Methods

Because of the nonlinear nature of the resulting partial differential equations, a time marching nonlinear finite element scheme with a set of nonlinear iterations for each time step was implemented to simultaneously solve both equations (5.14) and (5.18) for the unknown functions $u(r,t)$ and $d(r,t)$, subject to the boundary conditions (5.19) and the initial conditions (5.20). The infinitely long cylindrical annulus was modeled to simulate a stent deployed inside a coronary artery (shown in Figure 4.2). The outer radius, $b$, was chosen to be 2.5 mm, and represents an average coronary artery. Considering a typical strut thickness of 100 $\mu$m, the inner radius, $a$, was chosen to be 2.4 mm. Pressure applied on the inner surface of the annulus, $p_1(t)$ in (5.19), was chosen to simulate the pressure during blood flow. A steady pressure field of 13 kPa ($\approx$ 98 mmHg) was superimposed with an oscillatory component with an amplitude of 2.75 kPa, yielding a systolic and diastolic pressure of 118 mmHg and 77 mmHg respectively. The frequency was taken to be 1 beat per time unit, simulating resting conditions.

The outer pressure, $p_2(t)$ in (5.19), representing the crushing action of the artery in the stent after deployment, was taken to be constant and with value of 202 kPa (2 atm). This value was based in a 3-D finite element model of the deployment of a metal stent in a hyperelastic coronary artery [BEa]. The mechanical properties chosen were for PLLA assuming a linear elastic behavior. PLLA is generally incompressible, so a Poisson ratio $\nu$ of 0.475 was chosen. The nondegraded Young's modulus $E_0$ in (5.11) was assumed to be 3.5 GPa, according to published data obtained from experimental results [DRa, GAb, LUa].

The domain of the problem was meshed with 1-D Lagrangian quadratic elements. Mesh convergence was verified and 20 elements were chosen, being the best compromise between precision and a fast computational running time.

This problem deals with two completely different time scales. The frequency of the oscillatory component of the inner pressure is one sinusoidal cycle per time unit. The degradation, on the other hand, varies over days, months, or even years. The time march was performed through 25 time units. Although any degradation over this time interval is completely nonphysical, the main purpose of this analysis was to understand the role of the degradation constants $\beta$ and $D(t)$ on the degradation process (cf. Eq. (5.11) and (5.13)). A time step of 0.05 seconds was chosen. This time step was enough to assure stability of the time marching scheme.

Two types of data were analyzed. From one standpoint, temporal evolutions of important quantities at a given point in the spatial domain were obtained, for example, the displacement of the inner surface of the annulus as a function of time, or the variation of the three nonzero components of the stress field at the outer surface. On the other hand, fields can be plotted along the entire domain at a given moment in time, and the evolution of the fields considered can be characterized.

As a starting point, a representative problem was studied with the degradation parameters taken to be $\beta = 5$ and $D(t) = 1$. Then, a parametric analysis was performed with regard to these two representative parameters. Three more cases were considered to study the effect of $\beta$, one lower ($\beta = 1$), and two greater ($\beta = 10$ and $\beta = 20$). The study on the influence of $D(t)$ in the degradation was done in two distinct steps. Firstly, three more constant $D(t)$ were considered, $D(t) = 2.5$, $D(t) = 5$, and $D(t) = 10$. Then, to have a greater insight on the role of $D(t)$ on the degradation mechanism, several shapes were considered and are shown in Figure 4.3. Table 4.1 shows schematically the cases considered.

To access the effects of the thickness of the annulus on the degradation process, two additional geometries besides the one considered in the representative case above were considered. The parameters of degradation, boundary conditions, and initial material properties were kept constant, and an aspect ratio, defined to be the ratio of the outer radius to the thickness of the annulus (i.e. $b/(b-a)$), was varied and is systematized in Table 4.2.

Lastly, to assess the influence of the applied loads on the degradation of the annulus, a different type of analysis was performed. The outer pressure was cycled between the inner pressure and 202 kPa considered in all of the previous cases. Four linear loading and unloading cycles over 100 time

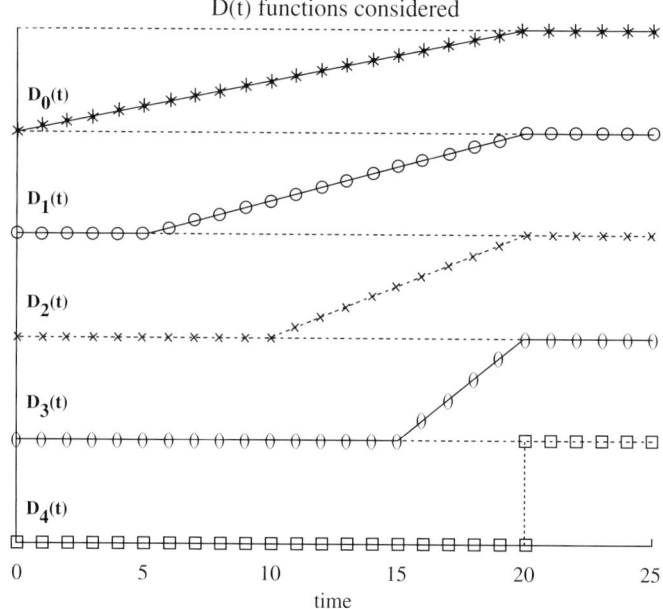

Figure 4.3. Shapes of several $D(t)$ considered. Ultimately, all will end with the same $D(t) = 1$. $D(t) = 0$ yields no degradation.

units were considered. Two different sets of degradation parameters were analyzed, $\beta = 5$, $D(t) = 1$ and $\beta = 10$, $D(t) = 2$.

## 4.7   Results

As time increases and degradation proceeds, the overall value of the radial displacement of the cylindrical annulus increases at a progressive rate, that is, the annulus creeps inwards when subjected to the same loads (Figure 4.4). For this initial representative case ($\beta = 5$ and $D(t) = 1$; cf. Table 4.1), the inner displacement increases 0.4 $\mu$m over 25 time units, a 15% increase when compared with the nondegraded radial displacement.

Due to the kinematical characteristics of this deformation and the strain field that it causes, a slightly inhomogeneous increase in degradation is observed (Figure 4.5). The inner part of the annulus is subjected to greater strains and therefore to a stronger degradation. This degradation has consequences on the material properties and the Young's modulus decreases steadily through the thickness as degradation and time increase. At the inner surface, where the degradation is more intense, a reduction of

| | $\beta$ | $D(t)$ |
|---|---|---|
| **Representative Case** | 5 | 1 |
| **Parametric Study** | | |
| *Influence of* $\beta$ | 1 | 1 |
| | 5 | |
| | 10 | |
| | 20 | |
| *Influence of* $D(t)$ | 5 | 1 |
| | | 2.5 |
| | | 5 |
| | | 10 |
| *Influence of the shape of* $D(t)$ | 5 | D0 |
| | | D1 |
| | | D2 |
| | | D3 |
| | | D4 |

Table 4.1. Summary of the cases considered for the parametric study of the degradation parameters $\beta$ and $D(t)$. The specific forms of the functions D0 through D4 are shown in Figure 4.2.

| | a | b | b/(b-a) |
|---|---|---|---|
| **Representative case (thin-walled annulus)** | 2.4 | 2.5 | 25 |
| **Medium thickness annulus** | 1.5 | 2.5 | 2.5 |
| **Large thickness annulus** | 0.5 | 2.5 | 1.25 |

Table 4.2. Summary of the cases considered for the parametric study of the degradation parameters $\beta$ and $D(t)$. The specific forms of the functions D0 through D4 are shown in Figure 4.2.

approximately 15% in the Young's modulus after 25 time units is observed (Figure 4.6).

The linearized strain field follows the same behavior as the displacement. The radial strain $\epsilon_{rr}(r,t)$ is positive and the hoop strain $\epsilon_{\theta\theta}(r,t)$ has a negative value (Figure 4.7). All the other components of the strain field tensor in this cylindrical coordinate system are zero. The two nonzero

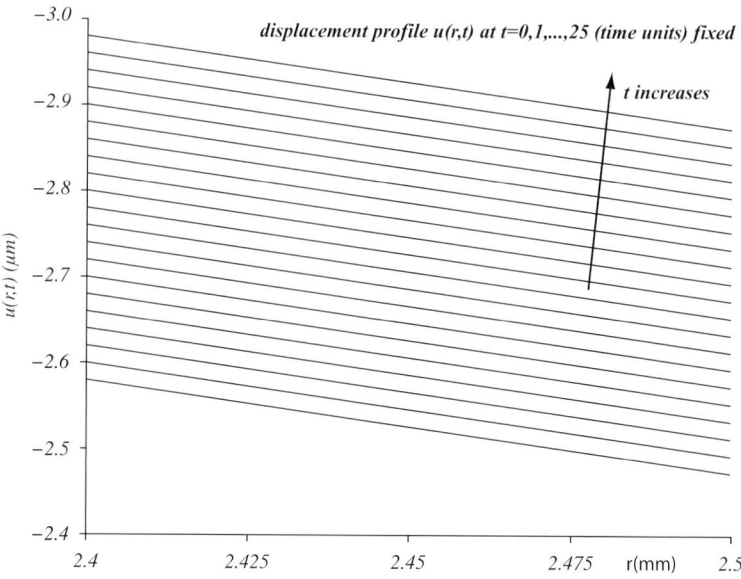

Figure 4.4. Displacement profiles $u(r,t)$ at $t = 0, 1, 2, \ldots, 25$ (time units). As time increases, the overall value of the displacement increases in absolute value with a progressive rate.

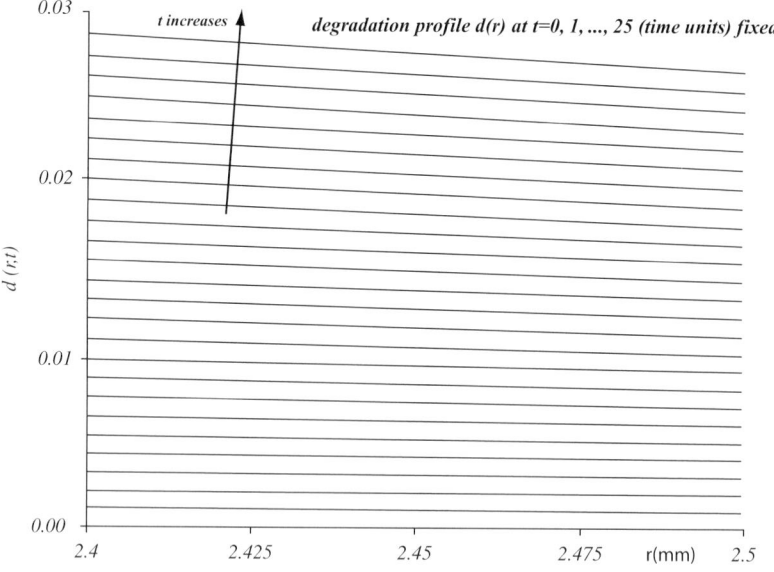

Figure 4.5. Degradation profiles $d(r,t)$ at $t = 0, 1, 2, \ldots, 25$ (time units). As time increases, the degradation increases. Degradation is nearly homogeneous, but is slightly more aggressive near the inner wall.

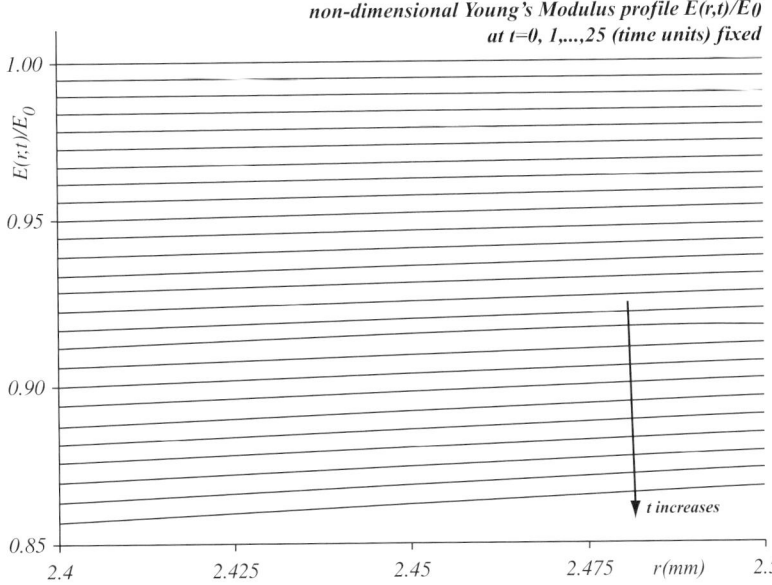

Figure 4.6. Nondimensional Young's modulus profile $E(r,t)/E_0$ at $t = 0, 1,$ $2, \ldots, 25$ (time units). The Young's modulus, initially constant throughout the thickness, shows radial dependence after the onset of degradation.

strains add up to approximately zero due to the near incompressibility of the material. Lastly, the values of the components of the linearized strain tensor are close to zero, thus the necessary requirement of small displacement gradients is verified.

Stress relaxation in the inner half of the annulus and an increase in the overall value of stress in the outer part are observed (Figures 4.8 and 4.9). Although both the hoop stress $\sigma_{\theta\theta}(r,t)$ and the axial stress $\sigma_{zz}(r,t)$ show significant differences from the classical linearized solution, stress relaxation and stress intensification are more relevant in the former than in the latter. An approximately 30 kPa relaxation in $\sigma_{\theta\theta}(r,t)$ at the inner surface and a similar increase in the outer part are observed (Figure 4.8). The axial stress departs from the constant through the thickness result obtained from the classical linearized solution, and the relaxation and intensification patterns are similar (Figure 4.9). However, the radial stress $\sigma_{rr}(r,t)$ shows little effect of degradation (Figure 4.10); due to the pressure boundary conditions, the whole radial stress profile is prescribed for all times.

### 4.7.1 On the Influence of the Load

Hysteresis loops are observed in the hoop stress $\sigma_{\theta\theta}(a,t)$ versus hoop strain $\epsilon_{\theta\theta}(a,t)$ diagram at the inner surface when the outer pressure is steadily

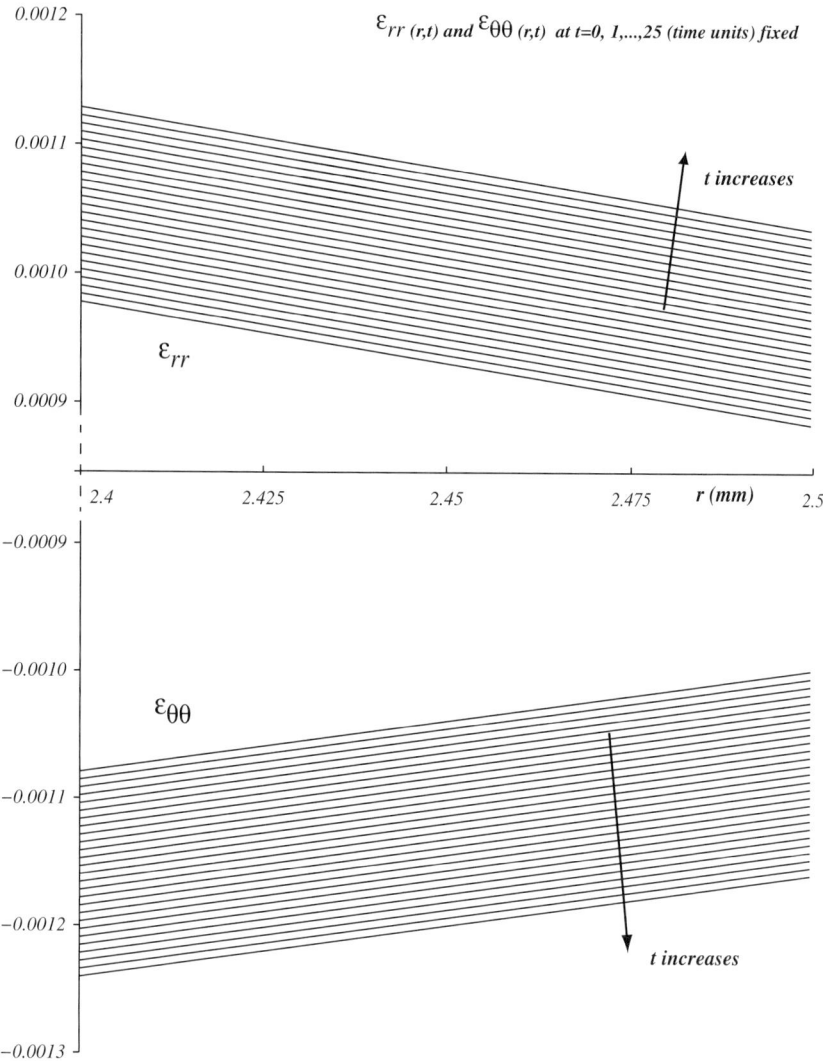

Figure 4.7. Strain field $\epsilon_{rr}(r,t)$ and $\epsilon_{\theta\theta}(r,t)$ at $t = 0, 1, 2, \ldots, 25$ (time units). The order of magnitude of the components of the infinitesimal strain tensor is small, therefore the assumption of linearized elasticity holds for this deformation.

cycled four times between the inner pressure and 202 kPa (Figure 4.11). Hysteresis is dependent on the degradation parameters, with the area spanned by the hysteresis loop increasing as degradation proceeds. Not only are the effects of degradation indistinguishable for the first cycle between

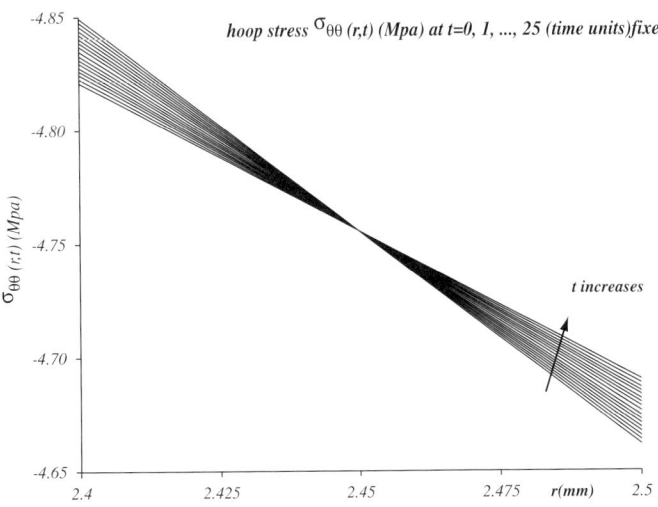

Figure 4.8. Hoop stress $\sigma_{\theta\theta}(r,t)$ at $t = 0, 1, 2, \ldots, 25$ (time units). Stress relaxation is observed in the inner half of the annulus. On the outer half, the stress increases in order to satisfy the linear momentum balance.

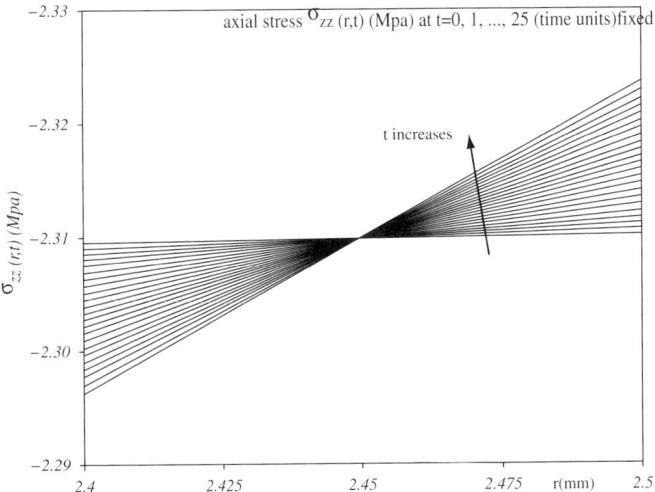

Figure 4.9. Axial stress $\sigma_{zz}(r,t)$ at $t = 0, 1, 2, \ldots, 25$ (time units). The nondegraded classical solution would yield a constant valued axial stress through all the annulus thickness. Due to degradation, relaxation occurs in the inner half.

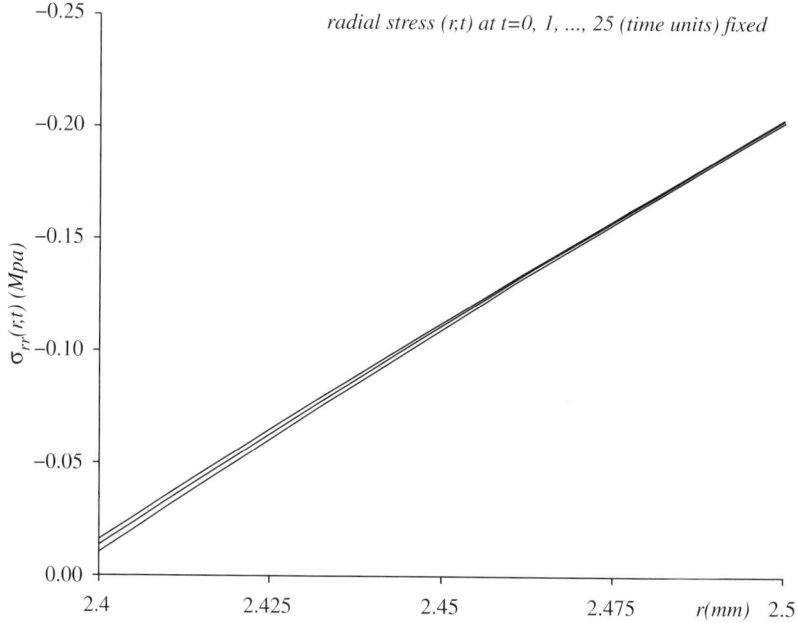

Figure 4.10. Radial stress $\sigma_{rr}(r, t)$ at $t = 0, 1, 2, \ldots, 25$ (time units). Due to BCs, $13 \pm 2.75$ kPa at the inner radius and 202 kPa in the outer radius, the whole profile is prescribed for all times.

the two sets of degradation constants, but also as degradation proceeds the curves deviate from each other. Lastly, no permanent set is induced due to the degradation imparted through this model; when the annulus is unloaded, it returns to its original configuration.

### 4.7.2 On the Influence of the Thickness of the Wall

The most important difference obtained through the different geometries referred to in Table 4.2 is the decrease of the homogeneity of the deformation. A deformation is said to be homogeneous if straight lines are mapped through the deformation into straight lines (or, if in a Cartesian coordinate system, the components of the deformation gradient are constants). In the case of a thin-walled annulus, the deformation is almost homogeneous, as can be inferred by the flatness of the displacement profile (Figure 4.12, first column). When the thickness of the annulus is increased, the deformation becomes more inhomogeneous, leading to a less flat radial displacement profile and consequently a more inhomogeneous degradation (Figure 4.12, second and third columns).

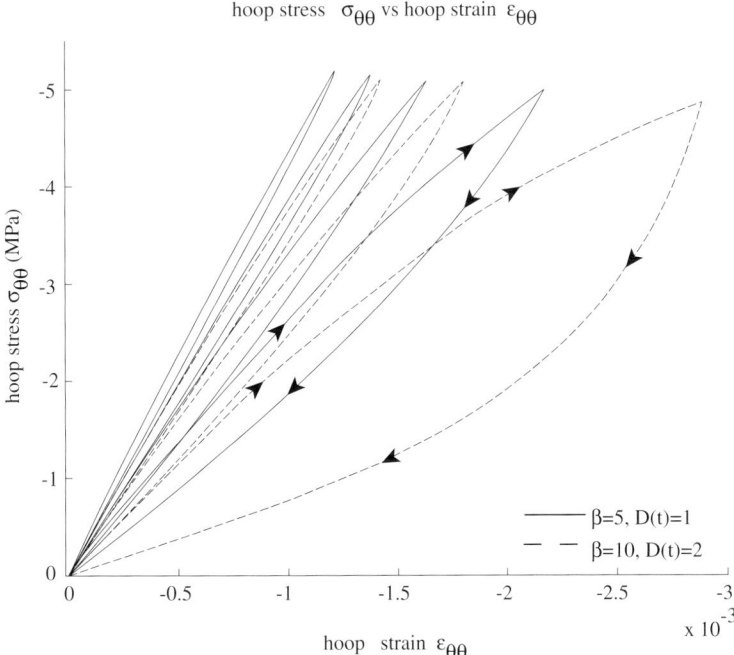

Figure 4.11. Hoop stress $\sigma_{\theta\theta}(a,t)$ versus hoop strain $\epsilon_{\theta\theta}(a,t)$ at the inner radius. The outer pressure $p_2(t)$ cycles four times between the inner pressure and 202 kPa. The loading/unloading curves are distinguished with arrows (shown only for the last cycle).

The radial displacement profile as well as its evolution show significant differences from the representative case: the nearly homogeneous deformation is lost, as well as the rate at which the annulus creeps inwards changes. An example of the former is the increase in the difference between inner and outer values of the radial displacement; that is, a 3% difference in between the inner and outer radial displacement for the representative case becomes 30% and almost three times greater in the other two cases, respectively; examples of the latter are the different "distances" that the displacement profiles (each profile is plotted at a fixed time) are from each other (Figure 4.12, first column). Similar results concerning the degradation and Young's modulus profiles are observed, but the main feature is the qualitative changes in the nature of each solution. An almost homogeneous degradation in the thin-walled annulus becomes markedly inhomogeneous as the thickness increases. Degradation and consequent depreciation clearly proceed in the outward direction, as can be observed in the significant differences between degradation and the Young's modulus at the inner and outer surface (Figure 4.12, second and third columns).

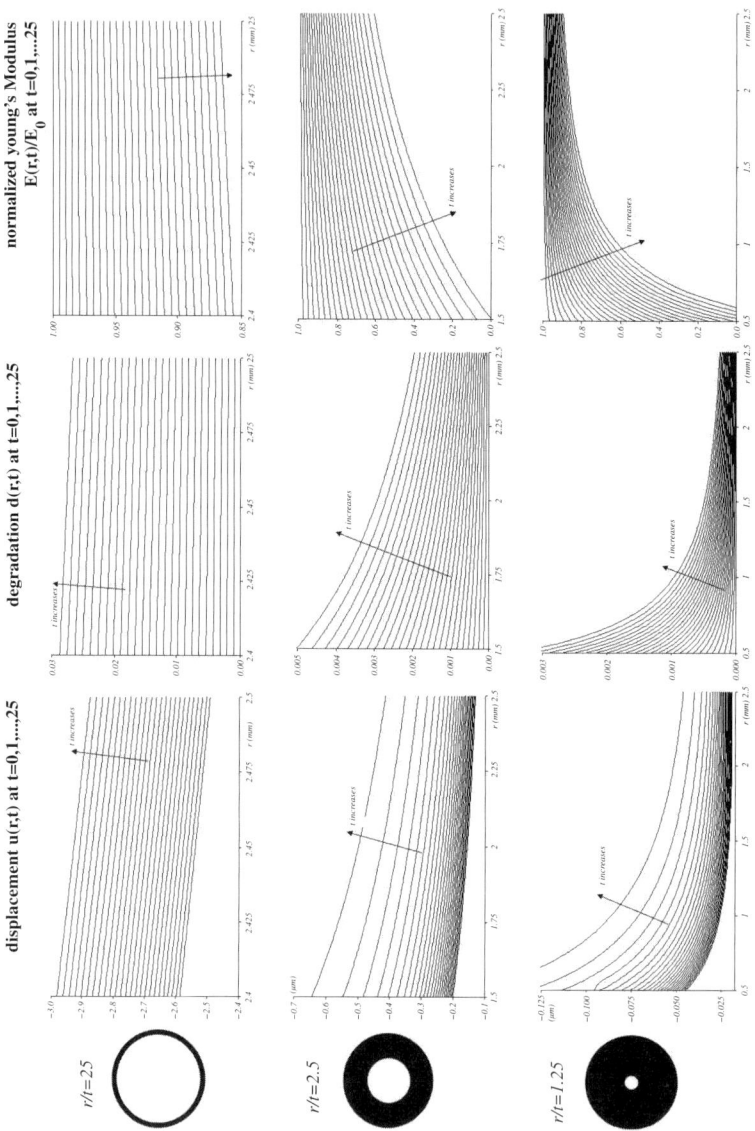

Figure 4.12. Influence of the wall thickness on the deformation, consequent degradation, and Young's modulus reduction. Profiles of displacement, degradation, and normalized Young's modulus for three different geometries are shown (the arrow means the increase in time). For a thin-walled cylinder such as a stent the degradation is almost homogeneous throughout the thickness. For the other two geometries, the inhomogeneity of the deformation is more intense. The inner and outer parts of the annulus are subjected to more different strain fields as the thickness increases and consequently greater asymmetries occur.

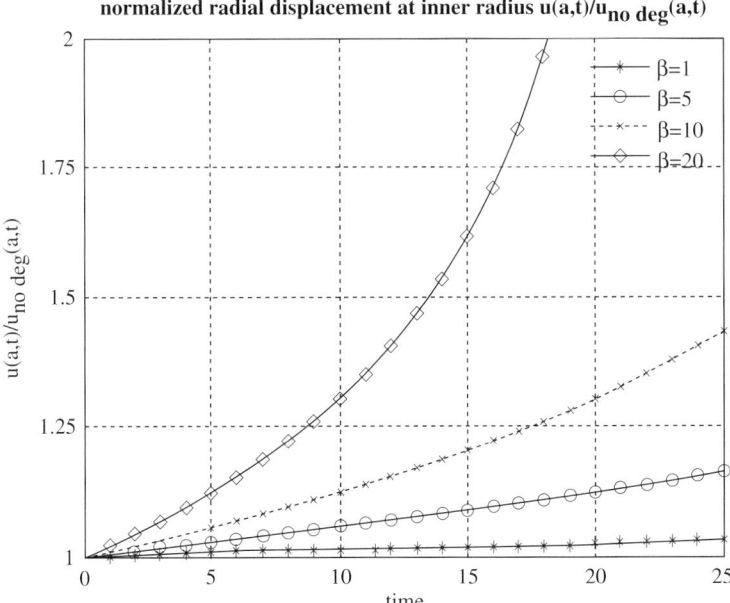

Figure 4.13.    Normalized displacement at the inner radius $u(a,t)/$ *unodeg*$(a,t)$ for several $\beta$. As $\beta$ increases, the impact of degradation on the mechanical properties is greater. Although the outer pressure remains constant, the inner surface creeps inwards.

### 4.7.3   On the Role of the Constant Governing the Mechanical Properties Reduction, $\beta$

The main characteristic of the constant governing the Young's modulus reduction $\beta$ is that when it increases in magnitude, so does the rate at which the material properties decrease due to a given degradation field (cf. Eq. (5.11)). Comparing several solutions with everything kept constant except the value of $\beta$, significant differences can be observed in the temporal evolution of the radial displacement, degradation, and Young's modulus at the inner surface (Figures 4.13, 4.14, and 4.15, respectively).

When $\beta = 20$ the annulus "collapses." This collapse is observed in all three figures: the displacement and degradation at the inner surface (the worst case scenario because the degradation proceeds from the inner surface) tend asymptotically to infinity whereas the Young's modulus reaches zero value. The numerical method does not converge, the computation is interrupted, and the assumption of classical elasticity does not hold anymore. On the other hand, a qualitative comparison between all the cases shown in the evolution of the Young's modulus profile (Figure 4.15) reveals

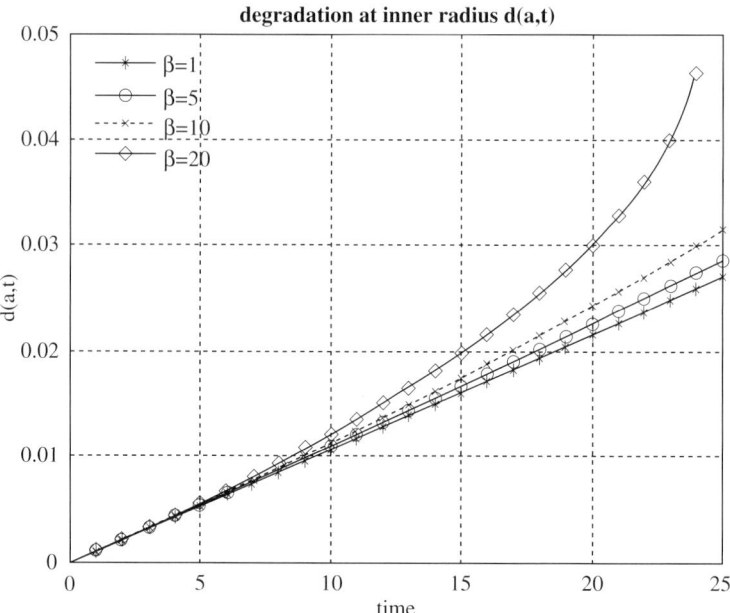

Figure 4.14.    Degradation at the inner radius $d(a,t)$ for several $\beta$. The degradation increases asymptotically and ultimately will lead to the collapse.

that the impact in the mechanical properties and the likelihood of collapse is intimately related to the value of $\beta$. Lastly, $\beta$ also influences the whole degradation process, as a softer material under the same load will be subjected to slightly greater strains and consequently to a slightly greater rate of degradation.

### 4.7.4   On the Parameter of the Mechanical Degradation Governing Equation, $D(t)$

The evolutions of the degradation at the inner surface for different $D(t)$ considered in the parametric study of the influence of the degradation parameter are clearly different from each other (Figure 4.16) and from the ones previously obtained in the parametric study for where the degradation achieved by each deformation is very close for all of the considered cases (cf. Figure 4.14). As $D(t)$ is connected to the amount of degradation that a given strain field can cause (cf. Eq. (5.13)), it follows logically that different values of this function will yield distinct degradations under the same strain field.

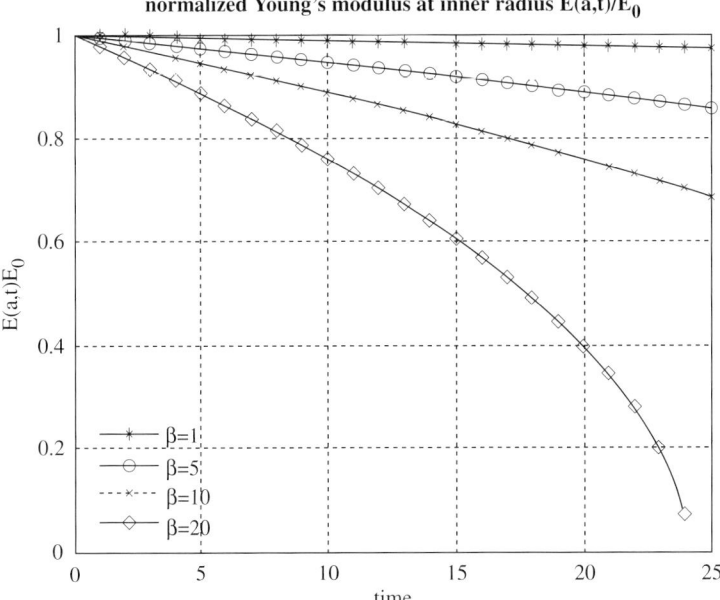

Figure 4.15. Normalized Young's modulus at the inner radius $E(a,t)/E_0$ for several $\beta$. As $\beta$ increases, the ability to support load is readily depreciated by the degradation field. In the solution corresponding to $\beta = 20$, $E(a,t)$ tends to zero before $t = 25$ time units and is responsible for the blowing up of the approximating method.

Increasing the value of $D(t)$ intensifies the degradation that a given strain can provoke and therefore makes collapse more likely. Two cases collapse within the time march considered (Figures 4.17 and 4.18).

### 4.7.5 On the Shape of $D(t)$

When $D(t)$ is considered as shown in Figure 4.3, it is observed that $D(t) = 0$ leads to no degradation and this function can be used as an activation criterion for the initiation of strain-induced degradation. The annulus starts to yield as soon as the onset of degradation takes place (Figure 4.19). Because the absolute value of $D(t)$ is directly related to the amount of degradation imparted by a given strain field, the smoothness of the transient regime in the degradation evolution is a direct consequence of the smoothness of each $D(t)$. D4, with an abrupt change from 0 to 1 at $t = 20$, provokes an abrupt change in the degradation of the annulus; however, with D0, where the increase is linear over the first 20 time units, the transition is

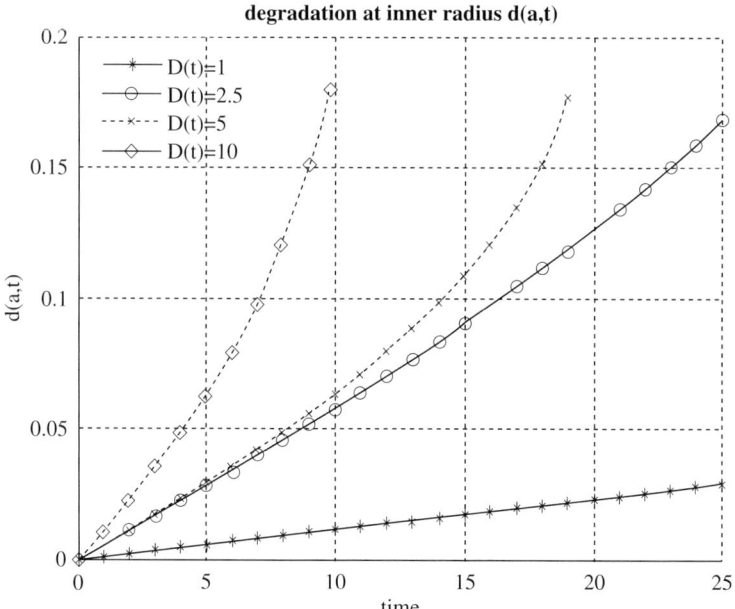

Figure 4.16. Degradation at inner radius $d(a,t)$ for several constant $D(t)$. Correspondingly, degradation increases at a faster rate for $D(t)$s of greater magnitude.

smoother (Figure 4.20). The functions tested ultimately end up with the value $D(t) = 1$ and it is observed that after a different transient period, degradation will eventually achieve approximately the same rate (slope in Figure 4.20), The time lag observed is a direct consequence of the different functions $D(t)$. Also, the annulus creeps inwards at an approximately constant rate and its material properties decrease in similar fashion for all the cases considered (slope in Figures 4.19 and 4.21).

## 4.8   Discussion

One particular case of the model for the mechanism of degradation shown in Figure 4.1 is the representative case taken as the initial result of the previous section. Data with the degradation parameters $\beta = 5$ and $D(t) = 1$ (cf. Figures 4.4 through 4.10), provide phenomenological support for the model. The annulus starts out nondegraded and the imposed loads in the inner and outer surfaces are responsible for a strain field through the balance

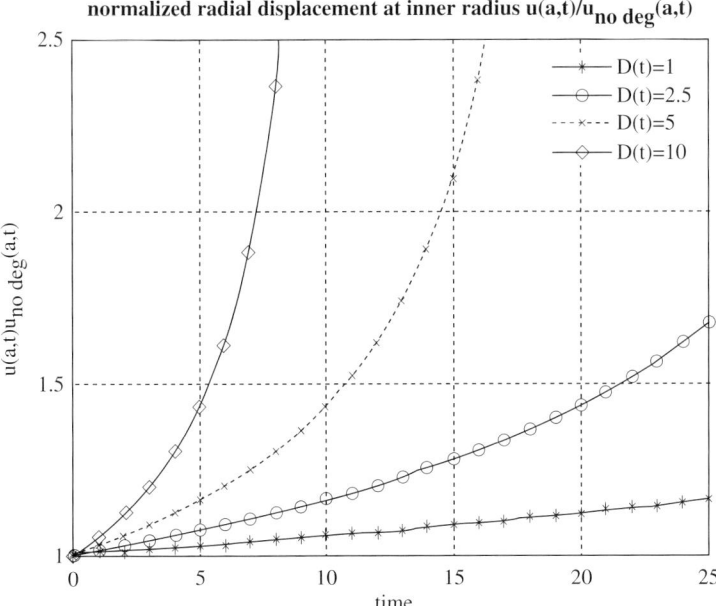

Figure 4.17. Normalized displacement at inner radius $u(a,t)/unodeg(a,t)$ for several $D(t)$. As the magnitude of $D(t)$ increases, the mechanical degradation induced by a strain field increases. Two cases $(D(t) = 5$ and $D(t) = 10)$ reach the point of collapse in 25 time units.

of linear momentum (cf. Eq. (5.10)). Simultaneously, a degradation field dependent on the strain field given by the equation governing the increase of degradation (5.13) comes into play. Both are related by the classical constitutive equation (5.12), where the Young's modulus is assumed to depend on the degradation field.

The overall features of the solution obtained as the representative case clearly show an increase in the degradation field and a reduction in the mechanical properties (cf. Figures 4.5 and 4.6, respectively). A decrease of approximately 15% of the value of the non-degraded Young's modulus is seen over the time march considered. However, it must be remarked that such fast degradation is a consequence of this particular choice of the degradation parameters. The main purpose for this choice was to obtain quick degradation, so one would be able to observe its effects with a feasible time march. Also, because the time scales associated with degradation and a cycle of blood pressure considered in this model are of the same order of magnitude, one would observe the influence, if any, of an oscillating pressure applied on the inner surface. Lastly, it must be

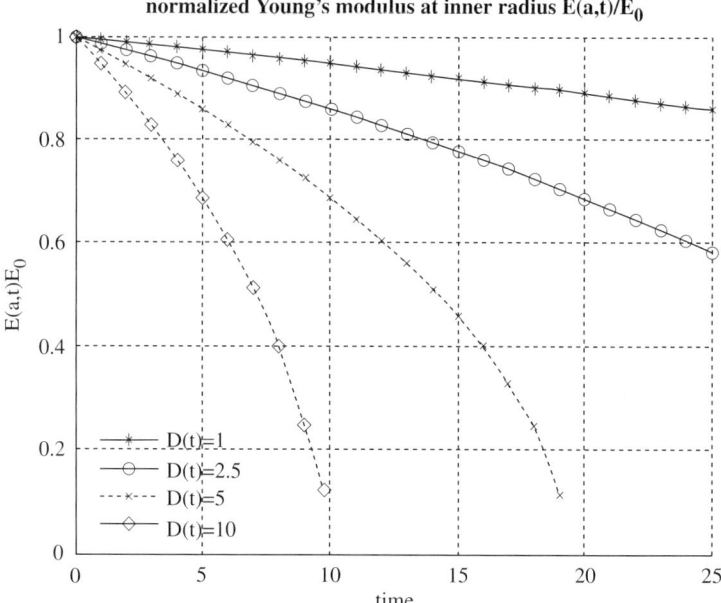

Figure 4.18. Normalized Young's modulus at inner radius $E(a,t)/E_0$ for several $D(t)$. A greater $D(t)$ would make a given strain yield a greater rate of increase of degradation.

remarked that in order to obtain a realistic model for describing a particular strain-degradable material, these parameters must be obtained from designed experiments.

The nonzero components of the stress tensor (Figures 4.8 through 4.10) have distinctive characteristics when compared with the nondegraded linear elastic case. The axial strain $\sigma_{rr}$ is completely prescribed by the boundary conditions. Stress relaxation of the hoop stress $\sigma_{\theta\theta}$ is observed in the inner half of the annulus caused by local softening of the material. The strain field is higher in magnitude over the inner half and therefore degradation is greater over this region. Because the degraded cylindrical annulus must withstand the same loading conditions, the stresses in the outer part increase in magnitude to compensate for the softening and relaxation in the inner part. The axial strain shows a similar relaxation behavior, although less intense, deviating from the constancy through the thickness obtained in the standard case of a nondegradable elastic infinite cylinder.

Due to the same degradation mechanism, the cylinder creeps inwards under a constant load. The displacement progressively increases with time and that is a clear sign of creep (Figure 4.4). Under the same crushing load, the displacement of the inner surface increases due to degradation

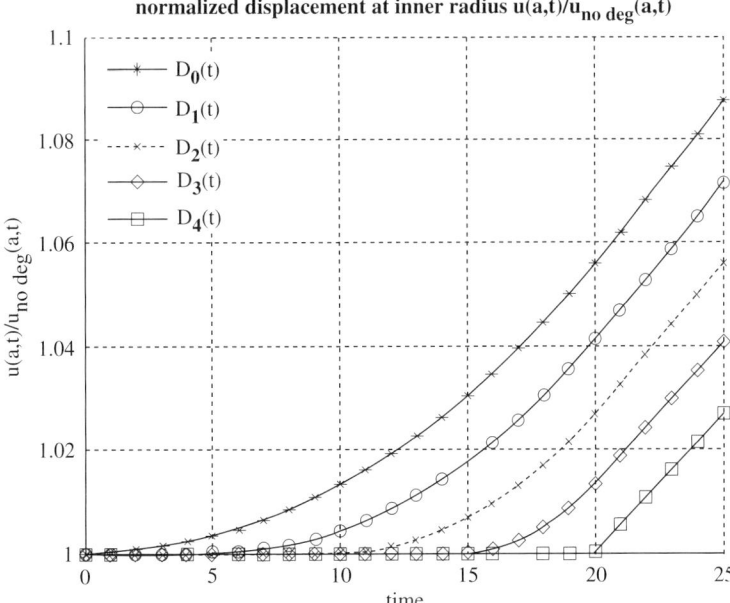

Figure 4.19. Normalized displacement at inner radius $u(a,t)/unodeg(a,t)$ for different functions $D(t)$. After a transient period, the cylinder creeps at an approximately constant rate.

and softening of the material. Thus the cylinder shrinks over time, inducing a higher strain field and consequently further degradation. On the other hand, when the evolution of the inner displacement is compared over time for different sets of degradation parameters (cf. Figures 4.13), it leads to such an intense degradation that the cylinder collapses within the time step considered and the numerical scheme does not converge.

Also, the material shows behavior similar to mechanical hysteresis when the outer pressure is cycled between the inner pressure and its constant value. Hysteresis is obviously dependent on the degradation, as the two cases shown in Figure 4.11 demonstrate. As the degradation proceeds, the loading and unloading curves are far apart, meaning that dissipation increases over time and over degradation.

All these three distinct features are inherent characteristics of a viscoelastic material. However, strain-induced degradation of the material considered here shows similar characteristics. The material is elastic if no degradation occurs, but when degradation is active, the material shows stress relaxation, creep, and hysteresis, but markedly different from the similar phenomena exhibited by viscoelastic materials (cf. Rajagopal and Wineman for similar results on aging viscoelastic materials: two different

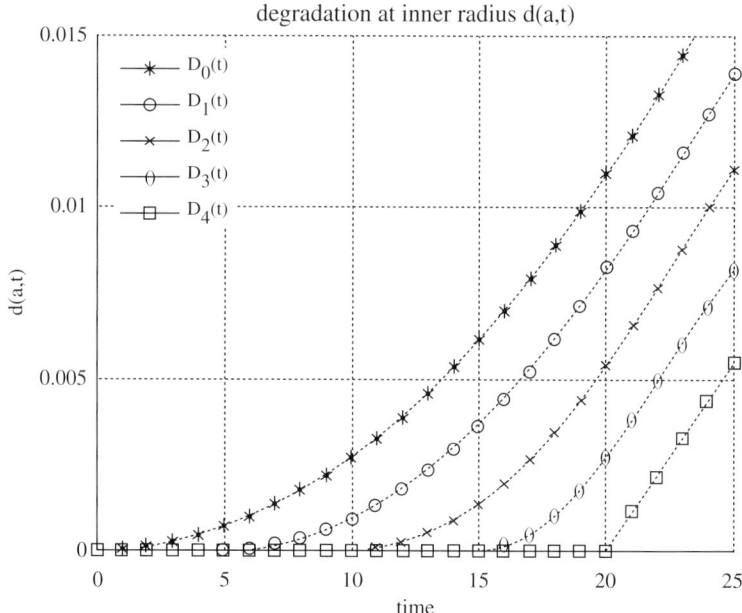

Figure 4.20. Degradation at inner radius $d(a, t)$ for different functions $D(t)$. The slope of each line will ultimately be the same and is related to the constant value that $D(t)$ assumes after $t = 20$.

components of the mechanisms of relaxation and creep were distinguished, one due to the inherent viscoelasticity of the material and the other due to aging processes [RAa]).

Another remarkable consequence of this degradation mechanism is its inhomogeneity. The equation governing degradation (5.13) states that degradation increases proportionally to the hoop strain. If two particles are subjected to different strains, consequently they will degrade to different extents. Their material properties, once the same before degradation, will vary with the current location. It is interesting how an initially homogeneous body becomes inhomogeneous after degradation induced by an inhomogeneous deformation.

Figure 4.12 shows qualitatively the differences among several inhomogeneous deformations. In the representative case, the profiles of displacement, degradation, and material properties reduction are almost straight lines. Each particle in the entire cross-section is subjected to an almost identical strain field and therefore the local degradation rates are similar and all particles of the body degrade approximately by the same amount. Thus, the differences in material properties are small. On the other hand, when the thickness of the cylinder is increased to obtain steeper hoop strain gradients

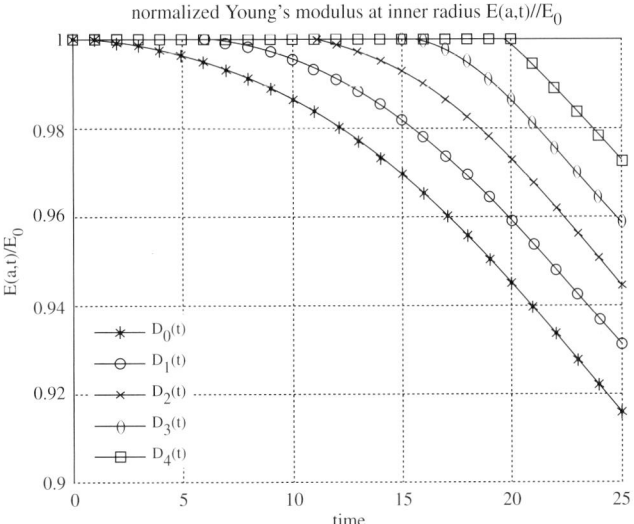

Figure 4.21. Normalized Young's modulus at inner radius $E(a,t)/E_0$ for different functions $D(t)$. The onset of degradation and consequent material properties reduction are directly related to $D(t)$ being nonzero.

throughout the thickness due to the inhomogeneity of the deformation, the degradation behavior of the body is completely different: (1) the flatness of the displacement profile is lost and the annulus is subjected to higher strains near the inside surface, (2) the degradation rate and the achieved degradation are greater in this region, (3) degradation clearly proceeds outwards, and (4) the differences in Young's modulus reduction obtained for different particles of the cylinder are enormous and the degree of inhomogeneity of the cylinder after some degradation is vast.

Degradation is driven by two distinct parameters that relate the two governing equations. Constant $\beta$ appearing in the constitutive equation (5.12) through (5.11) represents the impact of a given amount of degradation in the reduction of the Young's modulus. This relationship can have a more general form (space- and time-dependence, nonlinear in the degradation field, or even dependent on the diffusion of certain species), and (5.11) is just one particular description of the impact of the degradation in the constitutive equation. In this simplified model, the relationship is linear and $\beta$ is a constant (over space and time). The magnitude of $\beta$ is ultimately related to the amount of depreciation that a given degradation would produce (a negative $\beta$ would result in a material that strain hardens). Degradation is similar for the $\beta$s considered (Figure 4.14), but the corresponding Young's modulus reductions are clearly distinct (Figure 4.15).

The reason is that $D(t)$, which is responsible for the degradation due to a given strain field, is the same, whereas $\beta$ changes dramatically, provoking different Young's modulus reductions for approximately the same degradation field. Eventually, because of the softening of the material, the cylinder yields and degrades progressively at slightly different rates. When $\beta = 20$, the effect of degradation in the Young's modulus is so strong that collapse was observed, that is, a point where the numerical solution yields a Young's modulus close to zero. It must be noted that all the solutions will eventually collapse after some time. The value of $\beta$, together with $D(t)$ will ultimately decide when collapse occurs.

$D(t)$ influences the process through the equation governing degradation (5.1) and (5.13). Equation (5.13) is a particular form of the more general (5.1). The parametric study of the effect of the absolute value of $D(t)$ yielded similar effects when compared with the influence of $\beta$ (Figures 4.13 through 4.18). As expected, $D(t)$s of greater magnitude would motivate a stronger and faster degradation under the same strain field. Moreover, this parameter was chosen to be dependent on time in order to be possible to have activation criteria and temporal changes of the degradation rate. $D(t) = 0$ leads to no degradation and it can be used as an activation criterion (Figure 4.19). The rate of strain degradation will be closely related to this function. All five cases tend approximately to the same rate of degradation because all the $D(t)$s will ultimately have the value of 1. Young's modulus reduction and displacement at the inner surface show similar results (Figures 4.20 and 4.21).

As a final remark, it must be noted that this model is very simplistic in all its approximations: (1) a stent is not an infinite cylinder and this geometry was chosen in order to simplify the equations governing the process; (2) biodegradable polymers, such as PLLA, are not linear elastic materials and should be modeled as incompressible and fully nonlinear viscoelastic materials; (3) the particular forms of Eq. (5.13) and (5.11) describing the mechanism of degradation were chosen based on their simplicity, therefore the mechanism of strain-induced degradation is quite restrictive; and (4) mass balance must be met by taking $\rho$ constant in the linear momentum balance (5.9); thus this model describes the initial steps of the overall process when degradation occurs without significant effects of erosion or elimination.

## 4.9 Conclusions

As a first step towards a fully nonlinear model, biodegradable polymers were modeled as a class of linearized materials. From the phenomenological acceptable results obtained with this model, one concludes that strain-induced degradation can be modeled with a partial differential equation governing the rate of increase of degradation coupled with the equations of motion. The constitutive equation takes into account the effect of degradation of the mechanical properties.

The material with degradation considered here shows stress relaxation, creep, and hysteresis, thus a dissipative process governing this particular degradation mechanism. Inhomogeneous deformations lead to inhomogeneous degradations, hence the homogeneous cylinder considered here becomes inhomogeneous after the onset of degradation.

The efficacy of this approach was verified by solving this simple problem and the constitutive model will be integrated into a finite element software package in order to analyze more realistic and more complex stent geometries. Later, the model will be extended in such a way that it will be capable of describing the diffusion of a drug impregnated in the material reflecting the enhancement of the diffusion due to changes in the porosity due to degradation.

### Acknowledgments

This work was partially funded by the Portuguese FCT - Fundação para a Ciência e Tecnologia (SFRH/BD/17060/2004) and NIH grant R01 EB000115. A general thermodynamic framework for bodies capable of undergoing strain-induced degradation is being developed and will be published.

## 4.10   References

[AGa]  Agrawal, C.M., and Clark, H.G., Deformation characteristics of a bioabsorbable intravascular stent. *Invest. Radiol.*, **27** (1992), 1020–4.

[AGb]  Agrawal, C.M., Haas, K.F., Leopold, D.A., and Clark, H.G., Evaluation of poly(L-lactic acid) as a material for intravascular polymeric stents. *Biomaterials*, **13** (1992), 176–82.

[AHa]  Ahn, Y.K., Jeong, M.H., Kim, J.W., et al., Preventive effects of the heparin-coated stent on restenosis in the porcine model. *Catheter Cardiovasc. Interv.*, **48** (1999), 324–30.

[ALa]  Al Suwaidi, J., Berger, P.B., and Holmes, D.R., Jr., Coronary artery stents. *JAMA*, **284** (2000), 1828–36.

[ALb] Ali, S.A., Doherty, P.J., and Williams, D.F., Mechanisms of polymer degradation in implantable devices. 2. Poly(DL-lactic acid). *J. Biomed. Mater. Res.*, **27** (1993), 1409–18.

[ALc] Ali, S.A., Zhong, S.P., Doherty, P.J., and Williams, D.F., Mechanisms of polymer degradation in implantable devices. 1. Poly (caprolactone). *Biomaterials*, **14** (1993), 648–56.

[ARa] Ardissino, D., Cavallini, C., Bramucci, E., et al., Sirolimus-eluting vs. uncoated stents for prevention of restenosis in small coronary arteries: a randomized trial, *JAMA*, **292** (2004), 2727–34.

[ATa] Athanasiou, K.A., Agrawal, C.M., Barber, F.A., and Burkhart, S.S., Orthopaedic applications for PLA-PGA biodegradable polymers. *Arthroscopy J. Arthroscopic Related Surg.*, **14** (1998), 726–37.

[AXa] Axel, D.I., Kunert, W., Goggelmann, C., et al., Paclitaxel inhibits arterial smooth muscle cell proliferation and migration in vitro and in vivo using local drug delivery. *Circulation*, **96** (1997), 636–45.

[BAa] Babapulle, M.N., and Eisenberg, M.J., Coated stents for the prevention of restenosis: Part II. *Circulation*, **106** (2002), 2859–66.

[BEa] Bedoya, J., Meyer, C.A., Timmins, L.H., Moreno, M.R., and Moore, J.E., Jr., Effects of stent design parameters on artery wall mechanics (submitted).

[BEb] Bellenger, V., Ganem, M., Mortaigne, B., and Verdu, J., Lifetime prediction in the hydrolytic aging of polyesters. *Polym. Degradation Stability*, **49** (1995), 91–7.

[BEc] Bertrand, O.F., Sipehia, R., Mongrain, R., et al., Biocompatibility aspects of new stent technology. *J. Am. Coll. Cardiol.*, **32** (1998), 562–71.

[BIa] Bier, J.D., Zalesky, P., Li, S.T., Sasken, H., and Williams, D.O., A new bioabsorbable intravascular stent: in vitro assessment of hemodynamic and morphometric characteristics. *J. Interv. Cardiol.*, **5** (1992), 187–94.

[BLa] Blindt, R., Hoffmeister, K.M., Bienert, H., et al., Development of a new biodegradable intravascular polymer stent with simultaneous incorporation of bioactive substances. *Int. J. Artif. Organs*, **22** (1999), 843–53.

[BOa] Bose, S.M., and Git, Y., Mathematical modelling and computer simulation of linear polymer degradation: Simple scissions. *Macromol. Theor. Simul.*, **13** (2004), 453–73.

[BRa] Browarzik, D., and Koch, A., Application of continuous kinetics to polymer degradation. *J. Macromol. Sci. Pure Appl. Chem.*, **33A** (1996), 1633–41.

[BUa] Burkersroda, F.v., Schedl, L., and Gopferich, A., Why degradable polymers undergo surface erosion or bulk erosion. *Biomaterials*, **23** (2002), 4221–31.

[COa] Colombo, A., and Karvouni, E., Biodegradable stents : "Fulfilling the mission and stepping away." *Circulation*, **102** (2000), 371–3.

[CUa] Currier, J.W., and Faxon, D.P., Restenosis after percutaneous transluminal coronary angioplasty: have we been aiming at the wrong target? *J. Am. Coll. Cardiol.*, **25** (1995), 516–20.

[DEa] De Scheerder, I.K., Wilczek, K.L., Verbeken, E.V., et al., Biocompatibility of biodegradable and nonbiodegradable polymer-coated stents implanted in porcine peripheral arteries. *Cardiovasc. Intervent. Radiol.*, **18** (1995), 227–32.

[DEb] De Scheerder, I.K., Wilczek, K.L., Verbeken, E.V., et al., Biocompatibility of polymer-coated oversized metallic stents implanted in normal porcine coronary arteries. *Atherosclerosis*, **114** (1995), 105–14.

[DEc] DePalma, V.A., Baier, R.E., Ford, J.W., Glott, V.L., and Furuse, A., Investigation of three-surface properties of several metals and their relation to blood compatibility. *J. Biomed. Mater. Res.*, **6** (1972), 37–75.

[DIa] Di Mario, C., Griffiths, H., Goktekin, O., et al., Drug-eluting bioabsorbable magnesium stent. *J. Interv. Cardiol.*, **17** (2004), 391–5.

[DOa] Dotter, C.T., Transluminal angioplasty: A long view. *Radiology*, **135** (1980), 561–4.

[DOb] Douglas, J.S., Jr., King, S.B., 3rd, and Roubin, G.S., Influence of the methodology of percutaneous transluminal coronary angioplasty on restenosis. *Am. J. Cardiol.*, **60** (1987), 29B–33B.

[DRa] Drumright, R.E., Gruber, P.R., and Henton, D.E., Polylactic acid technology. *Adv. Mater.*, **12** (2000), 1841–6.

[DUa] Duda, S.H., Bosiers, M., Lammer, J., et al., Sirolimus-eluting versus bare nitinol stent for obstructive superficial femoral artery disease: The SIROCCO II trial. *J. Vasc. Interv. Radiol.*, **16** (2005), 331–8.

[DUb] Duda, S.H., Pusich, B., Richter, G., et al., Sirolimus-eluting stents for the treatment of obstructive superficial femoral artery disease: Six-month results. *Circulation*, **106** (2002), 1505–9.

[EDa] Edelman, E.R., and Rogers, C., Pathobiologic responses to stenting. *Am. J. Cardiol.*, **81** (1998), 4E–6E.

[FAa] Farb, A., Weber, D.K., Kolodgie, F.D., Burke, A.P., and Virmani, R., Morphological predictors of restenosis after coronary stenting in humans. *Circulation*, **105** (2002), 2974–80.

[FAb]  Faxon, D.P., Vascular stents. *Rev. Cardiovasc. Med.*, **2** (2001), 106–7.

[FIa]  Fischell, TA., Polymer coatings for stents. Can we judge a stent by its cover? *Circulation*, **94** (1996), 1494–5.

[FIb]  Fischman, D.L., Leon, M.B., Baim, D.S., et al., A randomized comparison of coronary-stent placement and balloon angioplasty in the treatment of coronary-artery disease. *N. Engl. J. Med.*, **331** (1994), 496–501.

[GAa]  Gallo, R., Padurean, A., Jayaraman, T., et al., Inhibition of intimal thickening after balloon angioplasty in porcine coronary arteries by targeting regulators of the cell cycle. *Circulation*, **99** (1999), 2164–70.

[GAb]  Garlotta, D., A literature review of poly(lactic acid). *J. Polym. Environ.*, **9** (2001), 63–84.

[GLa]  Glagov, S., Zarins, C.K., Masawa, N., Xu, CP., Bassiouny, H., and Giddens, D.P., Mechanical functional role of non-atherosclerotic intimal thickening. *Front. Med. Biol. Eng.*, **5** (1993), 37–43.

[GLb]  Glagov, S., Intimal hyperplasia, vascular modeling, and the restenosis problem. *Circulation*, **89** (1994), 2888–91.

[GOa]  Gopferich, A., and Langer, R., Modeling polymer erosion. *Macromolecules*, **26** (1993), 4105–12.

[GOb]  Gopferich, A., Mechanisms of polymer degradation and elimination. In: Domb, A.J., Kost, J., and Wiseman, D.M., Eds. **Handbook of Biodegradable Polymers**. Harwood Academic, Australia (1997), 451–71.

[GOc]  Gopferich, A., Polymer degradation and erosion: Mechanisms and applications. *Eur. J. Pharm. Biopharm.*, **4** (1996), 1–11.

[GRa]  Grabow, N., Martin, H., and Schmitz, K.P., The impact of material characteristics on the mechanical properties of a poly(L-lactide) coronary stent. *Biomed. Tech. (Berl.)*, **47** (2002), 503–5.

[GRb]  Grabow, N., Schlun, M., Sternberg, K., Hakansson, N., Kramer, S., and Schmitz, K.P., Mechanical properties of laser cut poly(L-lactide) micro-specimens: Implications for stent design, manufacture, and sterilization. *J. Biomech. Eng.*, **127** (2005), 25–31.

[GRc]  Grube, E., Gerckens, U., Muller, R., and Bullesfeld, L., Drug eluting stents: Initial experiences. *Z. Kardiol.*, **91** (2002), 44–8.

[GRd]  Grube, E., Silber, S., Hauptmann, K.E., et al., TAXUS I: Six- and twelve-month results from a randomized, double-blind trial on a slow-release paclitaxel-eluting stent for de novo coronary lesions. *Circulation*, **107** (2003), 38–42.

[GUa] Gutwald, R., Pistner, H., Reuther, J., and Muhling, J., Biodegradation and tissue-reaction in a long-term implantation study of poly (L-Lactide). *J. Mater. Sci. Mater. Med.*, **5** (1994), 485–90.

[HAa] Hawkins, W.L., Polymer degradation. In: Polymer degradation and stabilization. Springer-Verlag, Berlin (1984) 3–34.

[HAb] Hayashi, T., Biodegradable polymers for biomedical uses. *Prog. Polym. Sci.*, **19** (1994), 663–702.

[HEa] Heublein, B., Rohde, R., Kaese, V., Niemeyer, M., Hartung W., and Haverich, A., Biocorrosion of magnesium alloys: A new principle in cardiovascular implant technology? *Heart*, **89** (2003), 651–6.

[HIa] Hietala, E.M., Salminen, U.S., Stahls, A., et al., Biodegradation of the copolymeric polylactide stent. Long-term follow-up in a rabbit aorta model. *J. Vasc. Res.*, **38** (2001), 361–9.

[HOa] Holmes, D.R., Camrud, A.R., Jorgenson, M.A., Edwards, W.D., and Schwartz, R.S., Polymeric stenting in the porcine coronary artery model: Differential outcome of exogenous fibrin sleeves versus polyurethane-coated stents. *J. Am. Coll. Cardiol.*, **24** (1994), 525–31.

[HYa] Hyon, S.H., Jamshidi, K., and Ikada, Y., Effects of residual monomer on the degradation of DL-lactide polymer. *Polym. Int.*, **46** (1998), 196–202.

[ISa] Isotalo, T., Talja, M., Valimaa, T., Tormala, P., and Tammela, T.L., A bioabsorbable self-expandable, self-reinforced poly-L-lactic acid urethral stent for recurrent urethral strictures: Long-term results. *J. Endourol.*, **16** (2002), 759–62.

[IVa] Ivanova, T., Grozev, N., Panaiotov, I., and Proust, J.E., Role of the molecular weight and the composition on the hydrolysis kinetics of monolayers of poly(alpha-hydroxy acid)s. *Colloid Polym. Sci.*, **277** (1999), 709–18.

[JEa] Jeremias, A., Sylvia, B., Bridges, J., et al., Stent thrombosis after successful sirolimus-eluting stent implantation. *Circulation*, **109** (2004), 1930–2.

[JOa] Joshi, A., and Himmelstein, K.J., Dynamics of controlled release from bioerodible matrices. *J. Control. Release*, **15** (1991), 95–104.

[KAa] Kastrati, A., Dibra, A., Eberle, S., et al., Sirolimus-eluting stents vs. paclitaxel-eluting stents in patients with coronary artery disease: meta-analysis of randomized trials. *JAMA*, **294** (2005), 819–25.

[KAb] Kastrati, A., Hall, D., and Schomig, A., Long-term outcome after coronary stenting. *Curr. Control Trials Cardiovasc. Med.*, **1** (2000), 48–54.

[KAc]  Katti, D.S., Lakshmi, S., Langer, R., and Laurencin, C.T., Toxicity, biodegradation and elimination of polyanhydrides. *Adv. Drug Deliv. Rev.*, **54** (2002), 933–61.

[KHa]  Khang, G., Rhee, J.M., Jeong, J.K., et al., Local drug delivery system using biodegradable polymers. *Macromol. Res.*, **11** (2003), 207–23.

[KIa]  Kimura, T., Yokoi, H., Nakagawa, Y., et al., Three-year follow-up after implantation of metallic coronary-artery stents. *N. Engl. J. Med.*, **334** (1996), 561–6.

[LAa]  Labinaz, M., Zidar, J.P., Stack, R.S., and Phillips, H.R., Biodegradable stents: The future of interventional cardiology? *J. Interv. Cardiol.*, **8** (1995), 395–405.

[LAb]  Lambert, T.L., Dev, V., Rechavia, E., Forrester, J.S., Litvack F, and Eigler, N.L., Localized arterial wall drug delivery from a polymer-coated removable metallic stent. Kinetics, distribution, and bioactivity of forskolin. *Circulation*, **90** (1994), 1003–11.

[LAc]  Langer, R., Drug delivery and targeting. *Nature*, **392** (1998), 5–10.

[LAd]  Laufman, H., and Rubel, T., Synthetic absorbable sutures. *Surg. Gynecol. Obstet.*, **145** (1977), 597–608.

[LEa]  Lemos, P.A., Serruys, P.W., van Domburg, R.T., et al., Unrestricted utilization of sirolimus-eluting stents compared with conventional bare stent implantation in the "real world": The Rapamycin-eluting stent evaluated at Rotterdam Cardiology Hospital (RESEARCH) registry. *Circulation*, **109** (2004), 190–5.

[LEb]  Levenberg, S., and Langer, R., Advances in tissue engineering. In: **Current Topics in Developmental Biology**, Vol 61. Elsevier, San Diego (2004) 113.

[LIa]  Li, S.M., and McCarthy, S., Further investigations on the hydrolytic degradation of poly(DL-lactide). *Biomaterials*, **20** (1999), 35–44.

[LIb]  Li, S.M., and Vert, M., Morphological-Changes Resulting from the hydrolytic degradation of stereocopolymers derived from L-lactides and Dl-lactides. *Macromolecules*, **27** (1994), 3107–10.

[LIc]  Libby, P., Schwartz, D., Brogi, E., Tanaka, H., and Clinton, S.K., A cascade model for restenosis. A special case of atherosclerosis progression. *Circulation*, **86** (1992), 11147–52.

[LId]  Lincoff, A.M., Furst, J.G., Ellis, S.G., Tuch, R.J., and Topol, E.J., Sustained local delivery of dexamethasone by a novel intravascular eluting stent to prevent restenosis in the porcine coronary injury model. *J. Am. Coll. Cardiol.*, **29** (1997), 808–16.

[LIe]   Lincoff, A.M., Topol, E.J., and Ellis, S.G., Local drug delivery for the prevention of restenosis. Fact, fancy, and future. *Circulation*, **90** (1994), 2070–84.

[LIf]   Lipinski, M.J., Fearon, W.F., Froelicher, V.F., and Vetrovec, G.W., The current and future role of percutaneous coronary intervention in patients with coronary artery disease. *J. Interv. Cardiol.*, **17** (2004), 283–94.

[LUa]  Lunt, J., Large-scale production, properties and commercial applications of polylactic acid polymers. *Polym. Degradation Stability*, **59** (1998), 145–52.

[MAa]  Marx, S.O., Jayaraman, T., Go, L.O., and Marks, A.R., Rapamycin-Fkbp inhibits cell-cycle regulators of proliferation in vascular smooth-muscle cells. *Circulation Res.*, **76** (1995), 412–7.

[MIa]   Middleton, J.C., and Tipton, A.J., Synthetic biodegradable polymers as orthopedic devices. *Biomaterials*, **21** (2000), 2335–46.

[MIb]   Mintz, G.S., Hoffmann, R., Mehran, R., et al., In-stent restenosis: The Washington Hospital Center experience. *Am. J. Cardiol.*, **81** (1998), 7E–13E.

[MIc]   Mintz, G.S., Popma, J.J., Pichard, A.D., et al., Arterial remodeling after coronary angioplasty: A serial intravascular ultrasound study. *Circulation*, **94** (1996), 35–43.

[MOa]  Moore, J., Jr., and Berry, J.L., Fluid and solid mechanical implications of vascular stenting. *Ann. Biomed. Eng.*, **30** (2002), 498–508.

[MOb]  Morice, M.C., Serruys, P.W., Sousa, J.E., et al., A randomized comparison of a sirolimus-eluting stent with a standard stent for coronary revascularization. *N. Engl. J. Med.*, **346** (2002), 1773–80.

[Moc]   Moses, J.W., Leon, M.B., Popma, J.J., et al., Sirolimus-eluting stents versus standard stents in patients with stenosis in a native coronary artery. *N. Engl. J. Med.*, **349** (2003), 1315–23.

[MUa]  Muller, D.W., Ellis, S.G., and Topol, E.J., Experimental models of coronary artery restenosis. *J. Am. Coll. Cardiol.*, **19** (1992), 418–32.

[MUb]  Murphy, J.G., Schwartz, R.S., Edwards, W.D., Camrud, A.R., Vlietstra RE, and Holmes, D.R., Jr. Percutaneous polymeric stents in porcine coronary arteries. Initial experience with polyethylene terephthalate stents. *Circulation*, **86** (1992), 1596–604.

[MUc]  Murphy, J.G., Schwartz, R.S., Huber, K.C., and Holmes, D.R., Jr. Polymeric stents: Modern alchemy or the future? *J. Invasive. Cardiol.*, **3** (1991), 144–8.

[NGa]  Nguyen, K.T., Su, S.H., Sheng, A., et al., In vitro hemocompatibility studies of drug-loaded poly-(L-lactic acid) fibers. *Biomaterials*, **24** (2003), 5191–201.

[NGb] Nguyen, T.Q., and Kausch, H.H., GPC data interpretation in mechanochemical polymer degradation. *Int. J. Polym. Anal. Characterization*, **4** (1998), 447–70.

[NGc] Nguyen, T.Q., Kinetics of mechanochemical degradation by gel permeation chromatography. *Polym. Degradation Stability*, **46** (1994), 99–111.

[NUa] Nuutinen, J.P., Clerc, C., and Tormala, P., Theoretical and experimental evaluation of the radial force of self-expanding braided bioabsorbable stents. *J. Biomater. Sci. Polym. Ed.*, **14** (2003), 677–87.

[ONa] Ong, A.T., Serruys, P.W., and Aoki, J., et al., The unrestricted use of paclitaxel- versus sirolimus-eluting stents for coronary artery disease in an unselected population: One-year results of the Taxus-Stent Evaluated at Rotterdam Cardiology Hospital (T-SEARCH) registry. *J. Am. Coll. Cardiol.*, **45** (2005), 1135–41.

[OTa] Ottenbrite, R.M., Albertsson, and A.C., Scott, G., Discussion on degradation terminology. In: Vert, M., Feijen, J., Albertsson, A.C., Scott, G., Chiellini, E., Eds. **Biodegradable Polymers and Plastics**. The Royal Society of Chemisty, Cambridge (1992) 73–92.

[PAa] Palmaz, J.C., Balloon-expandable intravascular stent. *AJR Am. J. Roentgenol.*, **150** (1988), 1263–9.

[PEa] Peng, T., Gibula, P., Yao K.-d., and Goosen, M.F.A., Role of polymers in improving the results of stenting in coronary arteries. *Biomaterials*, **17** (1996), 685–94.

[PEb] Peuster, M., Wohlsein, P., Brugmann, M., et al., A novel approach to temporary stenting: Degradable cardiovascular stents produced from corrodible metal-results 6-18 months after implantation into New Zealand white rabbits. *Heart*, **86** (2001), 563–9.

[PIa] Pietrzak, W.S., Sarver, D.R., and Verstynen, M.L., Bioabsorbable polymer science for the practicing surgeon. *J. Craniofac. Surg.*, **8** (1997), 87–91.

[PIb] Pietrzak, W.S., Verstynen, M.L., and Sarver, D.R., Bioabsorbable fixation devices: Status for the craniomaxillofacial surgeon. *J. Craniofac. Surg.*, **8** (1997), 92–6.

[PIc] Pistner, H., Bendix, D.R., Muhling, J., and Reuther, J.F., Poly (L-Lactide) — A long-term degradation study invivo .3. Analytical characterization. *Biomaterials*, **14** (1993), 291–8.

[PId] Pistner, H., Gutwald, R., Ordung, R., Reuther, J., and Muhling, J., Poly(L-lactide) — A long-term degradation study in-vivo.1. Biological results. *Biomaterials*, **14** (1993), 671–7.

[RAa] Rajagopal, K.R., and Wineman, A.S., A note on viscoelastic materials that can age. *Int. J. NonLinear Mech.*, **39** (2004), 1547–54.

[ROa] Robaina, S., Jayachandran, B., He, Y., et al., Platelet adhesion to simulated stented surfaces. *J. Endovasc. Ther.*, **10** (2003), 978–86.

[ROb] Rogers, C., and Edelman, E.R., Endovascular stent design dictates experimental restenosis and thrombosis. *Circulation*, **91** (1995), 2995–3001.

[ROc] Roubin, G.S., Douglas, J.S., Jr., King, S.B., 3rd, et al., Influence of balloon size on initial success, acute complications, and restenosis after percutaneous transluminal coronary angioplasty. A prospective randomized study. *Circulation*, **78** (1988), 557–65.

[SAa] Saia, F., Marzocchi, A., and Serruys, P.W., Drug-eluting stents. The third revolution in percutaneous coronary intervention. *Ital. Heart J.*, **6** (2005), 289–303.

[SAb] Sarembock, I.J., LaVeau, P.J., Sigal, S.L., et al., Influence of inflation pressure and balloon size on the development of intimal hyperplasia after balloon angioplasty. A study in the atherosclerotic rabbit. *Circulation*, **80** (1989), 1029–40.

[SCa] Schakenraad, J.M., Hardonk, M.J., Feijen, J., Molenaar, I., and Nieuwenhuis, P., Enzymatic activity toward poly(L-lactic acid) implants. *J. Biomed. Mater. Res.*, **24** (1990), 529–45.

[SCb] Schatz, R.A., Baim, D.S., Leon, M., et al., Clinical-experience with the Palmaz-Schatz coronary stent - Initial results of a multicenter study. *Circulation*, **83** (1991), 148–61.

[SCc] Schatz, R.A., Introduction to intravascular stents. *Cardiol. Clin.*, **6** (1988), 357–72.

[SCd] Schnabel, W., **Polymer Degradation**. Macmillan, New York (1981).

[SCe] Schofer, J., Schluter, M., Gershlick, A.H., et al., Sirolimus-eluting stents for treatment of patients with long atherosclerotic lesions in small coronary arteries: Double-blind, randomised controlled trial (E-SIRIUS). *Lancet*, **362** (2003), 1093–9.

[SCf] Schwartz, R.S., Neointima and arterial injury: Dogs, rats, pigs, and more. *Lab. Invest.*, **71** (1994), 789–91.

[SCg] Schwartz, R.S., Pathophysiology of restenosis: Interaction of thrombosis, hyperplasia, and/or remodeling. *Am. J. Cardiol.*, **81** (1998), 14E–7E.

[SEa] Serruys, P.W., Dejaegere, P., Kiemeneij, F., et al., A comparison of balloon-expandable-stent implantation with balloon angioplasty in patients with coronary-artery disease. *N. Engl. J. Med.*, **331** (1994), 489–95.

[SEb]   Serruys, P.W., Emanuelsson, H., van der Giessen, W., et al., Heparin-coated Palmaz–Schatz stents in human coronary arteries. Early outcome of the Benestent-II Pilot Study. *Circulation*, **93** (1996), 412–22.

[SIa]   Siepmann, J., and Gopferich, A., Mathematical modeling of bioerodible, polymeric drug delivery systems. *Adv. Drug Delivery Rev.*, **48** (2001), 229–47.

[SIb]   Sigwart, U., Puel, J., Mirkovitch, V., Joffre, F., and Kappenberger, L., Intravascular stents to prevent occlusion and restenosis after transluminal angioplasty. *N. Engl. J. Med.*, **316** (1987), 701–6.

[SIc]   Silber, S., Hamburger, J., Grube, E., et al., Direct stenting with TAXUS stents seems to be as safe and effective as with predilatation. A post hoc analysis of TAXUS II. *Herz*, **29** (2004), 171–80.

[SId]   Siparsky, G.L., Voorhees, K.J., and Miao, F.D., Hydrolysis of polylactic acid (PLA) and polycaprolactone (PCL) in aqueous acetonitrile solutions: Autocatalysis. *J. Environ. Polym. Degradation*, **6** (1998), 31–41.

[STa]   Staab, M.E., Holmes, D.R., and Schwartz, R.S., Polymers. In: Sigwart U, Ed. **Endoluminal Stenting**. WB Saunders, London (1996), 34–44.

[STb]   Stone, G.W., Ellis, S.G., Cox, D.A., et al., One-year clinical results with the slow-release, polymer-based, paclitaxel-eluting TAXUS stent: The TAXUS-IV trial. *Circulation*, **109** (2004), 1942–7.

[SUa]   Su, S.H., Chao, R.Y., Landau, C.L., et al., Expandable bioresorbable endovascular stent. I. Fabrication and properties. *Ann. Biomed. Eng.*, **31** (2003), 667–77.

[TAa]   Tamada, J.A., and Langer, R., Erosion kinetics of hydrolytically degradable polymers. *Proceedings of the National Academy of Sciences of the United States of America*, **90** (1993), 552–6.

[TAb]   Tamai, H., Igaki, K., Kyo, E., et al., Initial and 6-month results of biodegradable poly-l-lactic acid coronary stents in humans. *Circulation*, **102** (2000), 399–404.

[TAc]   Tamai, H., Igaki, K., Tsuji, T., et al., A biodegradable poly-L-lactic acid coronary stent in the porcine coronary artery. *J. Interven. Cardiol.*, **12** (1999), 443–9.

[TAd]   Tammela, T.L., and Talja, M., Biodegradable urethral stents. *BJU Int.*, **92** (2003), 843–50.

[TAe]   Tanabe, K., Serruys, P.W., Grube, E., et al., TAXUS III Trial: In-stent restenosis treated with stent-based delivery of paclitaxel incorporated in a slow-release polymer formulation. *Circulation*, **107** (2003), 559–64.

[THa] Therasse, E., Soulez, G., Cartier, P., et al., Infection with fatal outcome after endovascular metallic stent placement. *Radiology,* **192** (1994), 363–5.

[THb] Thombre, A.G., Theoretical aspects of polymer biodegradation: Mathematical modeling of drug release and acid-catalyzed poly(othoester) biodegradation. In: Vert, M., Feijen, J., Albertsson, A.C., Scott, G., Chiellini, E., Eds. **Biodegradable Polymers and Plastics**. The Royal Society of Chemisty, Cambridge (1992) 214–25.

[TSa] Tsuji, T., Tamai, H., Igaki, K., et al., Biodegradable polymeric stents. *Curr. Interv. Cardiol. Rep.,* **3** (2001), 10–7.

[TSb] Tsuji, T., Tamai, H., Igaki, K., et al., Biodegradable stents as a platform to drug loading. *Int. J. Cardiovasc. Intervent.,* **5** (2003), 13–6.

[UNa] Unverdorben, M., Spielberger, A., Schywalsky, M., et al., A polyhydroxybutyrate biodegradable stent: Preliminary experience in the rabbit. *Cardiovasc. Intervent. Radiol.,* **25** (2002), 127–32.

[UUa] Uurto, I., Mikkonen, J., Parkkinen, J., et al., Drug-eluting biodegradable poly-D/L-lactic acid vascular stents: An experimental pilot study. *J. Endovasc. Ther.,* **12** (2005), 371–9.

[VAa] Valimaa, T., Laaksovirta, S., Tammela, T.L., et al., Viscoelastic memory and self-expansion of self-reinforced bioabsorbable stents. *Biomaterials,* **23** (2002), 3575–82.

[VAb] van der Giessen, W.J., Lincoff, A.M., Schwartz, R.S., et al., Marked inflammatory sequelae to implantation of biodegradable and non-biodegradable polymers in porcine coronary arteries. *Circulation,* **94** (1996), 1690–7.

[VAc] van der Giessen, W.J., Slager, C.J., Gussenhoven, E.J., et al., Mechanical features and in vivo imaging of a polymer stent. *Int. J. Card. Imaging.,* **9** (1993), 219–26.

[VAd] van der Giessen, W.J., Slager, C.J., van Beusekom, H.M., et al., Development of a polymer endovascular prosthesis and its implantation in porcine arteries. *J. Interv. Cardiol.,* **5** (1992), 175–85.

[VAe] van der Giessen, W.J., Vanbeusekom H.M.M., Vanhouten, C.D., Vanwoerkens, L.J., Verdouw, P.D., and Serruys, P.W., Coronary stenting with, polymer-coated and uncoated self-expanding endoprostheses in pigs. *Coronary Artery Disease,* **3** (1992), 631 40.

[VEa] Vert, M., Li, S., Garreau, H., et al., Complexity of the hydrolytic degradation of aliphatic polyesters. *Angew Makromol Chem.,* **247** (1997), 239–53.

[VEb] Vert, M., Li, S.M., Spenlehauer, G., and Guerin, P., Bioresorbability and biocompatibility of aliphatic polyesters. *J. Mater. Sci. Mater. Med.*, **3** (1992), 432–46.

[VEc] Vert, M., Aliphatic polyesters: Great degradable polymers that cannot do everything. *Biomacromolecules*, **6** (2005), 538–46.

[VIa] Virmani, R., Liistro, F., Stankovic, G., et al., Mechanism of late in-stent restenosis after implantation of a paclitaxel derivate-eluting polymer stent system in humans. *Circulation*, **106** (2002), 2649–51.

[WEa] Weir, N.A., Buchanan, F.J., Orr, J.F., and Dickson, G.R., Degradation of poly-L-lactide. Part 1: In vitro and in vivo physiological temperature degradation. *Proc Inst Mech Eng [H]*, **218** (2004), 307–19.

[WEb] Weir, N.A., Buchanan, F.J., Orr, J.F., Farrar, D.F., and Dickson, G.R., Degradation of poly-L-lactide: Part 2: Increased temperature accelerated degradation. *Proceedings of the Institution of Mechanical Engineers Part H-J. Eng. Med.*, **218** (2004), 321–30.

[WEc] Wentzel, J.J., Krams, R., Schuurbiers, J.C., et al., Relationship between neointimal thickness and shear stress after Wallstent implantation in human coronary arteries. *Circulation*, **103** (2001), 1740–5.

[WEd] Wentzel, J.J., Whelan, D.M., van der Giessen, W.J., et al., Coronary stent implantation changes 3-D vessel geometry and 3-D shear stress distribution. *J. Biomech.*, **33** (2000), 1287–95.

[WHa] Whelan, D.M., van Beusekom, H.M., and van der Giessen, W.J., Mechanisms of drug loading and release kinetics. *Semin. Interv. Cardiol.*, **3** (1998), 127–31.

[WIa] Williams, D.F., Biodegradation of surgical polymers. *J. Mater. Sci.*, **17** (1982), 1233–46.

[YAa] Yamawaki, T., Shimokawa, H., Kozai, T., et al., Intramural delivery of a specific tyrosine kinase inhibitor with biodegradable stent suppresses the restenotic changes of the coronary artery in pigs in vivo. *J. Am. Coll. Cardiol.*, **32** (1998), 780–6.

[YEa] Ye, Y.W., Landau, C., Willard, J.E., et al., Bioresorbable microporous stents deliver recombinant adenovirus gene transfer vectors to the arterial wall. *Ann. Biomed. Eng.*, **26** (1998), 398–408.

[YOa] Yoon, J.S., Jin, H.J., Chin, I.J., Kim, C., and Kim, M.N., Theoretical prediction of weight loss and molecular weight during random chain scission degradation of polymers. *Polymer*, **38** (1997), 3573–9.

[ZAa] Zahn, R., Hamm, C.W., Schneider, S., et al., Incidence and predictors of target vessel revascularization and clinical event rates of the sirolimus-eluting coronary stent (results from the prospective multi-center German Cypher Stent Registry). *Am. J. Cardiol.*, **95** (2005), 1302–8.

[ZHa]  Zhang, Y., Zale, S., Sawyer, L., and Bernstein, H., Effects of metal salts on poly(DL-lactide-co-glycolide) polymer hydrolysis. *J. Biomed. Mater. Res.*, **34** (1997), 531–8.

[ZIa]  Zidar, J., Lincoff, A., and Stack, R., Biodegradable stents. In: Topol, E.J., Ed. **Textbook of Interventional Cardiolology**. 2nd ed. WB Saunders, Philadelphia (1994), 787–802.

[ZIb]  Zilberman, M., Nelson, K.D., and Eberhart, R.C., Mechanical properties and in vitro degradation of bioresorbable fibers and expandable fiber-based stents. *J. Biomed. Mater. Res. B Appl. Biomater.*, **74** (2005), 792–9.

[ZIc]  Zilberman, M., Schwade, N.D., and Eberhart, R.C., Protein-loaded bioresorbable fibers and expandable stents: Mechanical properties and protein release. *J. Biomed. Mater. Res. B Appl. Biomater.*, **69** (2004), 1–10.

# 5

# Regulation of Hemostatic System Function by Biochemical and Mechanical Factors

K. RAJAGOPAL AND J. LAWSON

*Department of Surgery*
*Duke University Medical Center*
*Durham, NC 27710, USA*

ABSTRACT. The mammalian hemostatic system has evolved to accomplish the task of sealing defects in the cardiovascular system. Hemostasis occurs in and around a disruption in a vascular conduit through which blood normally flows, and is characterized by the localized formation of thrombus. Consequently, the process of hemostasis is influenced by: (1) the biochemical properties of the cellular and soluble components of the hemostatic system, counterregulatory networks, and the vascular conduit; (2) the local hemodynamic conditions, which regulate the influx and efflux of substrates, cofactors, and catalysts, and which also impose loads on the forming clot; and (3) the local mechanical properties of the vasculature. We review the components of the hemostatic and negative regulatory systems and their biochemical functions, and the roles that local hemodynamics play in the regulation of hemostasis.

## 5.1 Components of the Hemostatic System

The hemostatic system, prior to thrombus formation, is comprised of both platelets as well as water-soluble proteins that are in an inactive state (zymogens). Once thrombus formation is initiated, platelet recruitment and activation occur, and the coagulation zymogens, as well as counterregulatory zymogens, become proteolytically cleaved and thereby activated. The end result of these processes is the formation of thrombus that contains platelets and cross-linked (polymerized) fibrin. In addition, negative regulatory mechanisms exist that protect against excessive clot formation; these consist of anticoagulant factors that diminish clot formation, and fibrinolytic factors that degrade formed thrombus. In the following section, we discuss in detail the various components of the hemostatic system, as well as the regulation of their function.

### 5.1.1 Platelets

The principal cellular constituents of the hemostatic system are platelets. Similar to other blood cells, they are derived from bone marrow precursors (megakaryocytes [PAb]). However, there are features that distinguish platelet development substantially from erythroid, lymphoid, and myeloid development. A single nucleated megakaryocyte sheds multiple enucleated platelets as fragments [PAb]. In contrast, lymphoid and myeloid cells retain nuclei originating from their precursors and thus are capable of further mitoses, and individual enucleated erythrocytes develop from individual nucleated normoblast precursors.

Once shed from the marrow into the circulation, platelets may continue to circulate or be retained by the spleen. Under steady-state conditions, approximately two thirds of platelets circulate with a lifespan of seven to ten days, and one third are retained within the spleen. Consequently, the circulating platelet concentration is determined by:

1. The rate of platelet generation in the bone marrow

2. The rate of intravascular platelet destruction, or loss from the vasculature (i.e. bleeding)

3. The rate of platelet efflux (if any) from the spleen

4. The rate of splenic sequestration and destruction of circulating platelets (Figure 5.1).

Similarly, rates of production, rates of destruction, and rates of redistribution among physiological compartments, are the general determinants

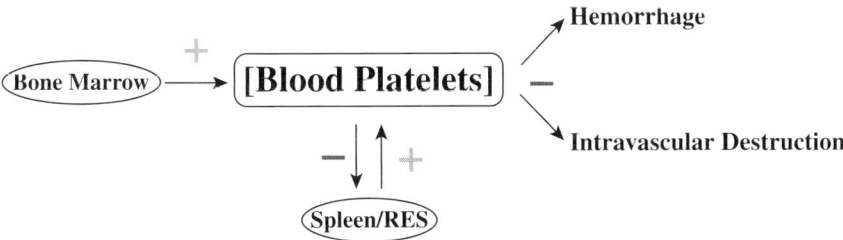

Figure 5.1. Diagram of variables affecting blood platelet concentrations (denoted by [ ]). As shown, platelets are generated in the bone marrow, and may be permanently lost from the circulation via either hemorrhage or intravascular destruction. Blood platelets may enter the spleen and reticuloendothelial system (RES), in which they may be sequestered or destroyed. Sequestered platelets may re-enter the circulation. Blood concentrations of all components are regulated similarly, by: (1) rates of production, (2) rates of destruction, and (3) rates of intercompartment redistribution. + and − denote positive and negative effects on [Blood Platelets].

of the concentrations of any biological agent (ions, macromolecules, and cellular constituents).

The molecular and cellular mechanisms involved in splenic regulation of circulating levels and function of blood cellular components are incompletely understood. However, it is known that several disease states are characterized by an augmented function of the spleen ("hypersplenism") in removal of circulating platelets, which in turn leads to a diminished blood platelet concentration (thrombocytopenia [GEa]). Furthermore, studies have demonstrated that the spleen may have a function as a platelet reservoir with a capacity to release sequestered platelets when thrombocytopenia secondary to platelet destruction or loss occurs [LEa, SCa].

Broadly, the function of platelets in hemostasis is, along with the injured endothelium, to provide a cellular scaffold on which the coagulation reactions proceed and fibrin is generated. In addition, platelets are a source of molecules critical to the process of coagulation (Table 5.1). These molecules are either expressed on the surface of platelets or are contained within one of two principal types of cytoplasmic granules [REa], dense or $\alpha$ granules.

The platelet surface molecules important in platelet aggregation are principally the glycoprotein GPIb, and the multiligand binding integrin GPIIb/IIIa. These are discussed in further detail. Platelet granules were originally described and classified based upon their electron microscopic appearance. Dense granules store adenosine diphosphate (ADP), serotonin, catecholamines, and $Ca^{2+}$. Activated platelets release stored ADP molecules that bind to platelet and endothelial ADP receptors, which are essential for platelet aggregation [FIa, HOb]; antagonism of ADP binding to its cognate receptor is the mechanism of action of specific clinically used drugs (e.g. clopidogrel) in inhibiting platelet aggregation [HOb, SAd]. $Ca^{2+}$ is a requisite cofactor for several of the coagulation reactions [CAb], and in the context of the regulation of local vascular tone, also functions as a vasoconstrictor [SHa]. Alpha granules store von Willebrand's factor (vWF), and Factors V and VIII.

vWF mediates platelet–endothelial adhesion and platelet–platelet adhesion. vWF is thought to bind to subendothelial collagen exposed by vascular injury, and also binds platelet surface glycoprotein GPIb [BOa, FUb], thereby forming a trimolecular complex that bridges the platelet and the injured vessel. In addition, platelets can directly bind subendothelial collagen via GPIa, an interaction potentiated by vWF interacting with GPIb [CRa]. Finally, GPIIb/IIIa is a platelet surface glycoprotein that binds multiple ligands, including fibrinogen, vWF, and the matrix proteins fibronectin and vitronectin, consequently mediating platelet–fibrin enmeshing and adhesion to the vessel wall [MEc]. Specific antagonists of this interaction ("GPIIb/IIIa inhibitors") are clinically utilized to inhibit platelet aggregation [LEb].

Upon vascular injury, a series of events occurs to recruit platelets to the site of injury, and once there, to retain and activate them. Processes directly concerning vascular responses to injury, and in particular, intercellular interactions, are discussed subsequently; we now turn to a review of the mechanisms underlying platelet recruitment.

In general, the mechanism employed to recruit cells to a particular site is that of chemotaxis, or movement along a concentration gradient of a soluble factor. Locally released agonists, many of which are classified as "chemokines," bind to cognate receptors on the target cell's surface, which induces signals that ultimately result in cytoskeletal reorganization and cellular movement. Chemokine receptors are members of the seventransmembrane (7TM) receptor superfamily, the single largest receptor superfamily (reviewed in [ROa]). 7TM receptors are traditionally understood to mediate signal transduction via coupling to heterotrimeric guanine nucleotide-binding (G) proteins, although recent studies have identified G protein-independent signal transduction via 7TM receptors (reviewed in [RAa]). Upon agonist binding to the receptor, the heterotrimeric G protein undergoes dissociation into a $G_\alpha$ subunit and a $G_{\beta\gamma}$ heterodimer. Different G proteins are defined by differing $G_\alpha$ subunits, and specific 7TM receptors preferentially couple to one or more distinct G proteins. The chemokine receptors generally couple to $G_\alpha i$ ("inhibitory" $G_\alpha$), activation of which leads to a reduction in cytoplasmic cyclic adenosine monophosphate (cAMP). Other nonchemokine chemoattractants are also ligands for 7TM receptors that couple to other $G_\alpha$ subunits; some are also ligands for receptors that are members of other classes, including receptor tyrosine kinases.

With respect to platelet recruitment, important chemoattractant factors include members of most classes of biological ligands: small molecules, lipids and lipid-derived compounds, and peptides and proteins. Examples of small molecule platelet chemoattractants are: ADP and other nucleotides [DIa], histamine, and serotonin. Lipids and related compounds derived from processing of membrane lipids also mediate platelet chemotaxis; important among these are platelet activating factor (PAF) (19) and thromboxane A2 [KHa]. Finally, peptides and proteins may be chemoattractants for platelets, notably collagen [LOc], platelet derived growth factor (PDGF) [WIa], and bacterial N-formylated peptides [CZa].

Platelet chemotaxis occurs in concert with platelet interactions with cellular components of both blood and the vessel wall, all of which together yield activated platelets, which constitute part of the thrombus. Platelet–cell interactions are reviewed subsequently.

### 5.1.2 Coagulation Factors

Thrombus formation has been viewed as principally involving a set of serially coupled enzymatic reactions; this is displayed in Figures 5.2 and 5.3. The general mechanism is that a zymogen factor is proteolytically cleaved into an enzymatically active factor, which in turn serves as a catalyst for proteolytic cleavage of another distinct coagulation factor. This process, by virtue of upstream reactions generating catalysts for downstream reactions, results in amplification of product formation (Figure 5.3). This generates a high sensitivity to thrombogenic stimuli, which provides an animal with rapid and efficacious prevention of further blood loss. From an evolutionary standpoint, it is reasonably postulated that trauma and resultant hemorrhage posed a more frequent threat to higher-order animals (particularly to prey species) than did thrombosis, and consequently, the development of a vigorous and effective response to injurious stimuli was of greater importance. The existence of negative regulatory networks to prevent inappropriate or excessive thrombus formation is important, as evidenced by disease states in which these pathways are deficient. However, it is also important to note that the negative regulatory mechanisms in the hemostatic system are generally activated by feedback, which creates a requirement for initiation of coagulation reactions for robust activation of anticoagulant and fibrinolytic pathways. These are discussed in the following subsections.

Conventionally, the coagulation cascade is split into two pathways, *extrinsic* and *intrinsic*. The extrinsic pathway is rapidly activated by tissue

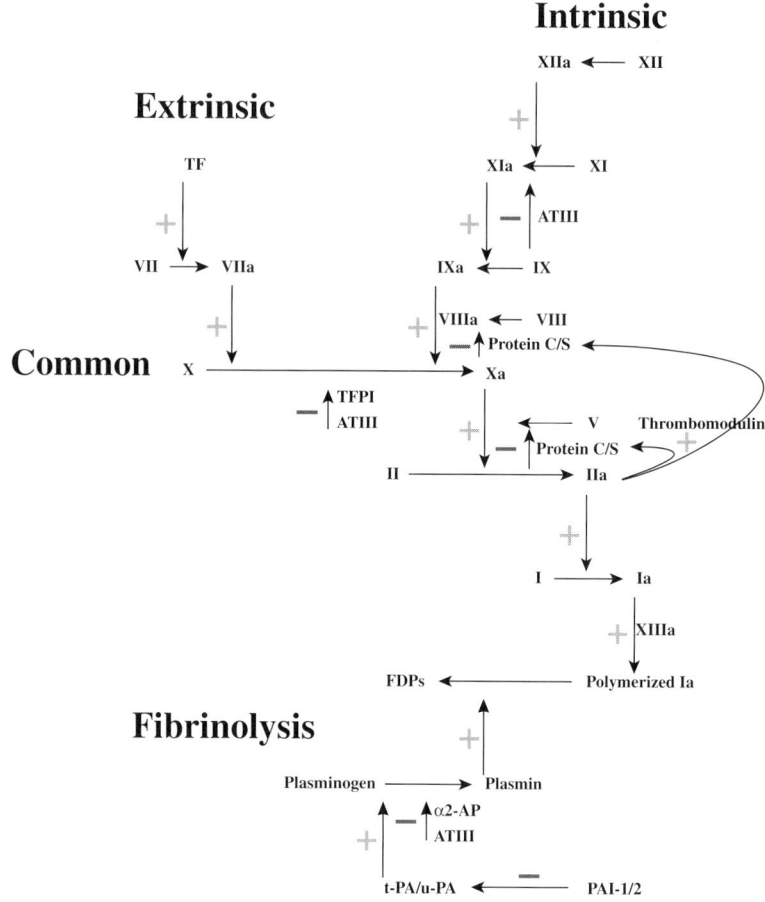

Figure 5.2. Diagram of the principal reactions of the intrinsic, extrinsic, and common pathways of coagulation, and counterregulatory (anticoagulant, fibrinolytic) mechanisms. + and − denote positive and negative effects on a reaction. Abbreviations used are noted in the text.

damage and the resultant release of tissue factor (TF). TF is a cell surface glycoprotein that binds the active form of Factor VII, VIIa, with high affinity [HIa]; this interaction is thought to substantially augment the activity of VIIa, and is requisite for optimal activation of the extrinsic pathway [KAa, OSa]. The conversion of VII to VIIa is substantially augmented by Factor Xa, an example of positive feedback [BRa, KIa]. VIIa in turn directly catalyzes the conversion of X to Xa [JEb], as well as IX to IXa [LAb], at which point the extrinsic and intrinsic pathways converge. Downstream of Xa, a common pathway serves to convert Factor I (fibrin) to Factor Ia (fibrin).

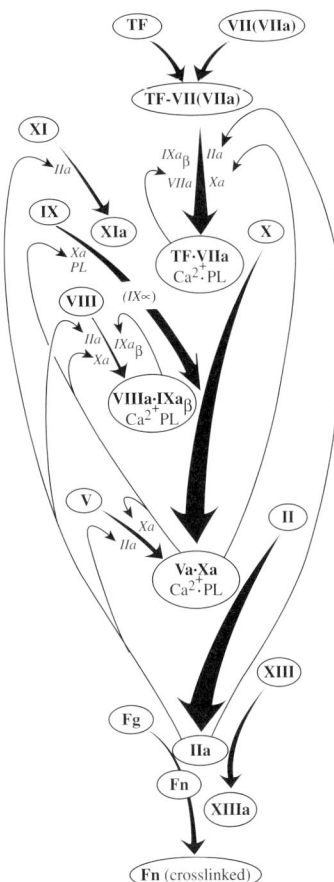

Figure 5.3. The hemostatic "vortex"; an alternate view of coagulation. This diagram illustrates the multiple levels of positive feedback/amplification inherent in the intrinsic, extrinsic, and common pathways of coagulation. Anticoagulant and fibrinolytic mechanisms are not depicted in this diagram. Abbreviations used: Fg, fibrinogen; Fn, fibrin; PL, phospholipid. Reproduced with permission from [Lawson, J.H. 1992. Tissue Factor-Dependent Hemostasis. Ph.D. Thesis, University of Vermont.].

Of note, Factors II, VII, IX, X, and Protein C (see next subsection) are all exclusively synthesized in the liver, and their activities require them being acted on by a Vitamin K-dependent carboxylase [SAb]. The liver also serves as a synthetic source for other coagulation factors, including Factors V and VIII (see below).

The intrinsic pathway is initiated by the conversion of Factor XII (Hageman Factor) to Factor XIIa upon exposure to negatively charged surfaces [HOa], which is positively regulated by prekallikrein/kallikrein. XIIa then catalyzes the $Ca^{2+}$-dependent conversion of Factor XI to Factor XIa [GRc]. XIa in turn catalyzes the $Ca^{2+}$-dependent conversion of Factor IX to Factor IXa [OSb]; this reaction is a point of cross-talk between the extrinsic and intrinsic pathways, because VIIa can catalyze the conversion of IX to IXa also [STa]. Finally, IXa and Factor VIIIa form a complex on phospholipid surfaces that catalyzes the conversion of X to Xa [VAa] in the presence of $Ca^{2+}$. This is one of two coagulation reactions (the second discussed in the next paragraph) in which there are five components: two active coagulation factors (IXa and VIIIa), cofactor ($Ca^{2+}$), surface (phospholipid), and substrate (X).

Once Xa is generated, a second five-component reaction converts Factor II (prothrombin) to Factor IIa (thrombin) [ESa]; the five components are Xa and Factor Va, $Ca^{2+}$, phospholipid, and prothrombin. In another example of positive feedback activation, thrombin cleaves and activates VIII and V, yielding VIIIa and Va [PIa]. Thrombin catalyzes the conversion of I (fibrinogen) to monomeric Ia (fibrin) [SHd, LAa]. Factor XIIIa, generated by the action of thrombin on Factor XIII, catalyzes fibrin polymerization [KAb].

### 5.1.3   Anticoagulant Factors

A distinguishing characteristic of higher-order organisms is the development of elaborate positive and negative regulatory networks that exert tight control over most processes. Hemostasis is an excellent example of this, with a balance existing between the tendency to form and maintain thrombus, and the tendency to not clot and destroy clot. Although vascular injury tips the balance in favor of thrombus formation, after time the balance shifts in the opposite direction, and thrombus destruction occurs. The counter-regulatory components regulating hemostasis may act either by preventing thrombus formation, as reviewed now, or by degrading existing thrombus.

Tissue factor pathway inhibitor (TFPI) is a potent protein inhibitor of extrinsic pathway-mediated coagulation. It contains three Kunitz-type serine protease inhibitor domains. Upon TF binding VII and conversion of VII to VIIa, TFPI binds to a quaternary complex of TF, VIIa, Xa (which, as discussed, is not only the product of the action of VIIa on X, but also promotes further production of VIIa from VII), and $Ca^{2+}$. This has biochemical sequelae:

1. The catalytic activity of formed Xa is inhibited

2. Factor VIIa/TF is inhibited [BRb]

Antithrombin III (ATIII) is, like TFPI, a protein serine protease inhibitor. It inhibits the catalytic activity of thrombin, as well as VIIa, Xa, IXa, and XIa [OLa]. The important polysaccharide anticoagulant heparin augments the inhibitory activity of ATIII (reviewed in [BJa]). Heparin-mediated activation of ATIII and resultant anticoagulation are clinically manifested by an elevated partial thromboplastin time (PTT), with a relatively unaffected prothrombin ratio/international normalized ratio (PT/INR).

Protein C is another clinically important anticoagulant protein, whose activity is induced upon thrombin generation. Thrombin binds and activates thrombomodulin, which in turn binds and activates Protein C [ESb]. Activated Protein C then interacts with the cofactor Protein S, and this complex inactivates VIIIa and Va [SAc].

### 5.1.4 The Fibrinolytic System

The second mechanism by which coagulation may be antagonized is via destruction of already formed thrombus, termed fibrinolysis. The best characterized fibrinolytic mechanism is that mediated by plasmin [SHb]. Plasmin is generated by proteolytic cleavage of the zymogen plasminogen, and is a serine protease like the coagulation factors [MAe]. Active plasmin then cleaves fibrin at multiple sites, yielding fibrin degradation products. Proteolysis of polymerized fibrin removes the structural framework of the thrombus, resulting in its dissolution.

The conversion of plasminogen to plasmin is a tightly regulated process, and requires a catalyst. Two classes of plasminogen activators, tissue plasminogen activator (t-PA) and urokinase plasminogen activator (u-PA), serve this function [DEb]. Both t-PA and u-PA display ubiquitous tissue expression patterns, and are synthesized and secreted by endothelial cells. The activities of t-PA and u-PA are negatively regulated by plasminogen activator inhibitors 1 and 2 (PAI-1 and PAI-2), polypeptides that bind to and inactivate them (reviewed in [COa]). Once plasmin has been generated, inhibition of plasmin as an antifibrinolytic mechanism becomes relevant. Several proteins inhibit plasmin, notably $\alpha$2-antiplasmin ($\alpha$2-AP; [MEa]), but also proteins with other functions in the coagulation system including ATIII [TEa].

## 5.2 Vascular Physiology in the Context of Hemostasis

The physiological response of the vascular conduit to injury is of critical importance to hemostasis. The functions of the vasculature within hemostasis can be classified as

1. Regulation of local hemodynamics (discussed later)

2. Regulation of the biochemical events of coagulation

In particular, the endothelium plays an essential role in modulating platelet activation, as well as the coagulation cascade. These processes are discussed below.

### 5.2.1 Endothelial Regulation of Local Hemodynamics

The direct effects of mechanical forces on the process of hemostasis are reviewed in detail subsequently. These forces, however, are physiologically influenced by the actions of several compounds liberated by endothelial cells, which in turn both positively and negatively regulate vascular smooth muscle contractile function, and thus local vascular resistance and impedance. Endothelial cells serve as sources of nitric oxide (NO) and NO carrier molecules, which stimulate vascular smooth muscle relaxation and vasodilation [IGa]. Another vasodilator agent released by endothelial cells includes prostaglandin I2 [HYa]. However, the endothelium also synthesizes and releases vasoconstrictor agents. These include endothelin-1, the most potent endogenous arteriolar vasoconstrictor identified to date (ET-1; [YAa]), and platelet activating factor [PAF; BAa].

### 5.2.2 Platelet–Endothelial Interactions

Like other blood cellular components, platelets undergo a process termed margination, whereby they move from the center channel of flowing blood to the boundary layer, thus coming into contact with the endothelium. The ability of different cell types to marginate has been shown to be dependent upon size [DOb], and is also promoted by chemoattractants (e.g. [OLb]); platelets are much smaller than leukocytes, and marginate to a lesser extent under normal conditions. Marginated platelets then bind endothelial cells through various receptor–ligand interactions, and along with chemotaxis, undergo activation. This moves the platelet from a quiescent state from the standpoint of hemostasis, to an active one. Several changes are observed, including:

1. Adhesion to the endothelium, other platelets, and fibrin, in part due to upregulation of pre-existing adhesion molecules

2. Release of the contents of platelet granules [MUb]

3. Platelet morphologic changes [BOc]

Some of these events have been briefly reviewed in the previous section. We now discuss them in detail, with particular emphasis on endothelial interactions.

Adhesion to the endothelium is predominantly mediated by high-affinity interactions between receptor–ligand pairs. Important among these receptors are GPIb and GPIIb/IIIa, briefly discussed earlier. GPIb is a member of the leucine-rich repeat group family [CLa], and GPIIb/IIIa is a member of the best-characterized family of adhesion receptors, the integrins [SHc]. Members of the leucine-rich repeat family include the Toll-like receptors, which are critical regulators of innate immunity [SCd]. The roles of other family members are less well characterized, but this class of cell surface receptors is present in primitive Metazoa, suggesting important and conserved functions. GPIb, along with GPIX and GPV, bind vWF released from the endothelium or platelet granules [RUa]. Additionally, the GPIb complex binds thrombin [HAb].

Similarly evolutionarily conserved are integrins, which are heterodimeric proteins comprised of $\alpha$ and $\beta$ subunits that may bind to cell surface or matrix ligands. GPIIb/IIIa interacts with thrombospondin [NAa], a soluble cell surface-binding glycoprotein produced by platelets (stored in $\alpha$ granules) and endothelial cells. The matrix ligands and other noncell surface ligands to which GPIIb/IIIa binds include fibrin, vWF, fibronectin, and vitronectin [MEc]. Recently, crystallographic approaches have been used to show that integrins have a structure with a "head" and two "legs," with one leg each from the $\alpha$ and $\beta$ subunit [XIa]. In the case of GPIIb/IIIa, these structural features may explain the mechanism underlying the efficacy of clinically used fibrinogen mimetic GPIIb/IIIa inhibitors [XIa].

The high avidity-binding of GPIIb/IIIa to its ligands has made it an attractive target for antiplatelet therapies, and the efficacy of these therapies in acute coronary syndromes, where they are most commonly used [INa], is well established. In contrast, although antagonism of the GPIb/vWF interaction has been shown in animal models to have antithrombotic effects [YEa], clinically utilized agents have not been developed to date. It is unclear whether GPIb is of less importance than GPIIb/IIIa in vivo, but loss-of-function mutations in GPIb result in Bernard–Soulier syndrome [HAa], a bleeding disorder.

Platelet degranulation is comprised of release of $\alpha$ and dense granule contents into the local extracellular milieu. Because these granules contain procoagulant molecules, degranulation is a positive feedback/amplificatory mechanism with respect to clot formation. It is an important and tightly regulated and inducible process, triggered by agonist ligands (e.g. thrombin and ADP) binding to platelet receptors. Respectively, these include the protease-activated receptors PAR1 and PAR4, and the ADP receptor [GRb], all of which are members of the 7TM receptor superfamily.

Temporally associated with platelet activation are morphological changes. Generally, changes in cell morphology require cytoskeletal reorganization, and platelet cytoskeletal reorganization is observed subsequent to exposure to activating stimuli [FUc]. Cytoskeletal reorganization induces structural changes in the fluid plasma membrane, namely, the formation of membrane projections. Surface receptors cluster at these projections, increasing the probability, and therefore, the avidity, of platelet–platelet, platelet–endothelial, and platelet–coagulation factor interactions.

### 5.2.3  Endothelial Regulation of the Coagulation Cascade

In addition to regulating platelet functions, the endothelium is critically involved in regulation of the coagulation cascade. These effects are due to endothelial surface expression or release of a variety of procoagulant, anticoagulant, profibrinolytic, and antifibrinolytic molecules (some of which have been discussed above).

Endothelial cells are also an important reservoir for vWF, and the clinically used vasopressin analogue DDAVP stimulates release of vWF from the Weibel–Palade bodies of endothelial cells [ROb]. Additionally, they synthesize and store the cofactors for the five-component reactions, V and VIII, although the physiological significance of these functions is unclear. Low levels of Factor V synthesis have been observed in cultured endothelial cells [GIa]. Factor VIII gene expression has been demonstrated in both hepatocytes and hepatic endothelial cells [DOa]; furthermore, liver transplantation is curative for Hemophilia A, which is a deficiency in Factor VIII [BOb].

The endothelium is generally thought, under baseline conditions, to be in a net "anticoagulant" state; it utilizes all of the predominant anticoagulant mechanisms reviewed previously [CIa]. Endothelial cells express thrombomodulin, thereby promoting thrombin-dependent activation of Protein C [ESb]. In addition, endothelial cells express TFPI [WAb], release of which is promoted by heparin [GIb]. Last, the subendothelial matrix, and in particular, polyanionic glycosaminoglycans, mediate activation of ATIII [DEc].

Upon exposure to a thrombotic stimulus, endothelial cells promote not only platelet recruitment and aggregation as discussed, but they also promote the generation of fibrin [CIa]. Tissue factor is inducibly expressed on the endothelium [PAa], although whether the endothelium actually serves as the principal source of TF synthesis is unclear. This results in activation of Factor VII (i.e. the extrinsic pathway), which is likely the predominant mechanism under most circumstances in vivo with respect to thrombus formation. Finally, endothelial cells express receptors for thrombin (PAR-1 and PAR-2; [HAc]) and other thrombus components, signaling through which induces a variety of responses.

## 5.3   Mechanics and Effects on Hemostasis

Although there are important biochemical functions that blood constituents and the cells of the vasculature perform in hemostasis, these functions serve a mechanical purpose, namely, sealing vascular defects. Consequently, it stands to reason that the efficacy of the hemostatic process is governed by the intrinsic mechanical properties of the parts of the system. Furthermore, the extrinsic mechanical forces to which the components are subjected may, and likely do, also influence the process of hemostasis in vivo. The solid mechanics of the clot and the vascular conduit, and local hemodynamics, are discussed in this section.

### 5.3.1   Mechanical Properties of Blood and Clots

Thrombus formation can be viewed as liquid blood progressively increasing viscosity as a function of time postexposure to "thrombogenic" stimulus, the end product being a gellike clot. The specific mechanical properties of clotted versus unclotted blood are more complex, however, drawing a simple distinction highlights the concept that blood and the hemostatic thrombus are vastly different in terms of their mechanical properties. If thrombus and flowing blood were mechanically similar, and moreover, if thrombi did not adhere to injured vessel walls, vascular disruptions could never be sealed and hemostasis would never occur. We now survey various studies, both theoretical and experimental, on the mechanical properties of blood and thrombi.

In the context of continuum models, blood is often treated as a Newtonian liquid (e.g. [MIa, LOb, PRa]). Some experimental data suggest that this may be a reasonable approximation for blood flow in large conduits (reviewed in [BEa]). However, even in large or intermediate-size conduits with certain flow regimes, and in smaller blood vessels (small arteries/veins, arterioles/venules), numerous studies have demonstrated that blood exhibits non-Newtonian behavior (reviewed in [BEa, HAe, MEb, MAb, MOa]). Over short time scales of analysis, blood appears to display yield stress (the value of which is proportional to the hematocrit [MOc]). Additionally, blood exhibits shear-thinning behavior (reviewed in [BEa]; [AMa]). Shear-thinning may be in part due to the fact that erythrocyte aggregation is inversely related to shear rate [SCc, MUa]. Numerous studies, notably those of Thurston [THa], have also demonstrated that blood exhibits stress relaxation. Finally, in small conduits in which cellular constituent size is on the order of the diameter of the vessel, blood cannot be reasonably treated as an homogenized continuum, in as much as the cellular components and plasma are separated within small blood volumes.

In sum, blood is a multiconstituent material. Approaches to modeling its macroscopic behavior include mixture theories versus homogenized single-constituent fluid models; although mixture theories should be suitable for modeling blood behavior across a wide range of vessel sizes and related hemodynamic conditions, homogenized models are clearly inappropriate in small vessels in which the cellular size is on the order of the lumen size. Furthermore, non-Newtonian behavior of blood is demonstrable within intermediate-sized conduits, thereby affecting the choice of an appropriate homogenized model.

The clinical and experimental consequences of these non-Newtonian properties are unclear. However, comprehensive mathematical models of the hemostatic system or hemodynamics in general will require incorporating these features, not only for the purposes of completeness, but because of the difficulty in being able to determine a priori the effects of including or excluding these various features on the accuracy of the predictions of theories, which would only be tested thereafter. As a preliminary example, Anand et al. [ANa] have modeled unclotted blood as a generalized (viscosity and relaxation time are both shear rate-dependent in this model) Oldroyd-B fluid, with clots as fluids exhibiting much higher viscosities as a function of shear rate (another pertinent example can be found in [YEb]). Even in large vessels, the mixture properties of blood may be relevant. In regions of arterial flow instability, margination and "sedimentation" of cellular and other large particles such as lipoproteins occur. This is thought to be necessary for the development of atherosclerosis; numerous observational and experimental studies, notably those of Glagov and co-workers [KUa] have demonstrated that atheromas preferentially develop in regions of flow stagnation. These and other related hemodynamic studies are discussed in the following section.

Clotted blood is evidently biochemically and mechanically distinct from unclotted static or flowing blood. Several lines of evidence demonstrate that clots display behaviors most consistent with highly viscous, gellike liquids, as they contain a volume of trapped liquid plasma within a network of fibrin polymers and cellular material (platelets, erythrocytes, etc.). This structural organization results in interesting mechanical properties. Clot creep was experimentally demonstrated over 30 years ago [GEb; NEa; NEb], and stress relaxation has also been observed [JAa]. More recent experimental studies have shown flow of established clots in cylindrical conduits when exposed to pulsatile pressure gradients [HEa; THb]. In summary, experimental data suggest that thrombi are highly viscous viscoelastic liquids; over short time scales of analysis, they can be treated as viscoelastic solids. However, models of clots as highly viscous liquids can capture solidlike behaviors over short time scales well (with the exception of "catastrophic" phenomena such as traumatic clot disruption), whereas in contrast, models of clots as solids will fail to capture liquidlike behavior over long time scales.

The properties of individual constituents of unclotted and clotted blood contribute to their global behaviors. First, platelets are essential for normal clot strength. Highly enriched platelet preparations exist as gels [ALa]. Jen and McIntire [Jen, CJ, and McIntire, LV. The structural properties and contractile force of a clot. *Cell Motil.*, 1982, **2**(5):445–455.] have shown that a mixture of plasma and platelets forms clots with elastic moduli and contractile force per unit area approximately 10 and 15 times that of plasma alone, respectively. Furthermore, clots formed from plasma with functionally defective platelets from a patient with Glanzmann's thrombasthenia (see previous) displayed similar properties to clots formed from plasma alone. These findings demonstrate that the mechanical properties of the platelets themselves are insufficient to account for the effects that platelets have on the solid mechanics of the clot, because clots with biochemically dysfunctional platelets present have vastly different material properties than those with functional platelets.

Although erythrocytes are not generally thought of as serving essential functions in clot formation, they comprise a substantial portion of clot mass and volume [CAa]. In theory, clot strength is proportionally related to clot viscosity, and experimental data have demonstrated that the principal determinant of blood viscosity is the hematocrit; presumably, clot viscosity is also influenced in a similar fashion by the hematocrit, due to the presence of erythrocytes in clots. This hypothesis is supported by data that suggest that erythrocytes contribute to clot strength, in as much as the deformations required to achieve clot disruption are proportional to the hematocrit [RIa]. Erythrocytes, and their plasma membranes in particular, behave as linearly elastic solids over short time scales of analysis [EVa]. However, over longer experimental time scales, erythrocyte membranes have been found to demonstrate both yield stress [EVa] and stress relaxation [MAd].

In contrast to whole blood, plasma appears to behave as a Newtonian liquid [CHa] under most circumstances. It does not display stress relaxation or other time-dependent behaviors as does whole blood, but like whole blood, plasma may display shear-thinning [ERa]. This may be due to shear-mediated breakdown of large plasma polymers that are important factors influencing viscosity. Once plasma has coagulated, resulting fibrin polymers (after Factor XIIIa-mediated cross-linking) are responsible for much of the clot's mechanical properties. Recent work [COb] suggests that fibrin polymers display anisotropy, insofar as the strain in response to application of tensile stress has been found to be less than that observed in response to application of flexion stress. However, it is important to note that in such studies, as the fibrin strands exist within a thrombus (in this case, a plasma clot without cellular constituents, but with molecules other than fibrin present, as well as aqueous solvent), the mechanical properties assayed are not isolated properties of an individual (or even cross-linked set, after exposure to Factor XIIIa) fibrin strand, but rather, are influenced by the concentrations and properties of the other components of the thrombus.

### 5.3.2   Hemodynamics

Local hemodynamic factors have important effects on clot formation and destruction, in addition to the effects that they have on blood and clot mechanical properties. We now discuss hemodynamic forces in the context of the hemostatic system. These have been reviewed in detail in Anand et al. [ANa]. It is important to note that in many, if not most cases, it is unclear which specific mechanical factors are causative with respect to effects on thrombus formation, and as such it is currently difficult to distinguish effects of specific hemodynamic variables on the hemostatic system from those that are proportionally related.

Under steady-state condtions, the cardiac output and venous return are identical, and moreover, in isolated tissue/organ beds, the equality of arterial inflow rate and venous outflow rate holds true. However, substantial differences in the characteristics of arterial and venous flow exist. In the systemic and pulmonary arterial trees, blood flow is unsteady, and more specifically, is pulsatile. Systolic, mean, and diastolic arterial pressures generally far exceed venous pressure, although pulmonary arterial diastolic pressure is minimally higher than pulmonary venous pressures under normal circumstances. In contrast, blood flow is essentially steady in the systemic and pulmonary venous circuits. Furthermore, not only does maximum flow velocity in arterial systems far exceed flow velocity in venous systems, as is obvious, but mean flow velocity in arterial systems is generally greater than in venous systems [BEa]. These findings correlate with those observed for intracardiac chamber pressures and flows. The pressure-generating ventricles, which pressurize and eject during systole, and which depressurize and fill during diastole, are responsible for generating pulsatile arterial blood flow. In contrast, the atria are principally reservoirs at the outlet of their respective venous circuits.

These hemodynamic profiles that vary as a function of anatomic location have theoretically differential effects on clot formation, maintenance, and lysis, which are supported by experimental data and clinical studies. Low flow and wall shear rates (i.e. flow regimes associated with stagnation) have been shown to promote thrombus formation (e.g. [YAb]), hypothesized over a century ago by Virchow [VIa]. Conversely, high wall shear rates have been shown to exert antithrombotic effects [HAd]. Based on the differing hemodynamic profiles discussed above, it might be reasonably hypothesized that veins are more prone to thrombus formation than are arteries, which is true. The systemic venous system as a whole exhibits a greater propensity for thrombosis than does the systemic arterial system, and the pelvis and lower extremities are at particular risk due to the effects of gravity on low-pressure venous blood flow in the upright position [MAc]. These are the most common sites for deep venous thrombosis (DVT). In addition, venous valves, which facilitate unidirectional flow through the venous system, are sites at which secondary flows may develop and at which thrombosis may occur [KAc].

Additionally, certain sites in the arterial tree are prone to thrombus formation. In the systemic arterial system, branching points, aneurysms, and sites of atherosclerotic plaques (see below) are all regions of secondary flows with stagnation, in which clots commonly form. Pertinent experimental studies are discussed in further detail later.

But what are the molecular mechanisms that underlie, or at least the molecular phenomena that are associated with, the hemodynamic regulation of thrombus formation? The effects of hemodynamic forces on thrombus formation can be understood as effects on the components of a nascent clot. Platelet activation, adhesion, and aggregation are in fact promoted by high shear stresses [MAa, WOa]. Although this may appear paradoxical insofar as high shear stresses are usually associated with high flow rates and velocities at which stasis and resultant thrombosis should be unlikely, the inhibitory effects of high shear stresses and/or proportionally related variables (and conversely, the stimulatory effects of low shear stresses) on the endothelium and coagulation/antifibrinolytic factors [GRa] are more consistent with the findings that thrombi predominate in regions of low shear stresses.

High shear stresses are associated with enhanced generation of NO, and presumably, biological NO carriers such as S-nitrosothiols, which are antithrombotic [FRa] by virtue of antiplatelet effects and inhibition of coagulation factor synthesis. Clinical and experimental evidence also suggest that high shear stresses cause proteolysis of vWF [TSa], which may be mediated by the metalloprotease ADAMTS-13 ([DOc]; evidence suggests that deficiency in ADAMTS-13 underlies the pathogenesis of thrombotic thrombocytopenia purpura [SAa]) particularly in the case of high molecular weight vWF, thereby impairing platelet aggregation and coagulation. Shear stresses also promote fibrinolysis, at least in part due to induction of increased t-PA expression/secretion by the endothelium [DIb]; in contrast, elevated normal stresses have been shown to suppress t-PA expression/secretion [SJa]. In fact, elevated normal stresses in arterial hypertensive diseases such as pulmonary hypertension [LOa] have been associated with increased platelet activation and vWF levels. Also, animal studies of venous interposition grafts into the carotid arterial circulation have demonstrated that elevated circumferential wall stress ($\sigma_{\theta\theta}$) decreases endothelial expression of thrombomodulin [SPa], generating a locally prothrombotic milieu due to functional deficiency of active protein C.

In aggregate, several lines of evidence suggest that thrombus formation and maintenance are antagonized by shear stresses and promoted by normal stresses. However, intralumenal pressures are substantially higher in systemic or pulmonary arteries than they are in systemic or pulmonary veins, suggesting that wall normal stresses, and thus at least one contributing mechanical factor for promotion of thrombus formation, may be greater in arteries than in veins. Higher intralumenal pressures do not necessarily translate to elevated wall normal stresses in arteries, however, because artery lumens are generally smaller and artery walls generally thicker than venous lumens and walls. Finally, spatiotemporal gradients in both normal and shear stresses may also be important in modulating thrombus formation.

These data are consistent with extensive experimental work on the roles of hemodynamic loads in the development of atherosclerosis, notably initiated by Fry [LUa] and Glagov. The experimental model systems of Glagov and co-workers [ZAa], and others (e.g. [WEa, GAa]), have focused principally on atheroma formation in the carotid arterial system; recent work has also examined atherosclerosis-related hemodynamics in the coronary arterial circulation [RAb]. Atherosclerosis has long been recognized as a diffuse disease that can afflict any portion of the systemic arterial tree, but comprised of numerous discrete lesions. Initial experimental studies [KUa, ZAa] showed that in human carotid artery autopsy specimens and in a scale model of the human carotid arterial bifurcation, zones of secondary flows with recirculation (the outer wall of the internal carotid artery) were those found to be the predominant susceptible sites for atherosclerotic plaques in a corresponding set of cadaveric specimens. A so-called "oscillatory shear index" was developed in order to quantify the extent of flow recirculation/stagnation, and has been successfully correlated in several systems to risk of atheroma formation (reviewed in [KLa]). The mechanisms by which these secondary flows with stagnation cause atheroma formation are not clear, but experimental evidence supports two distinct, but not mutually exclusive processes:

1. The cells of the vessel wall (endothelial, vascular smooth muscle) respond directly to altered hemodynamic loads, that is mechanosensation (reviewed in [WAa]).

2. Transport of blood-borne components onto (binding to endothelial surface receptors) and into (infiltrating into the interstitium and contacting vascular smooth muscle cells) the vessel is increased due to augmented local dwelling of the components in question (e.g. [MOb]).

Of these, the latter mechanism may directly contribute to thrombus formation.

## 5.4 Developing Physiological Experimental Model Systems and Mathematical Models for Coagulation

The study of complex biological systems has historically required reductionist approaches to perform controlled experiments. However, although such models provide the ability to do tightly controlled studies, they have the disadvantage of being of little physiological relevance; the more reductionist and controllable a system is, the less physiologically accurate and relevant it and the results obtained from it are. Conversely, in vivo experimental model systems are complicated by the difficulty in controlling variables.

Modern experimental techniques, both ex vivo and in vivo, may be able to address these difficulties. Microfluidics technology [RUb] has recently been used to develop simple experimental "biomimetic" models, which may be capable of reproducing ex vivo the fluid mechanical conditions in vivo. However, such approaches contain assumptions regarding the biochemical behavior of the reactants, catalysts, and products, because biomimetic models use different molecules from those biologically involved as replacements. Thus, these models have the opposite problem of classical biochemical studies of coagulation, which neglect fluid mechanical conditions. Yet, more complex microfluidics systems using actual blood components have been developed [DEa], and data obtained from these systems may prove useful in understanding hemostasis in vivo, and in aiding in the construction of mathematical models with robust predictive power.

The most physiologically accurate experimental systems in which to study hemostasis are in vivo (reviewed in [FUa]). Intravital microscopy [SCb] has been used to study the in vivo course of stimulus-induced thrombus formation. In studies published to date (notably [FAa]), rapid confocal imaging techniques combined with digital capture have been used, which yield temporally closely spaced three-dimensional images. Local hemodynamic variables can also be measured as a function of time, using standard pressure-transducing catheters and flow (e.g. Doppler) probes. Thus, intravascular thrombus formation and hemodynamic variables that influence thrombus formation can be both tracked as a function of time poststimulus. The stimuli used may be local endothelial injury using a laser coupled to the imaging system, or using agents such as ferric chloride ($FeCl_3$) that are caustic to the endothelium.

Comprehensive mathematical models of the hemostatic system have been lacking in general. This has been for several reasons. First, most models (e.g. [JOa]; a survey of such models, as well as others, is provided in Anand et al. [ANa]) have focused only on mechanically static biochemical systems. Second, even those models that incorporate fluid mechanical factors, such as that of Kuharsky and Fogelson [KUb], have neglected the direct effects of mechanical loading on the process of clot formation (and its maintenance and lysis), and only incorporate the effects of hemodynamic forces on the transport of reactants, catalysts, and products. Third, even those models that have incorporated at least some of the effects of mechanical loads (insofar as they affect component transport) on thrombus formation, also neglect the mechanical properties of blood and the forming clots. Fourth, these models assume spatially homogeneous distribution of reactants, catalysts, and products (as evidenced by the exclusive use of ordinary differential equations). Fifth and finally, the experimental data such as diffusion coefficients and rate constants, upon which all mathematical models rely, are generally obtained from reductionist in vitro experiments. Anand et al. [ANa] have recently developed a mathematical model of the hemostatic system that addresses many of the aforementioned problems, but is yet far from complete. It models blood as a non-Newtonian (generalized Oldroyd-B) liquid and the forming clot as such a liquid with much higher viscosity. A series of 25 coupled convection–reaction–diffusion partial differential equations are used to describe the flow, generation, and depletion of hemostatic system components. Thus the first through fourth concerns described previously are addressed at least in part. However, there are assumptions and oversimplifications inherent in the model. First, blood and the clot are both treated as homogenized continua; as discussed, a mixture theory model would be more physiologically accurate. Second, the model only includes the components of the extrinsic and common pathways, and neglects the intrinsic pathway. Third, the vascular conduit is assumed to be rigid, whereas native blood vessels are not. Fourth, the clot formation being modeled is that of a "plasma clot," which is devoid of cellular constituents (although platelets are included in the model, both as protein binding sites and as catalytic surfaces); this is clearly an incorrect oversimplification, because whole blood does not give rise to clots that selectively contain only plasma constituents. Nonetheless, such models provide a preliminary framework for the development of more comprehensive mathematical models of the hemostatic system that fully capture the salient biochemical and mechanical processes in vivo.

## 5.5   Conclusion

Gaining a full understanding of the functions of multicomponent biological systems in health and disease is dependent upon generating both mathematical and experimental models that capture as many biochemical and physical features as possible. The hemostatic system is one such complex multicomponent system. We are only beginning to develop mathematical models that are both rigorous and comprehensive in terms of the currently available experimental data, with the capacity to make predictions that can be tested in physiologically faithful experimental models. Equally important, pertinent experimental models are relatively nascent in their development. Such studies will, it is hoped, not only aid in our understanding of hemostasis and its regulation, but will also serve as a template for approaching other complex biological systems.

## 5.6   References

[ALa]   Altmeppen, J., Hansen, E., Bonnlander, G., Horch, R., and Jeschke, M., Composition and characteristics of an autologous thrombocyte gel. *J. Surg Res.*, **117** (2004), 202–207.

[AMa]   Amin, T. and Sirs, J., The blood rheology of man and various animal species. *Q. J. Exp. Physiol.*, **70** (1985), 37–49.

[ANa]   Anand, M., Rajagopal, K. and Rajagopal, K., A model incorporating some of the mechanical and biochemical factors underlying clot formation and dissolution in flowing blood. *J. Theor. Med.*, **5** (2003).

[BAa]   Baer, P. and Cagen, L., Platelet activating factor vasoconstriction of dog kidney. Inhibition by alprazolam. *Hypertension*, **9** (1987), 253–260.

[BEa]   Berne, R. and Levy, M., **Cardiovascular Physiology**. Mosby, St. Louis (1992).

[BJa]   Bjork, I. and Lindahl, U., Mechanism of the anticoagulant action of heparin. *Mol. Cell. Biochem.*, **48** (1982), 161–182.

[BOa]   Bockenstedt, P., Greenberg, J. and Handin, R., Structural basis of von Willebrand factor binding to platelet glycoprotein Ib and collagen. Effects of disulfide reduction and limited proteolysis of polymeric von Willebrand factor. *J. Clin. Invest.*, **77** (1986), 743–749.

[BOb]   Bontempo, F., Lewis, J., Gorenc, T., Spero, J., Ragni, M., Scott, J. and Starzl, T., Liver transplantation in hemophilia A. *Blood*, **69** (1987), 1721–1724.

[BOc] Boyle Kay, M. and Fudenberg, H., Inhibition and reversal of platelet activation by cytochalasin B or colcemid. *Nature*, **244** (1973), 288–289.

[BRa] Broze, G.J. and Majerus, P., Purification and properties of human coagulation factor VII. *J. Biol. Chem.*, **255** (1980), 1242–1247.

[BRb] Broze, G.J., Warren, L., Novotny, W., Higuchi, D., Girard, J. and Miletich, J., The lipoprotein-associated coagulation inhibitor that inhibits the factor VII-tissue factor complex also inhibits factor Xa: Insight into its possible mechanism of action. *Blood*, **71** (1988), 335–343.

[CAa] Cadroy, Y. and Hanson, S., Effects of red blood cell concentration on hemostasis and thrombus formation in a primate model. *Blood*, **75** (1990), 2185–2193.

[CAb] Carr, M. and Powers, P., Differential effects of divalent cations on fibrin structure. *Blood Coagul. Fibrinolysis*, **2** (1991), 741–747.

[CHa] Chien, S., Usami, S., Taylor, H., Lundberg, J. and Gregersen, M., Effects of hematocrit and plasma proteins on human blood rheology at low shear rates. *J. Appl. Physiol.*, **21** (1966), 81–87.

[CIa] Cines, D., Pollak, E., Buck, C., Loscalzo, J., Zimmerman, G., McEver, R., Pober, J., Wick, T., Konkle, B., Schwartz, B., et al., Endothelial cells in physiology and in the pathophysiology of vascular disorders. *Blood*, **91** (1998), 3527–3561.

[CLa] Clemetson, K., Platelet activation: Signal transduction via membrane receptors. *Thromb. Haemost.*, **74** (1995), 111–116.

[COa] Collen, D. and Lijnen, H., Fibrin-specific fibrinolysis. *Ann. N.Y. Acad. Sci.*, **667** (1992), 259–271.

[COb] Collet, J., Shuman, H., Ledger, R., Lee, S. and Weisel, J., The elasticity of an individual fibrin fiber in a clot. *Proc. Natl. Acad. Sci. USA*, **102** (2005), 9133–9137.

[CRa] Cruz, M., Chen, J., Whitelock, J., Morales, L. and Lopez, J., The platelet glycoprotein Ib-von Willebrand factor interaction activates the collagen receptor alpha2beta1 to bind collagen: Activation-dependent conformational change of the alpha2-I domain. *Blood*, **105** (2005), 1986–1991.

[CZa] Czapiga, M., Gao, J., Kirk, A. and Lekstrom-Himes, J., Human platelets exhibit chemotaxis using functional N-formyl peptide receptors. *Exp. Hematol.*, **33** (2004), 73–84.

[DEa] De Marco, L., Perris, R., Cozzi, M. and Mazzucato, M., Blood clotting in space. *J. Biol. Regul. Homeost. Agents*, **18** (2004), 187–192.

[DEb]  de Vries, C., Veerman, H., Blasi, F. and Pannekoek, H., Artificial exon shuffling between tissue-type plasminogen activator (t-PA) and urokinase (u-PA): A comparative study on the fibrinolytic properties of t-PA/u-PA hybrid proteins. *Biochemistry*, **27** (1988), 2565–2572.

[DEc]  Desai, U., Petitou, M., Bjork, I. and Olson, S., Mechanism of heparin activation of antithrombin. Role of individual residues of the pentasaccharide activating sequence in the recognition of native and activated states of antithrombin. *J. Biol. Chem.*, **273** (1998), 7478–7487.

[DIa]  Di Virgilio, F., Chiozzi, P., Ferrari, D., Falzoni, S., Sanz, J., Morelli, A., Torboli, M., Bolognesi, G. and Baricordi, O., Nucleotide receptors: an emerging family of regulatory molecules in blood cells. *Blood*, **97** (2001), 587–600.

[DIb]  Diamond, S., Eskin, S. and McIntire, L., Fluid flow stimulates tissue plasminogen activator secretion by cultured human endothelial cells. *Science*, **243** (1989), 1483–1485.

[DOa]  Do, H., Healey, J., Waller, E. and Lollar, P., Expression of factor VIII by murine liver sinusoidal endothelial cells. *J Biol. Chem.*, **274** (1999), 19587–19592.

[DOb]  Doerschuk, C., Downey, G., Doherty, D., English, D., Gie, R., Ohgami, M., Worthen, G., Henson, P. and Hogg, J., Leukocyte and platelet margination within microvasculature of rabbit lungs. *J. Appl. Physiol.*, **68** (1990), 1956–1961.

[DOc]  Dong, J., Cleavage of ultra-large von Willebrand factor by ADAMTS-13 under flow conditions. *J. Thromb. Haemost.*, **3** (2005), 1710–1716.

[ERa]  Ernst, E., The non Newtonian properties of plasma. *Haematologica*, **67** (1982), 321–322.

[ESa]  Esmon, C., Owen, W. and Jackson, C., A plausible mechanism for prothrombin activation by factor Xa, factor Va, phospholipid, and calcium ions. *J. Biol. Chem.*, **249** (1974), 8045–8047.

[ESb]  Esmon, N., Owen, W. and Esmon, C., Isolation of a membrane bound cofactor for thrombin–catalyzed activation of protein C. *J. Biol. Chem.*, **257** (1982), 859–864.

[EVa]  Evans, E. and La Celle, P., Intrinsic material properties of the erythrocyte membrane indicated by mechanical analysis of deformation. *Blood*, **45** (1975), 29–43.

[FAa]  Falati, S., Gross, P., Merrill-Skoloff, G., Furie, B. and Furie, B., Real-time in vivo imaging of platelets, tissue factor and fibrin during arterial thrombus formation in the mouse. *Nat. Med.*, **8** (2002), 1175–1181.

[FIa]   Figures, W., Scearce, L., Wachtfogel, Y., Chen, J., Colman, R. and Colman, R., Platelet ADP receptor and alpha 2-adrenoreceptor interaction. Evidence for an ADP requirement for epinephrine-induced platelet activation and an influence of epinephrine on ADP binding. *J. Biol. Chem.*, **261** (1986), 5981–5986.

[FRa]   Freedman, J., Frei, B., Welch, G. and Loscalzo, J., Glutathione peroxidase potentiates the inhibition of platelet function by S-nitrosothiols. *J. Clin. Invest.*, **96** (1995), 394–400.

[FUa]   Furie, B. and Furie, B., Thrombus formation in vivo. *J. Clin. Invest.*, **115** (2005), 3355–3362.

[FUb]   Furlan, M., Von Willebrand factor: Molecular size and functional activity. *Ann. Hematol.*, **72** (1996), 341–348.

[FUc]   Furman, M., Gardner, T. and Goldschmidt-Clermont, P., Mechanisms of cytoskeletal reorganization during platelet activation. *Thromb. Haemost.*, **70** (1993), 229–232.

[GAa]   Gambillara, V., Chambaz, C., Roy, S., Stergiopulos, N. and Silacci, P., Plaque-prone hemodynamic impairs endothelial function in pig carotid arteries. *Am. J. Physiol. Heart Circ. Physiol.* (2006) epub:epub.

[GEa]   George, J., el-Harake, M. and Raskob, G., Chronic idiopathic thrombocytopenic purpura. *New Engl. J. Med.*, **331** (1994), 1207–1211.

[GEb]   Gerth, C., Roberts, W. and Ferry, J., Rheology of fibrin clots. II. Linear viscoelastic behavior in shear creep. *Biophys. Chem.*, **2** (1974), 208–217.

[GIa]   Giddings, J., Jarvis, A. and Hogg, S., Factor V in human vascular endothelium and in endothelial cells in culture. *Thromb. Res.*, **44** (1986), 829–835.

[GIb]   Giraux, J., Tapon-Bretaudiere, J., Matou, S. and Fischer, A., Fucoidan, as heparin, induces tissue factor pathway inhibitor release from cultured human endothelial cells. *Thromb. Haemost.*, **80** (1998), 692–695.

[GRa]   Grabowski, E. and Lam, F., Endothelial cell function, including tissue factor expression, under flow conditions. *Thromb. Haemost.*, **74** (1995), 123–128.

[GRb]   Graff, J., Klinkhardt, U. and Harder, S., Pharmacodynamic profile of antiplatelet agents: Marked differences between single versus costimulation with platelet activators. *Thromb. Res.*, **113** (2004), 295–302.

[GRc] Griffin, J. and Cochrane, C., Mechanisms for the involvement of high molecular weight kininogen in surface-dependent reactions of Hageman factor. *Proc. Natl. Acad. Sci. USA*, **73** (1976), 2554–2558.

[HAa] Hagen, I. and Solum, N., Further studies on the protein composition and surface structure of normal platelets and platelets from patients with Glanzmann's thrombasthenia and Bernard-Soulier syndrome. *Thromb. Res.*, **13** (1978), 845–855.

[HAb] Hagen, I., Brosstad, F., Gogstad, G., Korsmo, R. and Solum, N., Further studies on the interaction between thrombin and GP Ib using crossed immunoelectrophoresis. Effect of thrombin inhibitors. *Thromb. Res.*, **27** (1982), 549–554.

[HAc] Hamilton, J., Nguyen, P. and Cocks, T., Atypical protease-activated receptor mediates endothelium-dependent relaxation of human coronary arteries. *Circ. Res.*, **82** (1998), 1306–1311.

[HAd] Hashimoto, S., Maeda, H. and Sasada, T., Effect of shear rate on clot growth at foreign surfaces. *Artif. Organs*, **9** (1985), 345–350.

[HAe] Haynes, R. and Burton, A., Role of the non-Newtonian behavior of blood in hemodynamics. *Am. J. Physiol.*, **197** (1959), 943–950.

[HEa] Henderson, N. and Thurston, G., A new method for the analysis of blood and plasma coagulation. *Biomed. Sci. Instrum.*, **29** (1993), 95–102.

[HIa] Higashi, S., Matsumoto, N. and Iwanaga, S., Molecular mechanism of tissue factor-mediated acceleration of factor VIIa activity. *J. Biol. Chem.*, **271** (1996), 26569–26574.

[HOa] Hojima, Y., Cochrane, C., Wiggins, R., Austen, K. and Stevens, R., In vitro activation of the contact (Hageman factor) system of plasma by heparin and chondroitin sulfate E. *Blood*, **63** (1984), 1453–1459.

[HOb] Hollopeter, G., Jantzen, H., Vincent, D., Li, G., England, L., Ramakrishnan, V., Yang, R., Nurden, P., Nurden, A., Julius, D., et al., Identification of the platelet ADP receptor targeted by antithrombotic drugs. *Nature*, **409** (2001), 202–207.

[HYa] Hyman, A., Chapnick, B., Kadowitz, P., Lands, W., Crawford, C., Fried, J. and Barton, J., Unusual pulmonary vasodilator activity of 13,14-dehydroprostacyclin methyl ester: comparison with endoperoxides and other prostanoids. *Proc. Natl. Acad. Sci. USA*, **74** (1977), 5411–5415.

[IGa] Ignarro, L., Nitric oxide as a unique signaling molecule in the vascular system: a historical overview. *J. Physiol. Pharmacol.*, **53** (2002), 503–514.

[INa] Investigators, T.P.T., Inhibition of platelet glycoprotein IIb/IIIa with eptifibatide in patients with acute coronary syndromes. The

PURSUIT Trial Investigators. Platelet Glycoprotein IIb/IIIa in Unstable Angina: Receptor Suppression Using Integrilin Therapy. *New Engl. J. Med.*, **339** (1998), 436–443.

[ISa]  Issekutz, A. and Ripley, M., The effect of intravascular neutrophil chemotactic factors on blood neutrophil and platelet kinetics. *Am. J. Hematol.*, **21** (1986), 157–171.

[JAa]  Janmey, P., Amis, E. and Ferry, J., Rheology of fibrin clots. VI. Stress relaxation, creep, and differential dynamic modulus of fine clots. *J. Rheol.*, **27** (1983), 135–153.

[JEa]  Jesty, J. and Nemerson, Y., Purification of Factor VII from bovine plasma. Reaction with tissue factor and activation of Factor X. *J. Biol. Chem.*, **249** (1974), 509–515.

[JOa]  Jones, K. and Mann, K., A model for the tissue factor pathway to thrombin. II. A mathematical simulation. *J. Biol. Chem.*, **269** (1994), 23367–23373.

[KAa]  Kalafatis, M., Egan, J., van 't Veer, C., Cawthern, K. and Mann, K., The regulation of clotting factors. *Crit. Rev. Eukaryot. Gene Expr.*, **7** (1997), 241–280.

[KAb]  Kanaide, H. and Shainoff, J., Cross-linking of fibrinogen and fibrin by fibrin-stablizing factor (factor XIIIa). *J. Lab. Clin. Med.*, **85** (1975), 574–597.

[KAc]  Karino, T. and Motomiya, M., Flow through a venous valve and its implication for thrombus formation. *Thromb. Res.*, **36** (1984), 245–257.

[KHa]  Khirabadi, B., Foegh, M., Goldstein, H. and Ramwell, P., The effect of prednisolone, thromboxane, and platelet-activating factor receptor antagonists on lymphocyte and platelet migration in experimental cardiac transplantation. *Transplantation*, **43** (1987), 626–630.

[KIa]  Kirchhofer, D., Eigenbrot, C., Lipari, M., Moran, P., Peek, M. and Kelley, R., The tissue factor region that interacts with factor Xa in the activation of factor VII. *Biochemistry*, **40** (2001), 675–682.

[KLa]  Kleinstreuer, C., Hyun, S., Buchanan, J.J., Longest, P., Archie, J.J. and Truskey, G., Hemodynamic parameters and early intimal thickening in branching blood vessels. *Crit. Rev. Biomed. Eng.*, **29** (2001), 1–64.

[KUa]  Ku, D., Giddens, D., Zarins, C. and Glagov, S., Pulsatile flow and atherosclerosis in the human carotid bifurcation. Positive correlation between plaque location and low oscillating shear stress. *Arteriosclerosis*, **5** (1985), 293–302.

[KUb]  Kuharsky, A. and Fogelson, A., Surface-mediated control of blood coagulation: the role of binding site densities and platelet deposition. *Biophys. J.*, **80** (2001), 1050–1074.

[LAa]  Laki, K., The action of thrombin on fibrinogen. *Science*, **114** (1951), 435–436.

[LAb]  Lawson, J. and Mann, K., Cooperative activation of human factor IX by the human extrinsic pathway of blood coagulation. *J. Biol. Chem.*, **266** (1991), 11317–11327.

[LEa]  Lee, E. and Schiffer, C., Evidence for rapid mobilization of platelets from the spleen during intensive plateletpheresis. *Am. J. Hematol.*, **19** (1985), 161–165.

[LEb]  Lefkovits, J., Plow, E. and Topol, E., Platelet glycoprotein IIb/IIIa receptors in cardiovascular medicine. *New Engl. J. Med.*, **332** (1995), 1553–1559.

[LOa]  Lopes, A., Maeda, N., Aiello, V., Ebaid, M. and Bydlowski, S., Abnormal multimeric and oligomeric composition is associated with enhanced endothelial expression of von Willebrand factor in pulmonary hypertension. *Chest*, **104** (1993), 1455–1460.

[LOb]  Lou, Z., Yang, W. and Stein, P., Errors in the estimation of arterial wall shear rates that result from curve fitting of velocity profiles. *J. Biomech.*, **26** (1993), 383–390.

[LOc]  Lowenhaupt, R., Silberstein, E., Sperling, M. and Mayfield, G., A quantitative method to measure human platelet chemotaxis using indium-111-oxine-labeled gel-filtered platelets. *Blood*, **60** (1982), 1345–1352.

[LUa]  Lutz, R., Cannon, J., Bischoff, K., Dedrick, R., Stiles, R. and Fry, D., Wall shear stress distribution in a model canine artery during steady flow. *Circ. Res.*, **41** (1977), 391–399.

[MAa]  Maalej, N., Holden, J. and Folts, J., Effect of shear stress on acute platelet thrombus formation in canine stenosed carotid arteries: An in vivo quantitative study. *J. Thromb. Thrombolysis*, **5** (1998), 231–238.

[MAb]  Mann, K., Deutsch, S., Tarbell, J., Geselowitz, D., Rosenberg, G. and Pierce, W.S., An experimental study of Newtonian and non-Newtonian flow dynamics in a ventricular assist device. *J. Biomech. Eng.*, **109** (1987), 139–147.

[MAc]  Markel, A., Manzo, R., Bergelin, R. and Strandness, D.J., Pattern and distribution of thrombi in acute venous thrombosis. *Arch. Ann.* **127** (1992), 305–309.

[MAd] Markle, D., Evans, E. and Hochmuth, R., Force relaxation and permanent deformation of erythrocyte membrane. *Biophys. J.*, **42** (1983), 91–98.

[MAe] Mattler, L. and Bang, N., Serine protease specificity for peptide chromogenic substrates. *Thromb. Haemost.*, **38** (1977), 776–792.

[MEa] Menoud, P., Sappino, N., Boudal-Khoshbeen, M., Vassalli, J. and Sappino, A., The kidney is a major site of alpha(2)-antiplasmin production. *J. Clin. Invest.*, **97** (1996), 2478–2484.

[MEb] Merrill, E. and Pelletier, G., Viscosity of human blood: Transition from Newtonian to non-Newtonian. *J. Appl. Physiol.*, **23** (1967), 178–182.

[MEc] Merten, M., Pakala, R., Thiagarajan, P. and Benedict, C., Platelet microparticles promote platelet interaction with subendothelial matrix in a glycoprotein IIb/IIIa-dependent mechanism. *Circulation*, **99** (1999), 2577–2582.

[MIa] Misra, J. and Choudhury, K., Effect of initial stresses on the wave propagation in arteries. *J. Math. Biol.*, **18** (1983), 53–67.

[MOa] Mo, L., Yip, G., Cobbold, R., Gutt, C., Joy, M., Santyr, G. and Shung, K., Non-Newtonian behavior of whole blood in a large diameter tube. *Biorheology*, **28** (1991), 421–427.

[MOb] Moore, J.J., Ku, D., Zarins, C. and Glagov, S., Pulsatile flow visualization in the abdominal aorta under differing physiologic conditions: implications for increased susceptibility to atherosclerosis. *J. Biomech. Eng.*, **114** (1992), 391–397.

[MOc] Morris, C., Smith, C.N. and Blackshear, P.J., A new method for measuring the yield stress in thin layers of sedimenting blood. *Biophys. J.*, **52** (1987), 229–240.

[MUa] Murata, T. and Secomb, T., Effects of shear rate on rouleau formation in simple shear flow. *Biorheology*, **25** (1988), 113–122.

[MUb] Mustard, J., Perry, D., Kinlough-Rathbone, R. and Packham, M., Factors responsible for ADP-induced release reaction of human platelets. *Am. J. Physiol.*, **228** (1975), 1757–1765.

[NAa] Nachman, R. and Leung, L., Complex formation of platelet membrane glycoproteins IIb and IIIa with fibrinogen. *J. Clin. Invest.*, **69** (1982), 263–269.

[NEa] Nelb, G., Gerth, C., Ferry, J. and Lorand, L., Rheology of fibrin clots. III. Shear creep and creep recovery of fine ligated and coarse unligated clots. *Biophys. Chem.*, **5** (1976), 377–387.

[NEb] Nelb, G., Kamykowski, G. and Ferry, J., Rheology of fibrin clots. V. shear modulus, creep, and creep recovery of fine unligated clots. *Biophys. Chem.*, **13** (1981), 15–23.

[OLa] Olson, S., Swanson, R., Raub-Segall, E., Bedsted, T., Sadri, M., Petitou, M., Herault, J., Herbert, J. and Bjork, I., Accelerating ability of synthetic oligosaccharides on antithrombin inhibition of proteinases of the clotting and fibrinolytic systems. Comparison with heparin and low-molecular-weight heparin. *Thromb. Haemost.*, **92** (2004), 929–939.

[OLb] Olson, T., Singbartl, K. and Ley, K., L-selectin is required for fMLP-but not C5a-induced margination of neutrophils in pulmonary circulation. *Am. J. Physiol. Regul. Integr. Comp. Physiol.*, **282** (2002), R1245–R1252.

[OSa] Osterud, B., Factor VII and haemostasis. *Blood Coagul. Fibrinolysis*, **1** (1990), 175–181.

[OSb] Osterud, B., Laake, K. and Prydz, H., The activation of human factor IX. *Thromb. Diath. Haemorrh.*, **33** (1975), 553–563.

[PAa] Parry, G. and Mackman, N., Transcriptional regulation of tissue factor expression in human endothelial cells. *Arterioscler. Thromb. Vasc. Biol.*, **15** (1995), 612–621.

[PAb] Patel, S., Hartwig, J. and Italiano, J.J., The biogenesis of platelets from megakaryocyte proplatelets. *J. Clin. Invest.*, **115** (2005), 3348–3354.

[PIa] Pieters, J., Lindhout, T. and Hemker, H., In situ-generated thrombin is the only enzyme that effectively activates factor VIII and factor V in thromboplastin-activated plasma. *Blood*, **74** (1989), 1021–1024.

[PRa] Prosi, M., Perktold, K., Ding, Z. and Friedman, M., Influence of curvature dynamics on pulsatile coronary artery flow in a realistic bifurcation model. *J. Biomech.*, **37** (2004), 1767–1775.

[RAa] Rajagopal, K., Lefkowitz, R. and Rockman, H., When 7 transmembrane receptors are not G protein-coupled receptors. *J. Clin. Invest.*, **115** (2005), 2971–2974.

[RAb] Ramaswamy, S., Vigmostad, S., Wahle, A., Lai, Y., Olszewski, M., Braddy, K., Brennan, T., Rossen, J., Sonka, M. and Chandran, K., Comparison of left anterior descending coronary artery hemodynamics before and after angioplasty. *J. Biomech. Eng.*, **128** (2006), 40–48.

[REa] Rendu, F. and Brohard-Bohn, B., The platelet release reaction: Granules' constituents, secretion and functions. *Platelets*, **12** (2001), 261–273.

[RIa] Riha, P., Wang, X., Liao, R. and Stoltz, J., Elasticity and fracture strain of whole blood clots. *Clin. Hemorheol. Microcirc.*, **21** (1999), 45–49.

[ROa] Rockman, H., Koch, W. and Lefkowitz, R. Seven-transmembrane-spanning receptors and heart function. *Nature*, **415** (2002), 206–212.

[ROb] Rosenberg, J., Foster, P., Kaufman, R., Vokac, E., Moussalli, M., Kroner, P. and Montgomery, R., Intracellular trafficking of factor VIII to von Willebrand factor storage granules. *J. Clin. Invest.*, **101** (1998), 613–624.

[RUa] Ruggeri, Z., De Marco, L., Gatti, L., Bader, R. and Montgomery, R., Platelets have more than one binding site for von Willebrand factor. *J. Clin. Invest.*, **72** (1983), 1–12.

[RUb] Runyon, M., Johnson-Kerner, B. and Ismagilov, R.F., Minimal functional model of hemostasis in a biomimetic microfluidic system. *Angew Chem. Int. Ed. Engl.*, **43** (2004), 1531–1536.

[SAa] Sadler, J., von Willebrand factor: Two sides of a coin. *J. Thromb. Haemost.*, **3** (2005), 1702–1709.

[SAb] Sadowski, J., Esmon, C. and Suttie, J., Vitamin K-dependent carboxylase. Requirements of the rat liver microsomal enzyme system. *J. Biol. Chem.*, **251** (1976), 2770–2776.

[SAc] Salem, H., Broze, G., Miletich, J. and Majerus, P., Human coagulation factor Va is a cofactor for the activation of protein C. *Proc. Natl. Acad. Sci. USA*, **80** (1983), 1584–1588.

[SAd] Savi, P. and Herbert, J., Clopidogrel and ticlopidine: P2Y12 adenosine diphosphate-receptor antagonists for the prevention of atherothrombosis. *Semin. Thromb. Hemost.*, **31** (2005), 174–183.

[SCa] Schaffner, A., Augustiny, N., Otto, R. and Fehr, J., The hypersplenic spleen. A contractile reservoir of granulocytes and platelets. *Arch. Int. Med.*, **145** (1985), 651–654.

[SCb] Schmid-Schonbein, G., Usami, S., Skalak, R. and Chien, S., The interaction of leukocytes and erythrocytes in capillary and postcapillary vessels. *Microvasc. Res.*, **19** (1980), 45–70.

[SCc] Schmid-Schonbein, H., Gallasch, G., von Gosen, J., Volger, E. and Klose, H., Red cell aggregation in blood flow. I. New methods of quantification. *Klin. Wochenschr.*, **54** (1976), 149–157.

[SCd] Schuster, J. and Nelson, P., Toll receptors: An expanding role in our understanding of human disease. *J. Leukoc. Biol.*, **67** (2000), 767–773.

[SHa] Shapira, N., Schaff, H., White, R. and Pluth, J., Hemodynamic effects of calcium chloride injection following cardiopulmonary bypass: response to bolus injection and continuous infusion. *Ann. Thor. Ann.*, **37** (1984), 133–140.

[SHb] Sherry, S., Fibrinolysis. *Ann. Rev. Med.*, **19** (1968), 247–268.

[SHc]  Shimaoka, M. and Springer, T., Therapeutic antagonists and conformational regulation of integrin function. *Nat. Rev. Drug Discov.*, **2** (2003), 703–716.

[SHd]  Shulman, S., Herwig, W. and Ferry, J., The conversion of fibrinogen to fibrin. V. Influence of ionic strength and thrombin concentration on the effectiveness of certain reversible inhibitors. *Arch. Biochem.*, **32** (1951), 354–358.

[SJa]  Sjogren, L., Doroudi, R., Gan, L., Jungersten, L., Hrafnkelsdottir, T. and Jern, S., Elevated intraluminal pressure inhibits vascular tissue plasminogen activator secretion and downregulates its gene expression. *Hypertension*, **35** (2000), 1002–1008.

[SPa]  Sperry, J., Deming, C., Bian, C., Walinsky, P., Kass, D., Kolodgie, F., Virmani, R., Kim, A. and Rade, J., Wall tension is a potent negative regulator of in vivo thrombomodulin expression. *Circ. Res.*, **92** (2003), 41–47.

[STa]  Stern, D., Drillings, M., Kisiel, W., Nawroth, P., Nossel, H. and LaGamma, K., Activation of factor IX bound to cultured bovine aortic endothelial cells. *Proc. Natl. Acad. Sci. USA*, **81** (1984), 913–917.

[TEa]  Telesforo, P., Semeraro, N., Verstraete, M. and Collen, D., The inhibition of plasmin by antithrombin III-heparin complex in vitro in human plasma and during streptokinase therapy in man. *Thromb. Res.*, **7** (1975), 669–676.

[THa]  Thurston, G., Viscoelasticity of human blood. *Biophys. J.*, **12** (1972), 1205–1217.

[THb]  Thurston, G. and Henderson, N., Impedance of a fibrin clot in a cylindrical tube: Relation to clot permeability and viscoelasticity. *Biorheology*, **32** (1995), 503–520.

[TSa]  Tsai, H., Sussman, I. and Nagel, R., Shear stress enhances the proteolysis of von Willebrand factor in normal plasma. *Blood*, **83** (1994), 2171–2179.

[VAa]  van Dieijen, G., Tans, G., Rosing, J. and Hemker, H., The role of phospholipid and factor VIIIa in the activation of bovine factor X. *J. Biol. Chem.*, **256** (1981), 3433–3442.

[VIa]  Virchow, R. Uber den Fasenstoff: V. phlogose und thrombose im gefabsystem. In **Gesammelte Abhandlungen zur wissenschaftlichen Medicin**. Verlag v. Meidinger, Sohn and Corp., Frankfurt am Main (1856).

[WAa]  Wang, J. and Thampatty, B., An introductory review of cell mechanobiology. *Biomech. Model Mechanobiol.*, **5** (2006), 1–16.

[WAb] Warn-Cramer, B., Almus, F. and Rapaport, S., Studies of the factor Xa-dependent inhibitor of factor VIIa/tissue factor (extrinsic pathway inhibitor) from cell supernates of cultured human umbilical vein endothelial cells. *Thromb. Haemost.*, **61** (1989), 101–105.

[WEa] Wells, D., Archie, J.J. and Kleinstreuer, C., Effect of carotid artery geometry on the magnitude and distribution of wall shear stress gradients. *J. Vasc. Ann.*, **23** (1996), 667–678.

[WIa] Wissler, R., Update on the pathogenesis of atherosclerosis. *Am. J. Med.*, **91** (1991), 3S-9S.

[WOa] Wootton, D. and Ku, D., Fluid mechanics of vascular systems, diseases, and thrombosis. *Annu. Rev. Biomed. Eng.*, **1** (1999), 299–329.

[XIa] Xiao, T., Takagi, J., Coller, B., Wang, J. and Springer, T. Structural basis for allostery in integrins and binding to fibrinogen-mimetic therapeutics. *Nature*, **432** (2004), 59–67.

[YAa] Yanagisawa, M., Kurihara, H., Kimura, S., Tomobe, Y., Kobayashi, M., Mitsui, Y., Yazaki, Y., Goto, K. and Masaki, T., A novel potent vasoconstrictor peptide produced by vascular endothelial cells. *Nature*, **332** (1988), 411–415.

[YAb] Yasuda, T., Sekimoto, K., Taga, I., Funakubo, A., Fukui, Y. and Takatani, S., New method for the detection of thrombus formation in cardiovascular devices: Optical sensor evaluation in a flow chamber model. *ASAIO J.*, **51** (2005), 110–115.

[YEa] Yeh, C., Wang, W., Hsieh, T. and Huang, T., Agkistin, a snake venom-derived glycoprotein Ib antagonist, disrupts von Willebrand factor-endothelial cell interaction and inhibits angiogenesis. *J. Biol. Chem.*, **275** (2000), 18615–18618.

[YEb] Yeleswarapu, K., Antaki, J., Kameneva, M. and Rajagopal, K., A mathematical model for shear-induced hemolysis. *Artif. Organs*, **19** (1995), 576–582.

[ZAa] Zarins, C., Giddens, D., Bharadvaj, B., Sottiurai, V., Mabon, R. and Glagov, S., Carotid bifurcation atherosclerosis. Quantitative correlation of plaque localization with flow velocity profiles and wall shear stress. *Circ. Res.*, **53** (1983), 502–514.

# 6

# Mechanical Properties of Human Mineralized Connective Tissues

R. De Santis, L. Ambrosio,

*IMCB Institute for Composite and Biomedical Materials*
*National Research Council*
*I-80125 Napoli, Italy*

F. Mollica,

*University of Ferrara*
*Department of Engineering*
*I-44100 Ferrara, Italy*

P. Netti, and L. Nicolais

*University of Naples "Federico II"*
*Department of Materials and Production Engineering*
*I-80125 Napoli, Italy*

ABSTRACT.   Experimental work has to be tightly linked with modeling. In fact, successful modeling requires firstly comparison with experiments in order to verify its predictions or to set its range of validity. Secondly, experiments measuring the mechanical properties of tissues are needed as input to calibrate mechanical models of organs that can be used to run simulations *in silico*. In this chapter we wish to provide a comprehensive literature review covering the mechanical characterization of hard tissues, in particular compact bone, trabecular bone and dentine.

Elastic and strength properties of such tissues are carefully reviewed, together with fracture mechanics properties when available. Properties obtained from different measurement devices are gathered and compared with each other. The anisotropy and the inelasticity of the mechanical properties of hard tissues is particularly stressed.

## 6.1   Introduction

Over the past half century, research on the evaluation of the mechanical properties of mineralized connective tissues (hard tissues) has been extremely active. It remains predominantly a fundamental research field, motivated by two principal objectives. The first is to achieve a deeper knowledge of the relation between the tissue structure and the mechanical properties in order to develop diagnostic tools. The second is to improve the design of prosthetic devices used as hard tissue substitutes.

The reliability and the significance of mechanical measurements upon hard tissues are dependent on a wide range of factors. These include anisotropy, time-dependent properties (e.g. viscoelasticity), the patient's sex and age, the organ from which the specimen is taken, its site and its adaptation, and response to the loads, diseases, or traumas experienced. Also, specimen conditioning (e.g. storage conditions and temperature) has a major influence upon mechanical properties. The difficulty in obtaining specimens of suitable dimensions for testing is also a significant issue.

### 6.1.1   Mechanical Testing

Mechanical properties have been assessed through several direct and indirect measurements of stress and strain.

Testing using dynamometers is the basic direct method for estimating the mechanical properties of tissue specimens loaded in tension, compression, bending, shear, or torsion. A dynamometer consists of an actuator mounted on a frame. Loading cells and displacement sensors are used to measure the load and the deformation, respectively. This analysis can be carried out in static or dynamic conditions thus providing direct information on the static, viscoelastic, and fatigue behavior of materials. This investigation technique has been widely performed on macroscopic and microscopic specimens, the latter having cross-sectional areas of a few square millimeters. This essentially characterizes standard mechanical and micromechanical testing [ERa, DOa].

Ultrasonic testing provides information on the elastic properties of materials by measuring the longitudinal and shear speeds of sonic waves propagating into the material. The equipment used principally comprises a piezoelectric transducer and a sensor. Although this technique does not provide mechanical strength information, it is especially useful in order to detect the anisotropy of materials by determining the elastic constants. In fact, compared to classical mechanical testing, it is not destructive, thus the same specimen can be used for further analysis in different directions. Moreover, ultrasonic testing—which measures resonant frequencies, rather than wave speed—allows the assessment of mechanical anisotropy from one measurement at a single orientation of the specimen. A further advantage of ultrasonic testing is its tolerance of a diversity of specimen geometry and size. Specimen shapes need not be particularly refined, and very small specimens may be tested [SKa].

Indentation testing has mainly been used to measure the hardness of materials, from which the Young's modulus can be derived. An indenter is applied to the specimen, whose hardness is calculated by measuring the load and penetration depth. Compared to the mechanical investigation techniques outlined earlier, testing with an indenter provides data indicative of surface properties rather than bulk properties. This method is particularly attractive for the simplicity of the equipment, and also because it enables the mapping of properties as a function of the site. This technique has further evolved over recent decades with the emergence of mechanical analysis by atomic force microscopy (AFM). This basically consists of using an indenter tip, with dimensions of the order of a few atoms, and a laser to detect the interference between the tip and the substrate material. With AFM, it is possible to identify variation in local hardness at a nanoscale level [LAb, CHa]. Scanning acoustic microscopy (SAM) is a very promising tool for nondestructive investigations of the elastic properties of a material at a scale resolution level similar to optical microscopy. The image contrast and fringes are related to both superficial and bulk elastic properties of the material [YUa, KAb].

In the last two decades fracture mechanics testing has been used to evaluate the fracture toughness of hard tissues by applying linear elastic fracture mechanics (LEFM). Fracture toughness plays a decisive role in hard tissue functionality by determining the level to which the material can be stressed in the presence of cracks, or, equivalently, the magnitude of cracking that can be tolerated at a given stress level. The major effort has concentrated on propagation of the crack using a variety of precracked specimens: three-point bending specimens, compact tension specimens (CT), single edge-notched specimens (SEN), center-notched cylindrical specimens

(CNC) and compact sandwich specimens. The CT geometry has proved to be the most useful for studying Mode I fracture mechanics of hard tissues [DOa, DEb].

### 6.1.2 Imaging

Mechanical properties of hard tissues are strongly related to the structural organization of the constituent materials. Optical or light microscopy is the first technique used to analyze small objects, their images being magnified through convex lenses, and the polarizing microscope is largely used for the examination of thin sections of materials. A method of particular value in examining living tissues is phase-contrast microscopy, which detects the phase differences due to the variations in the refractive index of material structures by imaging light intensity variations.

Through electron microscopy, sharper images, which better distinguish the hierarchical organization of living tissues, are observed below the resolution level of the optical microscopy with a greater depth of focus. This technique detects the interactions between an electron beam and the material. Images are formed either directly, by focusing the electron beam that passes through a thin specimen (transmission electron microscopy, TEM), or indirectly by using the information carried by secondary electrons or X-rays (scanning electron microscopy, SEM). Specimens are often coated with heavy metals. Environmental SEM has recently been developed. This has the advantage of eliminating artifacts due to the staining process, which modify the structure. Moreover, this imaging technique is particularly attractive for living tissues, because the hydrated state of the specimen can be preserved [BOd, BRa]. Topographic imaging of hard tissue at a very low resolution may also be obtained using AFM and SAM [LAb, CHa].

The X-ray computed microtomography (X-ray $\mu$CT) is a 3-D microscopic technique developed over the past 20 years and it is a miniaturized version of the well-known computer axial tomography method used for medical imaging. This new method of microimage analysis has gained attention in experimental biomechanics due to its ability to measure accurately the mineral content of tissues. The result of the X-ray $\mu$CT is in the form of a three-dimensional image reconstructed on the basis of multisliced planar tomographs taken at a fine angular pitch along the rotational axis. Moreover, compared to the previous microscopy techniques, the 3-D structure organization of specimens is detected in a nondestructive manner [WEb, DEa].

### 6.1.3 Structure–Property Relationship

Attaining knowledge of the relationship of the structure of living hard tissue to the physical and mechanical properties it exhibits is one of the principal objectives of this field of research. Along with this, the characterization of interfaces, such as those of synthetic devices with biological tissues, is a topic of extreme importance. For instance, a sufficiently precise identification of the structure–property relationships could facilitate diagnosis in certain clinical conditions. Unfortunately, as of today, the available data are rather poor but more refined structure–property correlations towards increasing levels of precision are an actual research target. In particular, structural correlations are sought with macroscopic quantities that are easily measurable. The measurement of tissue density probably represents the first notable attempt adopting this strategy. However, this quantity is not completely satisfactory because, at a macroscopic level, density alone is not adequate to distinguish tissue anisotropy. The investigation of the constituent materials' organization and distribution within the tissue therefore remains a compelling need. As a model is defined, its calibration passes through in vitro and in vivo testing. The main function of a hard tissue is to sustain and transmit loads, therefore mechanical testing is extremely important in order to validate mechanical models. Accordingly, density and microstructural organization information—obtained from microcomputed tomography imaging—combined with homogenization methods and finite element modeling (FEM), can be extremely useful to determine the material properties of a tissue. Thus the potential of all these methods, when used in a combined way, to characterize the mechanical behavior of the tissue and relate this to the disease state suggests great promise for its use in clinical diagnostic investigations [HOa, NIc].

In the following sections, the mechanical properties of human hard tissues—such as trabecular bone, compact bone, and dentine—are reviewed. Particular emphasis is placed upon the measurement of tissue anisotropy.

### 6.1.4 Hierarchical Structures in Hard Tissue

Figure 6.1 presents the hierarchical structure of a hard tissue related to imaging and to mechanical characterization methods and capabilities.

The main constituents of hard tissues are the extracellular matrix (i.e. collagen, apatite, and water), which mainly provides mechanical support, and cells (viz. osteocytes in bone and odontoblast in dentine) which, among various tasks, continuously control and adapt the extracellular matrix with the aim of achieving an optimal performance. It is considered that the mechanosensitivity of cells is related to the processes intimately connected with the mechanical behavior of the extracellular matrix [YOa] The extracellular matrix of bone consists of the soft and ductile collagen fibrils, reinforced by the stiff and brittle apatite crystals. Water is located within the fibrils and within the triple-helical molecules at a subnanostructure level. The collagen fibrils—about 100 nm in diameter—are held together

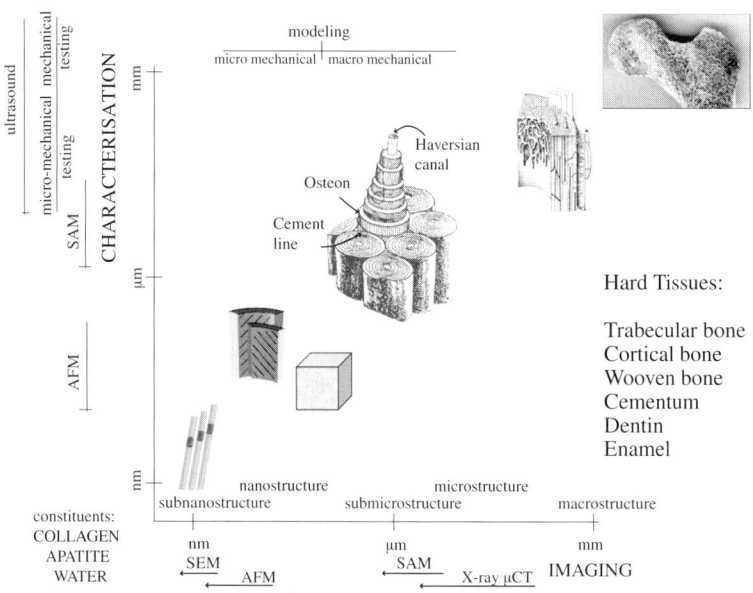

Figure 6.1. Hierarchical structure of hard tissues and characterization capabilities of mechanical and imaging techniques.

by a nonfibrillar organic matrix that acts as a glue. Less than 1% by weight of this glue would be enough to provide a link between the mineralized fibrils adequate to confer the known toughness properties to bone [THa, CUc, FAb]. Because the elastic properties of collagen and hydroxyapatite are about 1.5 GPa and 110 GPa, respectively, this is the range of stiffness one has to expect for hard tissues. Indeed, in the human body, enamel, basically constituted by apatite prismatic crystals, is the stiffest hard tissue, showing a Young's modulus of about 100 GPa. On the other hand, spongy (or trabecular) bone presents the lowest Young's modulus values. However, trabecular bone is a porous material and experimental values of elastic modulus have to be interpreted as apparent values. In fact, true trabecular tissue properties suggest an elastic modulus between 10 GPa and 20 GPa [WEa, ZId, SIa].

### 6.1.5    Elastic Properties of Individual Trabeculae

A notable account of the mechanical properties of individual trabeculae is already available [KEa]. Early investigations focused on direct measurements, obtained through buckling and bending tests on individual trabeculae. These reported mean elastic modulus values of 8.69 GPa [RUa] and 7.8 GPa [MEa]. More recently, through microtensile tests on long, uniform, rodlike trabeculae from human tibia, a mean value in tension of 10.4 GPa has been published. In contrast, ultrasonic measurements of similar individual trabeculae suggest an elastic modulus of 14.8 GPa [RHa].

Differences between direct mechanical testing and the wave propagation techniques are partially related to artifacts that affect direct mechanical measurements. Moreover, bone is a viscoelastic material, therefore strain rate dependence of the measured properties is to be expected. In general, the elastic modulus is an increasing function of strain rate. Consequently, continuous wave experiments give higher values of the elastic modulus because the strain rate is much higher than those performed by traditional dynamometers.

Higher values of the Young's modulus are recorded using the AFM nanoindentation technique. Individual trabeculae from the proximal femur are found to be significantly subject-dependent. The elastic modulus range is between 6.9 and 15.9 GPa, with an average value of 11.4 GPa [ZYa]. Elsewhere, the elastic modulus derived from nanoindentation of individual trabeculae from lumbar vertebrae is estimated to be 19.4 GPa [RHa].

In addition, SAM investigations have also been carried out on individual trabeculae from lumbar vertebrae. An average elastic modulus of 7.47 GPa is obtained for trabeculae from young individuals. This value decreases by about 30% for middle-aged and older people [ZHa]. Using trabecular material from the distal femur, a higher Young's modulus value (17.50 GPa) can be derived from SAM. Consistency with AFM investigations is also observed [TUa]. Similar average values of the Young's modulus obtained through SAM are later reported [BUa].

It is interesting to note that experimental results on individual trabeculae and single osteons, obtained in the same testing condition, generally provide higher values for the Young's modulus of osteons. Therefore, studies focusing on the mechanical properties of bone material at a microscale level divide into two groups. The first one assumes that trabeculae and osteons can be treated in the same manner, whereas the second one distinguishes trabeculae from osteons, on the basis of their different mechanical properties.

### 6.1.6 Elastic Properties of Single Osteons

The mechanical behavior of a single osteon is an intriguing research field, its study providing structural insights at both the macro and nano level. In fact, given the properties of a single osteon, the mechanical performance of a whole bone could be obtained and this can be useful also to validate averaging models against mechanical properties measured in a traditional way.

The effects of lamellar orientation on the tensile properties of single osteons were observed long ago. Osteons whose lamellae are mainly oriented in the longitudinal direction (dark osteons) are characterized by an elastic modulus of 12 GPa and a strength of 120 MPa, and an elastic modulus of 5.5 GPa and a strength of 102 MPa is found for osteons with lamellae oriented in intermediate directions [ASc]. The same authors later recorded a lower stiffness in compression [ASb] becoming even lower in bending [ASa]. The confidence in the strength values reported here is merited, because the load is constant in each cross-section during testing. The same, though, does not hold for strain, whose measurement is affected by the dynamometer compliance, thus leading to systematic errors in the measurement of the elastic modulus. Recently, nanoindentation tests in the longitudinal direction of cortical tibia provided elastic modulus values of 22.4 GPa and 25.7 GPa for the osteon and the interstitial lamellae, respectively [ROb].

Similar values have been measured in lamellae of dry cortical bone from the femoral neck; moreover, using low depth indents, thick lamellae results in stiffer rather than thin lamellae, therefore suggesting a longitudinal orientation of mineralized fibers in thick lamellae [HEa]. Comparable AFM indentation results have been obtained on the cortical bone from the femoral mid-shaft, the reported values being 24.36 GPa and 20.34 GPa for thick and thin lamellae, respectively [XUb]. However, it is important to notice that indentation measurements on cortical bone depend on the indentation depth, on the environmental conditions, and specimen preparation [HEa, XUb]. Moreover, the indentation modulus is determined somehow arbitrarily, that is, by assuming a certain value for the Poisson ratio. It is notewnorthy, however, that SAM investigation of cortical bone from the proximal femur also suggests a difference of about 20% between the acoustic impedance of lamellae with high and low reflectivity (thick and thin lamellae, respectively). The highest impedance values are obtained for the innermost lamellae, whereas the mean impedance of secondary osteons is higher than primary osteons [RAb]. The outermost lamellae of osteons are always more compliant [BUa] and the average elastic modulus derived from SAM investigations on cortical bone from the mid-shaft of the femur in the longitudinal and transverse directions are 20.55 GPa and 14.91 GPa, respectively [TUa].

A strong correlation between crystal features (essentially, shape and size) and the mechanical properties of osteons would be expected, though. Bone is more brittle in older and osteoporotic animals, where large size crystals are observed, whereas it is stronger and tougher in young animals, where a broader crystal size distribution is reported. Crystal size increases in osteoporotic and in aging bones [BOc]. Through X-ray investigations, structure distortions pictured as orientation of crystals have been observed in both longitudinal and alternate osteons cyclically loaded at a high stress level. Higher distortions are detected for longitudinal osteons. This supports the hypothesis that this type of osteon is more prone to buckling, and suggests that the thicker crystallites that are between the collagen fibers are more easily unbound from the matrix as a consequence of buckling [ASd].

The elastic behavior of osteons has also been predicted by several composite models. Common inputs are the elastic modulus of collagen (range 1.0 GPa to 1.5 GPa), the elastic modulus of apatite crystals (range 110 GPa to 150 GPa), and the Poisson ratio of 0.3. The analysis of the mechanical response of composite materials involves investigations on the micro- and macroscale level [NIc]. Using a cross-ply model of the osteon and an homogenization procedure for the properties of lamellae, a value of the elastic modulus of 22 GPa in the longitudinal direction is obtained at a mineral concentration of 0.6 [LEa]. However, the same concentration leads to very low values in the transverse direction. Incorporating the three partial porosities of bone [COc] related to Haversian canals, Volkman's canaliculi and lacunae, the typical elastic orthotropic constants of bone are derived using an hydroxyapatite content of 35% and an overall porosity of 8% [SEa]. Instead, modeling at the submicrostructural level the single lamella and using routines developed for a random network of cellulose fibers, nonuniform strains and clustering of aligned fibrils are displayed under load simulations [JAb]. However, model results of compact bone are still misleading because there are no reliable data regarding the properties of the constituent materials and interfaces [RHb]; moreover, at the macroscale level, the nondestructive imaging of the structure through micro-CT is not as powerful as it revealed to be for trabecular bone.

## 6.2 Trabecular Bone

Trabecular, cancellous, or spongy bone is abundant in the inner regions of the epiphysis of long bones, in the mandibular bone, within vertebral bones, and in flat and irregular bones. Therefore, this material represents the core of bony organs. The constituent materials are collagen, apatite, and marrow, differently organized across sites according to the specific adaptation process. The trabeculae are organized in very complex networks, which constitute an open porosity occupied by the bone marrow. The similarity between these organizations and engineering foam materials classifies trabecular bone within cellular solids, whose microstructural features affect the mechanical behavior. The microscopic level of the hierarchical organization of trabecular bone distinguishes rods and plates as the basic elements of the cellular structure.

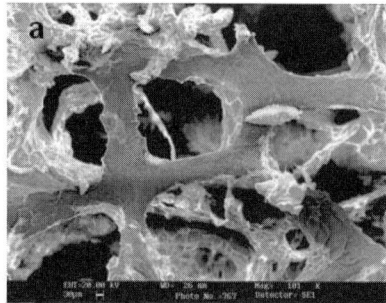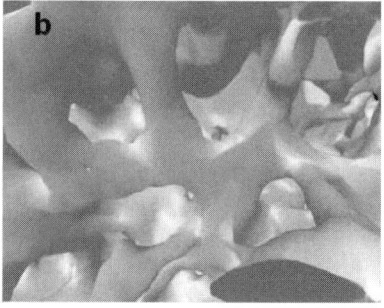

Figure 6.2. Trabecular bone network of a 50-year-old man: (a) SEM image (courtesy of Prof. D. Ronca, Orthopedic Clinic, II University of Naples, Italy); (b) 3-D reconstruction of micro X-ray CT (courtesy of Prof. S. Rengo, School of Dentistry, University of Naples Federico II, Naples, Italy).

### 6.2.1 Tibial Trabecular Bone

Trabecular bone obtained from human tibia has been used extensively, because this tissue is very abundant in the proximal tibial epiphysis thus providing the possibility to fabricate macroscopic specimens used for standard mechanical testing. Moreover, the proximal tibial epiphysis is important because of the prosthesis that is often required to alleviate certain pathologies of the knee. Figures 6.2a and b show the structural organization of trabeculae inside the subchondral trabecular bone of a 50-year-old man, depicted through SEM and X-Ray $\mu$CT, respectively.

This tissue is obtained from the weight-bearing area of proximal tibia, as depicted in Figure 6.3. A different mechanical behavior has been commonly observed between the medial and the lateral condyle region, and a strong anisotropy is generally recognized [KEb]. The relationship between trabecular orientation and mechanical properties in the longitudinal direction of the tibia (direction 3 of Figure 6.3) has been a matter of concern since about 1980. A complete map of the Young's modulus in this direction ($E_3$) related to the subchondral bone plate shows a marked site dependency [GOa]; in agreement with biomechanical models of femur–knee–tibia, the medial region is stiffer than the lateral region. A maximum peak value of 433 MPa is distinguished in the medial region. The mechanical method employed is the compression test on cylinders of 7 mm × 10 mm. A more precise average value of $E_3$ is later obtained by using compression and tension tests and measuring local deformations through an extensometer with 20-mm gauge length; mean values of 485 MPa and 483 MPa measured in compression and tension, respectively, have been obtained suggesting no statistical difference in the Young's modulus along direction 3 according

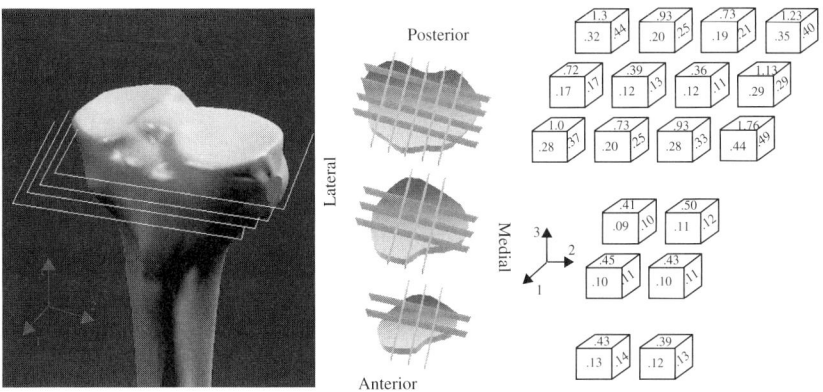

Figure 6.3. 3-D reconstruction of proximal tibia. Anisotropy in the Young's modulus of trabecular bone [RHd]. Values are expressed in GPa.

to the mode of loading [ROa]. More recently, optimized protocols for machining macroscopic specimens, obtained from the medial and the lateral regions, have been used. These employ 5-mm gauge length extensometers, thus preventing edge effects of previous investigations. In this case average values of $E_3 = 1091$ MPa (S.D.634) and $E_3 = 1068$ MPa (S.D.840) have been obtained in compression and tension, respectively [MOa]. Compression measurements of $E_3$ generally reveal that trabecular bone is stiffer in the medial condyle than in the lateral condyle region [GOa, DIa].

It is evident that a strong variation of $E_3$ according to the site is to be found also along the longitudinal direction. On the other hand, by using the ultrasonic investigation technique and smaller specimens, the Young's modulus anisotropy of trabecular bone has been measured through a complete 3-D mapping of the tibial epiphysis [RHd] as depicted in Figure 6.3. Also, these findings are in qualitative agreement with earlier measurements through penetration tests [HVa].

Unfortunately, the agreement among material properties measured using a variety of equipment setups, specimen geometry, environment, and mechanical conditioning is in the best cases qualitative rather than quantitative. If data obtained in tension, compression, and bending of trabecular bone from a specific site of the tibia are compared, an inconsistency of the results is generally observed. The statistical difference becomes more evident as the set of mechanical data includes micromechanical tests. In particular, a value of the Young's modulus of 4.59 GPa is obtained through microbending tests [CHb], whereas microtensile and ultrasonic tests on individual dry trabeculae provide an elastic modulus of 10.4 GPa and 14.8 GPa, respectively [RHa]. In contrast, the Young's moduli derived from nanoindentation tests on trabecular bone are estimated to be 19.6 GPa and 15.0 GPa in the longitudinal and transverse directions, respectively [RHb], thus in a range similar to that of compact bone. It is clear, then, that the Young's modulus of trabecular bone is strongly dependent on the specimen dimensions. Macroscopic specimens can be regarded as a porous structure delimited by a trabecular network. Accordingly, mechanical measurements need to take into account the effective volume fraction of bone. Indeed, bone density is the parameter most commonly used to determine the variations of the Young's modulus, which is assumed to depend on density through a power law. However, the precise form of this relationship and the fact that the modulus would be a function of density alone is controversial: the trabecular architecture is known to play a crucial role in the properties of the material [KEb] and density cannot take this into account. Instead, mechanical testing on microscopic specimens (i.e. individual trabeculae) and nanoindentation measurements provide the true elastic modulus of bone, which indeed produces results very close to the value measured for compact bone. Efforts to account for the structural anisotropy of trabecular bone are described by a general theory relating the elasticity tensor to the fabric tensor [COd, COb]. Age-dependence variations in the Young's modulus are generally described by a nonlinear relationship: this property increases up to about 45 years and significantly decreases after 60 years [DIa].

As with the Young's modulus, the strength of trabecular bone also depends on the loading direction, the site, and the age. The strength is higher in the longitudinal direction, and tension measurements provide yield stress values lower than in compression [MOa]. The trabecular bone strength can vary by an order of magnitude across the tibia sites, spanning between 1 and 10 MPa, and the medial subchondral bone region appears to be tougher than the lateral region [DIa]. A power law relation between strength and density is generally observed [KEb]. Again, this relationship is controversial because the trabecular architecture plays a fundamental role in determining bone strength. The strength dependence on age follows a similar trend observed for the Young's modulus. Table 6.1 presents the yield and the ultimate stresses and strains of trabecular bone from the tibia, depending on age [MOa, DIa].

### 6.2.2 Trabecular Bone from the Vertebral Body

The vertebral body is another site from where trabecular bone tissue is often harvested in order to determine its mechanical properties. This tissue is particularly abundant in the vertebrae from the lumbar region of the spine. The knowledge of the mechanical behavior of trabecular bone from this site of the body is particularly important because of various pathologies that

| Age(Years) | 16–39 | 40–59 | 60–83 |
|---|---|---|---|
| Compression | | | |
| Ultimate Stress [MPa] | 10.6 | 9.86 | 7.27 |
| Ultimate Strain [mm/mm] | 2.48 % | 2.12 % | 2.05 % |
| Yield Stress [MPa] | | 5.83 | |
| Yield Strain [mm/mm] | | 0.73 % | |
| Tension | | | |
| Yield Stress [MPa] | | 4.50 | |
| Yield Strain [mm/mm] | | 0.65 % | |

Table 6.1. Yield, ultimate stresses, and strains of trabecular bone from tibia.

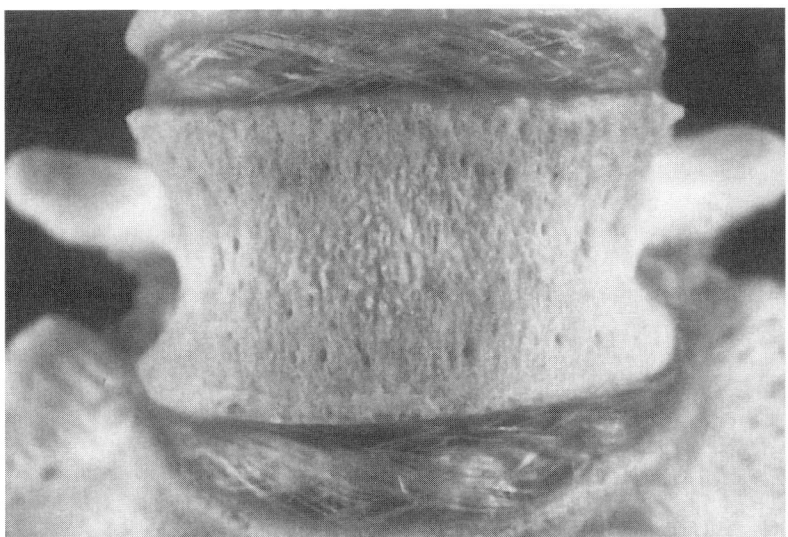

Figure 6.4. Optical image of a vertebra and the intervertebral discs. A main pattern of trabeculae orientation along the longitudinal axis is evident.

can affect the vertebral bodies, such as osteoporosis. Another important point is that the prosthetic devices that are often required to solve a spinal injury or disease will be in contact with this tissue.

Figure 6.4 shows an optical image of a vertebra; the intervertebral disc fibers are also shown. A main pattern of trabeculae orientation along the longitudinal axis is evident.

Figure 6.5 shows the 3-D image of a lumbar vertebra. The bone's image along a coronal section from an L3 vertebra of a 35-year-old man is provided. A trabecular network, principally oriented along the longitudinal direction, is rendered clearly visible through optical microscopy.

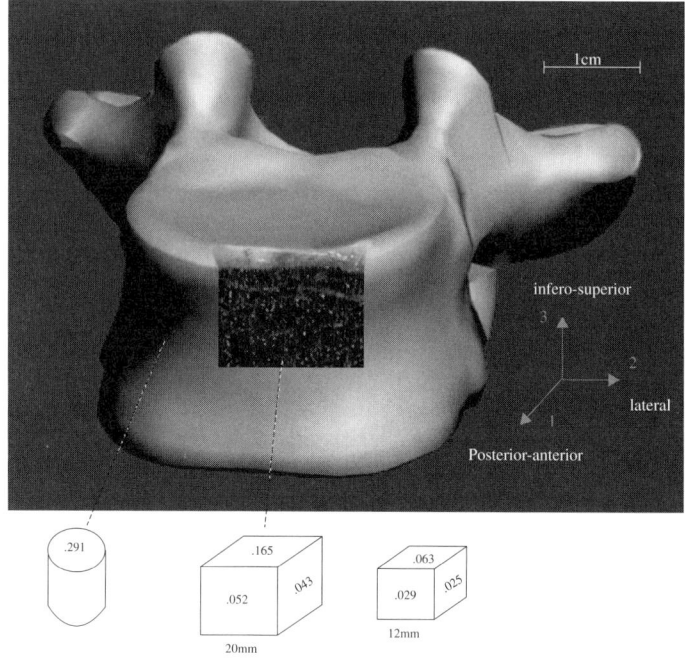

Figure 6.5. 3-D imaging of a lumbar vertebra showing Young's modulus anisotropy and site dependence. Values are expressed in GPa [NIb, AUa].

Compression, bending, and torsion are the loads that act on the spine through complex articulations between vertebrae and intervertebral discs [CAd, CAe]. The orientation of trabeculae along the axes of the spine suggests the adaptation to a compressive stress field transmitted through each vertebra. Trabeculae align along the direction of the functional principal stresses, as hypothesized by Wolff's law [WOa, COe, HUa].

Uniaxial compression tests on vertebral trabecular bone along the cephalo-caudal or superior–inferior direction (direction 3 of Figure 6.5) have been approached using cylindrical and cubic specimens. This direction is perpendicular to the articular end-plates [CAe] and the related properties are thus expected to be higher than those measured in the other directions. A mean value of 67 MPa for the elastic modulus of bone from the central part of the first lumbar vertebra in 15–87-year-old individuals has been measured [KEb, MOb]. Later, investigations in compression and in tension on bone specimens of 8-mm diameter, obtained from the lateral regions of thoracic and lumbar vertebrae, produced mean values of 291 MPa and 301 MPa, respectively [KOb]. Data reprocessed by using a 5-mm gauge length extensometer implied compression and tension mean values of 344 MPa and 349 MPa, respectively [MOa].

As with bone from the tibia, a marked anisotropy is also observed for vertebral trabecular bone.

Figure 6.5 reports the elastic modulus, determined in compression on cubic specimens from lumbar vertebrae, according to the posterior–anterior, the lateral, and inferior–superior directions (direction 1, 2, and 3, respectively, of Figure 6.5). Again, a variation between results obtained on cubic specimens with sides 20 mm [NIb] and 12 mm [AUa] is clearly evident. Not only do the mean values differ, but also the anisotropy—that is, the ratio between the modulus in the principal direction and the modulus along the orthogonal direction—is different. This inconsistency is due to the mechanical set-up and to bone quality. Furthermore, specimen conditioning can have a marked effect on the mechanical properties. In these studies, the specimens of 20-mm length were stored frozen at $-20°C$, while those of 12-mm length were kept in a saline solution at $4°C$. The site from where the tissue is harvested is also of significance.

As expected, the Young's modulus determined through ultrasonic tests gives significantly higher values of the elastic modulus. The wave speed measurements are mainly related to the trabecular struts, and the mechanical sound signal is faster than the strain rates induced with dynamometers. A mean value of 9.98 GPa is derived from ultrasonic tests [NIb] whereas the mechanical anisotropy is described by speed values of 1545 m/s, 1540 m/s, and 1979 m/s in directions 1, 2, and 3, respectively, in Figure 6.5 [NIa]. It is important to observe that mechanical compression studies on bone anisotropy generally use a single specimen to detect the property for all three loading directions. Therefore, using a low strain level, it is assumed that the material behavior is not affected by previous mechanical tests. However, measurements of the elastic modulus through destructive tests performed after the investigation at low strain levels suggest an increase of the modulus of about 25% [AUa].

In contrast, local measurements using the nanoindentation technique provide higher values of the elastic modulus. An advantage here is that the data can be directly related to specific areas of the trabecular material. The derived Young's moduli obtained on specimens with a thickness of 3 mm from the mid-sagittal plane of the L1 vertebra are 19.4 GPa and 15.0 GPa in the longitudinal and transverse directions, respectively [RHc]. Surprisingly, a detailed map of the elastic properties suggests that the material is stiffer in the transverse, rather than the longitudinal, direction. Measurements taken at a microstructural level also reflect the local variation in the extent of mineralization [ROb].

Table 6.2 summarizes the mechanical properties derived from destructive tests.

| Method | Y. Stress (MPa) | Y. Strain (mm/mm) | U. Stress (MPa) | U. Strain (mm/mm) | Ref. |
|---|---|---|---|---|---|
| Compression | 1.92 | 0.84 % | 2.23 | 1.45 % | Kopperdahl, |
| Tension | 1.75 | 0.78 % | 2.23 | 1.59 % | Keaveny, 98 |
| Compression | 2.02 | 0.77 % | | | Morgan, |
| Tension | 1.72 | 0.70 % | | | Keaveny 01 |
| Compression | | | 1.3 | 2.9 % | Augat et al. 98 |

Table 6.2. Ultimate properties of trabecular bone from vertebra.

### 6.2.3  Trabecular Bone from the Femur

The femoral head is one of the sites where the mechanical properties of trabecular bone are of great importance. Bone fractures at this location are a common injury especially in old people. The hip joint bears and transmits very high loads, amplified by the offset between the acetabular axis and the femoral axis. Several investigations focus on the mechanical properties of this bone, in order to improve preventive, therapeutic, and prosthetic strategies. Very complex patterns of stress distribution in the whole femoral proximal epiphyses and its single head are documented [VAc, APa].

Compression tests, using 12 mm in diameter cylindrical specimens, give the difference in the elastic moduli values in the anterior–posterior direction for inferior and superior sites on the head (Figure 6.6). As indicated by radiographs of healthy and osteoporotic femoral heads [VAc], a higher porosity is found in the inferior region, where the compressive modulus is five times lower. Surprisingly, no statistical differences are observed between the elastic properties of normal and osteoarthritic bones [BRb].

Compression tests on 8-mm cubic specimens from the region immediately inferior to the epiphyseal line, oriented along the superior–inferior direction (the primary compressive axis) provide a higher value of the elastic modulus (Figure 6.6). Again, no difference is observed in the modulus between tissues from cadavers and fractured hips [CIa, HOb]. It is clear from these results that the elastic modulus depends on the region and direction of loading and the strong mechanical anisotropy is already well evident. The elastic modulus of macroscopic trabecular specimens reaches values up to 1 GPa [CIa].

The elastic modulus measurements, averaged in the whole proximal femur, suggest values of 0.13 GPa, 0.06 GPa, and 0.05 GPa in the axial, sagittal, and coronal direction, respectively [AUa, MAa], thus indicating a certain degree of anisotropy. The lower observed values depend also on specimen conditioning (i.e. defatting procedures). The Young's modulus measured in compression on cylindrical specimens from the bearing areas oriented perpendicular to the articular surface of the femoral head (Figure 6.4) suggests differences between the weight-bearing and partial weight-bearing head site [DEc].

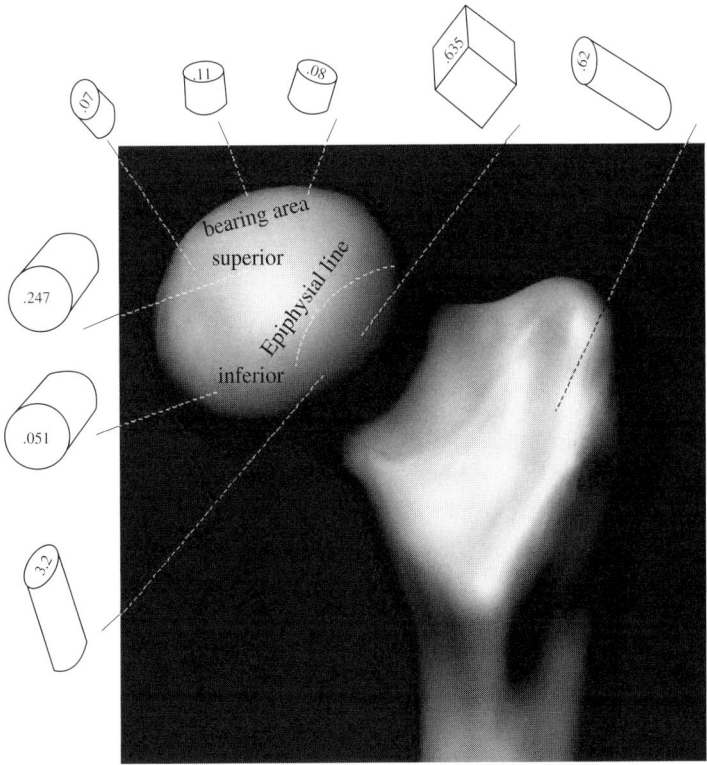

Figure 6.6. 3-D imaging of a femur showing Young's modulus anisotropy and site dependence of trabecular bone. Values are expressed in GPa [VAc, CIa].

Of course, the differences in the measured elastic property are also related to the mechanical preconditioning. Higher values of the elastic modulus are measured in the linear region of the stress–strain curves [CIa, HOb], or after a dynamic creep conditioning at low load levels [AUa, MAa].

Using a 5-mm gauge length extensometer and specimens oriented in the main direction of the trabecular network from the femoral neck, frozen and low strain conditioned bone shows moduli of 3.23 GPa and 2.7 GPa in compression and tension, respectively. The greater trochanter has an elastic modulus of about 0.6 GPa in both compression and tension-loading mode [MOa]. In contrast, nanoindentation measurements of trabecular bone from the femoral neck give a value of 11.4 GPa for the tissue modulus [ZYa]. Indeed, micro-FEM investigations use a uniform Young's modulus value for the trabecular bone material of between 10 and 18 GPa [BAd, ULa], in order to match the macroscopic apparent tissue modulus derived

| Site | Dir. | Meth. | Y. Stress (MPa) | Y. Strain (mm/mm) | U. Stress (MPa) | U. Strain (mm/mm) | Ref. |
|------|------|-------|-----------------|-------------------|-----------------|-------------------|------|
| Neck | IS | C | 17.45 | 0.85 % | | | Morgan |
| Neck | IS | T | 10.93 | 0.61 % | | | Keaveny 01 |
| Prox. Femur | IS | C | | | 2.9 | 3.2 % | Augat et al. 98 |
| Dist. Femur | IS | C | | | 2.2 | 2.9 % | |
| Sup. Head | AP | C | | | 2.35 | | Brown et al. 02 |
| Inf. Head | AP | C | | | 0.56 | | |
| Prox. Femur | IS | C | | | 2.5 | | Majumdar et al. 98 |

Table 6.3.    Mechanical properties derived from destructive tests. (C = compression;  T = tension;  AP = anterior–posterior;  IS = inferior–superior.)

from experimental investigations. Table 6.3 summarizes the mechanical properties derived from destructive tests.

### 6.2.4   Trabecular Bone from the Mandible

This site is of enormous importance for the outcome of prosthetic reconstructions with dental implants. The principal orientations of the trabecular network are affected by the teeth: mandible joint and the adaptation and remodeling consequent to chewing.

Compression tests on cubic specimens, whose sides are smaller than 5 mm, harvested from human edentulous mandible [OMa] show that a remarkable anisotropy exists in the premolar-incisal regions of the mandible arch (Figure 6.7). Compression tests on cylindrical specimens with similar dimensions from frozen mandibles aged between 56 and 90 years suggests an increase of the elastic modulus in the inferior-superior direction and a further decrease in the molar region [MIb]. It is also observed that the elastic modulus derived from unconfined or unconstrained compression tests lead to values that are about 40% lower than those detectable through constrained tests. An anisotropy in the elastic modulus of trabecular bone from the mandible condyle is apparent from compression tests on cylindrical specimens. Modulus values of 0.431 GPa and 0.127 GPa are exhibited in the superior–inferior and medio–lateral directions, respectively [GIa].

### 6.2.5   Anisotropy in the Elastic Modulus of Trabecular Bone

Based on the previous database of elastic modulus values it is possible to investigate the anisotropy of trabecular bone tissue according to the organ and the site. To do this the $E_3/E_2$ and $E_3/E_1$ ratios are considered. A first advantage of considering the anisotropy ratios is that a higher consistency

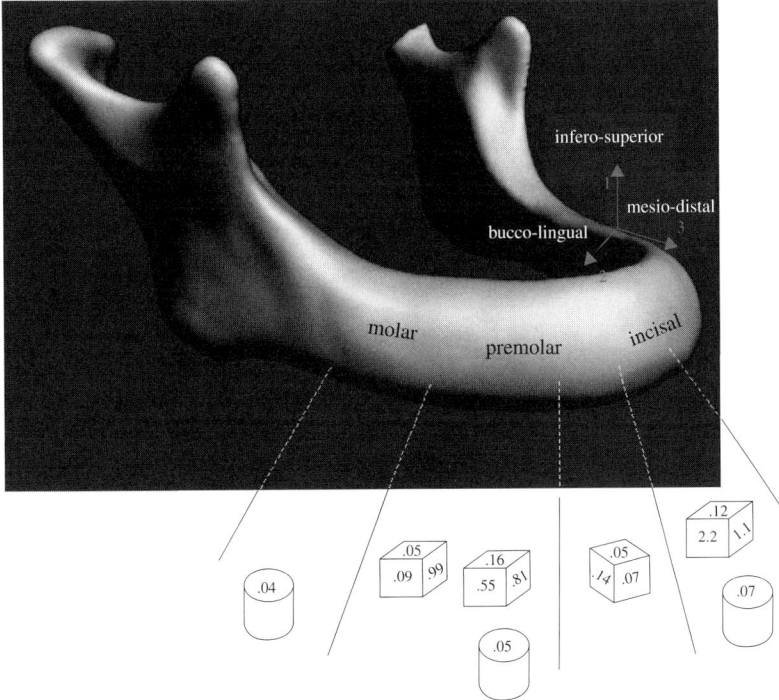

Figure 6.7. 3-D imaging of a mandible showing Young's modulus aniso-tropy and site dependence of trabecular bone. Values are expressed in GPa [OMa].

of the results is expected. These ratios are, of course, less sensitive to the specimen geometry and the mechanical testing setup.

It is interesting to observe from Table 6.4 that a higher anisotropy exists in the proximal tibia than in the proximal femur. Also, the vertebral trabec-ular bone displays a higher anisotropy than the proximal femur. However, compression tests on trabecular bone from the distal femur indicate an anisotropy ratio between the cephalo–caudal and medio–lateral direction of 3.5 [AUa]. Tibia, femur, and vertebra show that the highest elastic mod-ulus is obtained in the cephalo–caudal or superior–inferior direction, and the anisotropy ratios with respect to the orthogonal directions (directions 1 and 2 of Figures 6.2, 6.3 and 6.4) are similar. This suggests that these sites can be modeled as transversely isotropic materials as a first approximation. The mandible, however, displays a completely different anisotropy. In each site the elastic modulus is higher in the mesio–distal direction, that is, the axis of the mandibular arch. Moreover, a higher anisotropy is observed compared to the trabecular bone from other organs.

| Organ | Method | Site | c—Anisotropy | | |
|-------|--------|------|------|------|------|
| | | | $E_3/E_2$ | $E_3/E_1$ | |
| Tibia | US | Subchondral bone plate | 2.7–4.2 | 3.0–4.2 | Rho et al. 96 |
| | | Middle bone plate | 3.9–4.2 | 4.3–4.5 | |
| | | Deep bone plate | 3.0–3.1 | 3.2–3.3 | |
| Vertebra | MC | Central body L4 | 3.8 | 3.2 | Nicholson et al. 98 |
| | | Central body L1 | 2.5 | 2.2 | Augat et al 98 |
| | | Vertebral body | 1.8 | 1.8 | Hengsberger et al. 02 |
| | US | Central body | 1.3 | 1.3 | Nicholson et al. 98 |
| | | Vertebral body | 1.0 | 1.0 | Hengsberger et al. 02 |
| | NI | Axial trabeculae | 0.8 | — | Roy et al. 99 |
| Femur | MC | Proximal femur | 2.3 | 2.5 | Augat et al. 98 |
| | | Proximal femur | 2.5 | 2.2 | Majumdar et al. 98 |
| | | Femoral head* | 2.2 | 4.0 | Deligianni 91 |
| Mandible* | MC | Incisal | 2.0 | 18 | O'Mahoney et al. 00 |
| | | Canine | 2.0 | 2.8 | |
| | | First premolar | 1.5 | 5.1 | |
| | | Second premolar | 11.0 | 19.8 | |
| | | condyle | 3.4 | — | Giesen et al. 01 |

Table 6.4. Anisotropy ratios of trabecular bone from different organs. (MC = mechanical compression; US = ultrasound; NI = nanoindentation. *a local reference system is used with direction 3 being the osteon orientation.)

## 6.2.6 Viscoelasticity of Trabecular Bone

Time-dependent properties of trabecular bone have been observed, as the strain rate in static compression tests on human trabecular bone is varied. The ability of fresh vertebral bone to bear compressive stress in the superior–inferior direction is enhanced at high levels of strain rate [GAa]. A power law relating elastic modulus to strain rate is commonly observed. Assuming a cubic dependence of the elastic modulus from the apparent density, an exponent of 0.06 for specimens tested in confined compression [CAb] and an exponent of 0.05 [LIb] for tibial bone have been reported. An exponent of 0.1 is derived for vertebral bone [OUa]. The dependence of the elastic modulus from the strain rate is also strongly related to bone marrow, especially at high strain rates. At a strain rate of 10 s$^{-1}$, both the elastic modulus and strength measurements of bone with marrow are about five times higher than those of bone without marrow. This suggests that marrow is a dominant factor controlling the mechanical behavior of bone. However, at strain rates lower than 1 s$^{-1}$, the effect of bone marrow is far less evident. Moreover, it has been pointed out that the effects of viscous flow of bone marrow on mechanical properties depend upon boundary conditions. Trabecular bone in vivo is confined by cortical bone, with the marrow therefore acting as an incompressible fluid, redistributing the stress to the trabecular network and thus increasing bone strength and modulus. In contrast to this, in vitro testing is generally carried out in unconfined conditions where, at low strain rates, marrow can flow through the pores and exit the specimen. In vitro testing at high strain rates provides a better representation of confined conditions [CAb]. In order to describe the viscoelastic features of trabecular bone, stressed within physiological limits, a first approximation model is to consider trabeculae as elastic and the remaining material as a Maxwell body. This simplified model suggests that, during impact loading, the viscous marrow bears up to 30% of the applied load [KAa].

The effect of a fluid phase filling the bone porosity on the viscoelastic properties of human trabecular bone from the femoral head has also been detected through stress relaxation tests in compression [SCb, DEd]. By performing repeated stress relaxation tests at a fixed strain level, a higher stress decay is measured during the first test. However, no difference in the stress decay is observed through repetitive tests, if specimens are left to recover in Ringer's solution for 24 h. Moreover, a nonlinear viscoelastic behavior and a marked anisotropy for the relaxation constants are observed [DEd]. The viscoelastic properties of trabecular bone are more sensitive to storage conditions than is the elastic modulus. Defatting procedures lead to a 50% decrease in the viscoelastic energy dissipation [LIc]. Also, creep tests on trabecular bone from the femur suggest that time-dependent properties are a function of the site [ZIa]. Creep tests at low loading levels on trabecular bone from a vertebral body suggest a nonlinear viscoelastic behavior. Moreover, a full recovery of the strain is obtained only over a period 20 times longer than the duration of the creep test [YAa].

## 6.3 Cortical Bone

Cortical or compact bone constitutes the outer shell of bony organs such as the tibia, the femur, the vertebra, and the mandible. Osteons or the Haversian systems are the basic elements distinguished at a microscopic level. An osteon is generally an arrangement of concentric lamellae, each comprising mineralized collagen fibers oriented according to determined patterns, its shape approximating a cylinder. A more random organization of the mineralized collagen fibers distinguishes the woven bone (e.g. the bone in the callus). In long bones, the osteons are mainly oriented in the longitudinal direction and the peripheral osteons present their outer lamellae almost tangential to the external surface, forming, together with woven bone, a plywoodlike structure. This is a thick layer that represents, at a macroscopic level, the basic outer lamellar bone. This material is also the fundamental structural element of cementum, the material which coats the tooth's root and interfaces itself with periodontal ligaments in the alveolar bone. Orthogonal "plywood" structure is present in the outer lamellae of long bones [WEa]. The interface between osteons and bone matrix is the cement line, which represents a region of reduced mineralization [BUb]. The cement line provides a ductile interface that promotes crack initiation but slows crack growth.

Imaging of cortical bone through a polarized light microscope suggests that the appearance of lamellae depends on the collagen fibers' orientation. In dark osteons, the lamellae are almost orthogonal to the plane of section— that is, parallel to the osteon axis—whereas in light osteons, the fibers are oriented almost parallel to the plane of section [EVa]. It has also been observed that transversely oriented collagen fibers prevail in regions of compact bone adapted to high compression stress, whereas longitudinal fibers are abundant in regions corresponding to high tensile stresses [BRa].

### 6.3.1   Elastic Properties of Cortical Bone at a Macroscale Level

It is not surprising that almost all of the research on mechanical properties of hard tissues developed during the last century focused on cortical bone. Compared to trabecular bone, this tissue is dense, so measurement errors related to porosity are reduced. Moreover, the fabrication of trabecular bone specimens generates peripheral artifacts caused by the fracture of trabeculae along the specimen's edges. Instead, the human body presents several sources from where cortical bone can be harvested in order to obtain specimens for standard mechanical characterization at the continuum level. Classical mechanical testing (compression, tension, bending, torsion, etc.) on cortical bone suggests the use of specimens that contain several Haversian systems. For this, a minimum cross-sectional area of 4 mm$^2$ is recommended [REa]. Through uniaxial tests on cortical bone from the middle one-third of the diaphysis of fresh femur, an average value of 17.1 GPa has been recorded. No difference has been observed between the tensile and compression-loading modes. The mechanical anisotropy of cortical bone has been investigated using several techniques. With piezoelectric transducers and sensors, a fast decrease of the Young's modulus from about 17 GPa to 14 GPa has been observed through a slight decrease of the specimen orientation with respect to the longitudinal axis of the femur [BOb]. An inconsistency is observed between the experimental data and the predictions of composite models based upon a simple linear superposition principle of collagen and apatite. Subsequently, using a hierarchical material-structure composite model, the variation of the Young's modulus with specimen orientation is explained [KAc]. Because compact bone is quite stiff, physiological loads usually produce small deformations, thus linear elasticity can be used to describe the mechanical behavior of compact bone.

The general linear elastic relationship between stress and strain ($\sigma - \epsilon$) is given by the tensorial form of Hooke's law

$$\sigma_{ij} = \mathcal{C}_{ijkl}\epsilon_{kl},$$

where $\mathcal{C}_{ijkl}$ is the elastic modulus tensor (the stiffness matrix) characterized by 81 elements. However, cortical bone can usually be assumed to behave as an orthotropic material, therefore only 9 independent constants can be distinguished. The compliance tensor for these model is given by:

$$S_{ij} = \begin{pmatrix} \frac{1}{E_1} & \frac{-\nu_{21}}{E_2} & \frac{-\nu_{31}}{E_3} & 0 & 0 & 0 \\ \frac{-\nu_{12}}{E_1} & \frac{1}{E_2} & \frac{-\nu_{32}}{E_3} & 0 & 0 & 0 \\ \frac{-\nu_{13}}{E_1} & \frac{-\nu_{23}}{E_2} & \frac{1}{E_3} & 0 & 0 & 0 \\ 0 & 0 & 0 & \frac{1}{G_{23}} & 0 & 0 \\ 0 & 0 & 0 & 0 & \frac{1}{G_{31}} & 0 \\ 0 & 0 & 0 & 0 & 0 & \frac{1}{G_{12}} \end{pmatrix},$$

where $E_i, G_{ij}$, and $\nu_{ij}$ are the elastic moduli, the shear moduli, and the Poisson ratios, respectively. The directions 1, 2, and 3 are the transverse, radial, and longitudinal directions, respectively. The elastic constants of the stiffness matrix are given by

$$C_{11} = \frac{1 - \nu_{23}\nu_{32}}{E_2 E_3 \Delta}, \quad C_{22} = \frac{1 - \nu_{13}\nu_{31}}{E_1 E_3 \Delta}, \quad C_{33} = \frac{1 - \nu_{12}\nu_{21}}{E_1 E_2 \Delta},$$

$$C_{12} = \frac{\nu_{12} - \nu_{32}\nu_{13}}{E_1 E_3 \Delta}, \quad C_{13} = \frac{\nu_{13} - \nu_{12}\nu_{23}}{E_1 E_2 \Delta}, \quad C_{23} = \frac{\nu_{23} - \nu_{21}\nu_{13}}{E_1 E_2 \Delta},$$

$$C_{44} = G_{23}, \qquad C_{55} = G_{31}, \qquad C_{66} = G_{12},$$

$$\Delta = \frac{1 - \nu_{12}\nu_{21} - \nu_{23}\nu_{32} - \nu_{31}\nu_{13} - 2\nu_{21}\nu_{32}\nu_{13}}{E_1 E_2 E_3}.$$

Often, bone can be adequately described as a transversely isotropic material, because an axis of material symmetry can usually be recognized. Thus the number of independent constants reduces to five and the compliance tensor is characterized by the following relationships: $E_1 = E_2$, $\nu_{31} = \nu_{32}$, $\nu_{12} = \nu_{21}$, $G_{31} = G_{23}$, and $G_{12} = E_1/2(1+\nu_{12})$. Mechanical tests employing the strain gauge technique and continuous acoustic measurements have been the first methods to measure the mechanical anisotropy of human femoral cortical bone. Experimental mechanical testing aimed at determining five elastic constants are depicted in Table 6.5.

Ultrasonic data on cortical human bone from the tibia and mandible also support the hypothesis of an orthotropic anisotropy of cortical bone (Table 6.5). Nevertheless, the data presented in Table 6.5 are simplified,

| Year | 1975 | | 1984 | 1987 | 1996 | | | | 2003 | | |
|------|------|---|------|------|------|------|------|------|------|------|------|
| Organ | Fm | | Fm | Fm | Tb | | | | Mn | | |
| Site | | | | | An | Lt | Ps | Md | Al | Ib | Sy |
| Meth. | M | | U | M | U | | | | U | | |
| Mode | C | T | | T | | | | | | | |
| $E_1$ [GPa] | 18.2 | 17.7 | 20.0 | 22.5 | 20.9 | 20.6 | 21.1 | 21.2 | 25.2 | 23.0 | 21.1 |
| $E_2$ [GPa] | 11.7 | 12.8 | 13.4 | 13.4 | 11.5 | 11.9 | 12.3 | 12.9 | 12.7 | 16.0 | 16.8 |
| $G_{12}$ [GPa] | 3.3 | 3.3 | 6.23 | 6.23 | 5.5 | 5.7 | 5.8 | 6.1 | 6.8 | 7.2 | 7.4 |
| $\nu_{12}$ | 0.63 | 0.53 | 0.35 | 0.32 | 0.40 | 0.40 | 0.40 | 0.38 | 0.41 | 0.34 | 0.41 |

Table 6.5. For the sake of simplicity only four of the measured elastic constants are reported [REb, ASe, DAa, RHd, SCc]. Direction 1 stands for the axis of material symmetry. (Fm=femur; Tb=Tibiae; Mn=mandible; A=anterior; L=lateral; P=posterior; M=medial; Al=alveolar; I=inferior border; S=symphysis. C and T denote compression and tension, respectively. M and U denote mechanical and ultrasound testing, respectively.)

consistent with the case of transverse isotropy. Table 6.5 shows the dependence of the elastic constants on the organ and site. It is interesting to note that, compared to investigations on trabecular bone, local elastic properties measured by applying AFM and SAM techniques to single osteons provide slightly higher values [TUa] than traditional mechanical measurements, namely direct mechanical testing and ultrasound, on cortical bone.

A detailed map depicting the spatial distribution of the Young's modulus during bending of cortical bone tissue from the femoral midshaft along the longitudinal direction is shown in Figure 6.8 [CUa]. Specimens with a thickness of 1 mm were obtained through serial cutting of the femoral shaft with a diamond saw. A span of 30 mm is used for a three-point bending test. Elastic values range from 14.0 to 22.8 GPa, with an average value of 18.6 GPa.

### 6.3.2   Yield and Failure Properties of Cortical Bone

Yield and failure properties of materials are determined through direct mechanical testing according to the traditional tensile, compression, bending, and torsion tests. Inasmuch as it is impossible to determine exactly the linear elastic limit of a material, a yield point is defined as the intersection of the stress–strain curve and a straight line whose slope is the Young's modulus of the material and which is positioned at 0.2% strain offset from the origin. Ultimate properties refer to the stress and strain of the maximum stress point, which in the case of cortical bone coincides with the point of rupture of the stress–strain curve.

Table 6.6 shows yielding and failure properties of human compact bone from the femur. It is apparent that higher values of ultimate strength in tension and compression are observed along the longitudinal direction.

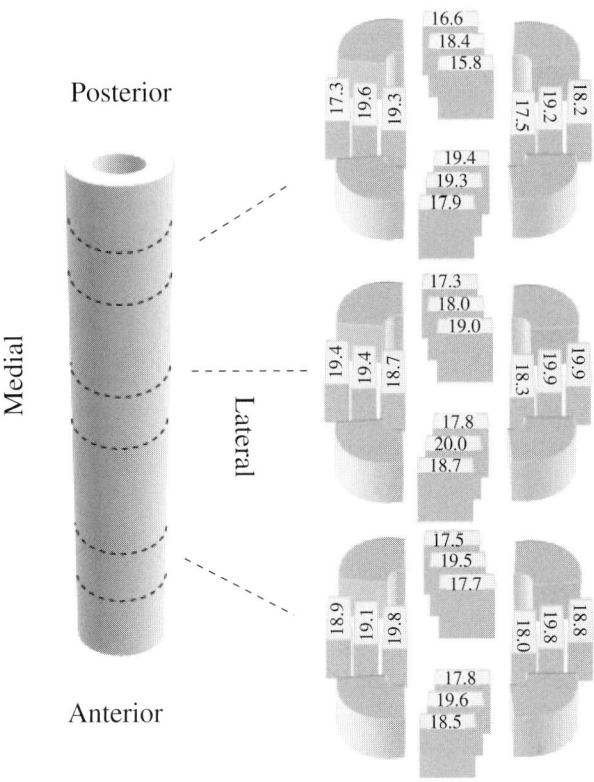

Figure 6.8. Map of the Young's modulus in bending of cortical bone tissue from the femoral midshaft along the longitudinal direction [CUa].

| Reference | Reilly and Burstain 75 | | | | | Bayractar04 | Curre04 | Jepsen97 |
|---|---|---|---|---|---|---|---|---|
| Test | Tension | | Compression | | Shear | Tension | Bending | Torsion |
| Direction | Long. | Trans. | Long. | Trans. | Trans. | Long. | Long. | |
| Y. Strain [mm/mm] | | | | | | 0.73 % | 1.41% | 1.3 % |
| Y. Stress [MPa] | | | | | | 107.9 | | 55.8 |
| U. Strain [mm/mm] | 3.8 % | 0.7 % | 1.9 % | 0.5 % | | | | 5.2 % |
| U. Stress [MPa] | 133 | 53 | 205 | 131 | 67 | | 208 | 74.1 |

Table 6.6. Yielding and failure properties of human compact bone from femur.

| Test | Property | Children | Young adult | Adult | Elderly | Ref. |
|------|----------|----------|-------------|-------|---------|------|
| T | Modulus [GPa] | | 15.69 | | 14.78 | Courtney et al 96 |
| T | Yield Strain [mm/mm] | | 0.378 % | | 0.356 % | |
| T | Yield Stress [MPa] | | 55.3 | | 49.3 | |
| T | Failure Stress [MPa] | | 81.6 | | 78.7 | |
| T | Failure Stress [MPa] | 114 | 122 | 102 | 86 | Natali Hart 02 |
| C | Failure Stress [MPa] | | 167 | 158 | | |
| B | Failure Stress [MPa] | 151 | 173 | 158 | 139 | |
| S | Failure Stress [MPa] | | 57 | 52 | 49 | |

Table 6.7. Effects of age on the mechanical properties of cortical bone from the human femur. (T=tension; C=compression; B=bending; S=Shear (torsion).)

| Direction | Children | Adult | Ref. |
|-----------|----------|-------|------|
| 0° | 12.8 | 17.5 | Hara et al. 98 |
| 30° | 9.9 | 14.3 | |
| 60° | 5.4 | 11.7 | |

Table 6.8. Effects of age on the elastic modulus of cortical bone from the mandible in bending.

The effects of age on mechanical properties of human cortical bone have also been investigated. Generally, these properties increase from the child (5–20 years) to the young adult (20–40 years), decrease slightly in the adult (40–60 years) and then decrease more markedly in the elderly person (60–80 years). Table 6.7 illustrates the effects of age on the mechanical properties of cortical bone from the human femur in tension compression, bending, and torsion, and Table 6.8 illustrates the same effects concerning the elastic modulus in bending of the mandible at various angles.

### 6.3.3 Viscoelasticity of Cortical Bone

The mechanical properties of human cortical bone depend on strain rate in the sense that a higher strain rate produces a higher elastic modulus, thus suggesting that cortical bone behaves as a viscoelastic material [MCa]. As with trabecular bone, the viscoelastic properties of cortical bone are strongly dependent upon the water phase. Dynamic mechanical analyses (DMA) at 1 Hz and 37°C on 3P-bending specimens of human cortical bone from the tibia (Table 6.9) suggest that water mainly affects the damping factor, and therefore the loss modulus, rather than the storage modulus [YAb]. The effect of interstitial fluid flow on hydraulic strengthening of bone has been analyzed through nonlinear models. It is suggested that a window of strain rates exists between $10^{-3}$ s$^{-1}$ and $10^{+3}$ s$^{-1}$. Within this window, the loss tangent is higher at low and high strain rates. Outside the boundaries of this strain rate range, there are no further strengthening effects on fluid flow resulting from recourse to higher or lower strain rates [LIa].

| Reference | | Yamashita 00 | Bargren 74 | | FanRho 02 | | |
|---|---|---|---|---|---|---|---|
| | Conditions | Bending | Compression | | AFM | | |
| Freq./Load | | 1 Hz | 5.2 Hz | 7.5 Hz | 10 $\mu$ N/s | $10^2 \mu$ N/s | $10^3 \mu$ N/s |
| | | | | | 18.6 | 26.5 | 32.2 |
| Storage | 0° wet | 9.4 | | | | | |
| Modulus | 0° dry | 9.4 | | | | | |
| [GPa] | 0° | | 15.8 | 16.3 | | | |
| | 30° | | 10.8 | 11.4 | | | |
| | 45° | | 12.5 | 12.4 | | | |
| Damping | 0° wet | 0.041 | | | | | |
| factor (tan $\delta$) | 0° dry | 0.029 | | | | | |

Table 6.9. Viscoelastic properties of cortical bone. Data from Bargren et al. 1974 are converted in standard units.

Anisotropy in viscoelastic properties is, of course, often observed. The data in Table 6.9 clearly illustrate that both the storage and loss moduli of cortical bone from the diaphysis of a human femur (from a woman, age 26 years) are greatest in the longitudinal direction [BAc]. Furthermore, both these moduli increase with increasing frequency. The increase in the elastic modulus has also been observed through AFM, performing viscoelastic measurements by changing the stress rate of the loading–unloading cycles of the tip indenter [FAa].

### 6.3.4 Fracture Mechanics

Bone undergoes microcracking as a result of repetitive daily stress while, simultaneously, the remodeling process occurs, establishing an equilibrium in the microcrack density within the living tissue [LAa]. The microstructure of bone anyway provides sites of discontinuity—including lacunae, canaliculi, blood vessels, and muscle insertions—all of which act as stress raisers [CUb]. The dimensions of intrinsic crack initiation sites are, for example, of the same order as those of the vascular spaces [BOa]. An in vitro comparison of the fracture mechanisms of human cortical bone suggests that the microcracks which form around a cracktip redistribute the stresses, increasing the toughness of the material [SIb, VAd].

Fracture toughness capability of Haversian bone is related to the osteonal structure. Similar examples among engineering materials are tough discontinuous fiber-reinforced laminates [NIc, NOb]. A ductile osteon–matrix interface—the cement line—promotes crack initiation [EVa], but slows crack propagation in compact bone [BUb] by blunting the crack tip and trapping it within the lamellar structure. A bridging effect of osteon is observed for cracks propagating in the osteon orientation [DEb].

Debonding of an osteon produces changes in the region of the Haversian canal wall adjacent to the crack, such as interruption of streaming potentials and nutrients. This debonding initiates remodeling and production of a new

| Site | Geometry | Direction | Speed | $K_c[MNm^{-3/2}]$ | References |
|------|----------|-----------|-------|-------------------|------------|
| Tibia | SEN | longitudinal | Fast | 2.2–4.6 | Bonfield Datta 76 |
| Tibia | CT | longitudinal | Slow | 2.1–4.7 | Behiry Bonfield 84 |
| Tibia | CT | longitudinal | Slow | 4.48 | Norman et al 91 |
| Tibia | CT | longitudinal | Slow | 4.05 | Norman et al. 95 |
| Tibia | CT | longitudinal | Slow | 2.08 | Norman et al. 96 |
| Tibia | CS (mode II) | longitudinal | Slow | 8.32 | Norman et al. 96 |
| Femur | CT | longitudinal | Slow | 1.6–2.5 | Vashishth et al. 97 |
| Femur | CT | longitudinal | Slow | 1.71 | Akkus et al. 00 |
| Femur | CT | transverse | Slow | 3.47 | Akkus et al. 00 |

Table 6.10. Stress intensity factor of human cortical bone.

secondary osteon, initially less calcified than the surrounding bone [SIb], which may act as a fiber reinforcement [BUb] preventing the accumulation of microdamage due to repetitive loading [MAd].

Micromechanical models of osteonal cortical bone suggest that the ratio of the modulus of the osteon to that of the interstitial bone controls crack propagation. Newly formed, low-stiffness osteons may toughen cortical bone by promoting crack propagation toward tougher osteons whereas stiff osteons repel the microcrack from more brittle osteons [GUa]. With aging, microdamage occurs more rapidly than the intrinsic repair process [SCa]. The effects of this accumulation of microcracks, which is higher in females than males [NOd], are reduced mechanical properties and increased fragility [COa, MAe]. Thus bone fracture has an increased occurrence within older persons because the microstructural organization of bone controls the relationship between loading conditions and fracture patterns [EVa, CLa]. Hence fracture toughness plays a decisive role in bone functionality by determining the level to which the material can be stressed in the presence of cracks, or, equivalently, the magnitude of cracking that can be tolerated at a given stress level.

Almost all of the literature data on fracture mechanics of cortical bone are based on the CT specimen and mode I condition. Table 6.10 shows the stress intensity factor of human cortical bone. In particular, it is suggested that fracture toughness of cortical bone in the osteonal direction is higher in shear than in tension [NOc]. Furthermore, fracture toughness in the transverse direction is higher than in the longitudinal direction [AKa].

### 6.3.5  Fatigue of Cortical Bone

Several fatigue investigations have been carried out on cortical bone through in vitro testing. It must be noted that the fatigue behavior of living bone is not known, and that the in vitro testing of cortical bone does not consider the remodeling process. Therefore, care has to be taken when transferring in vitro results to in vivo behavior of bone and failure predictions.

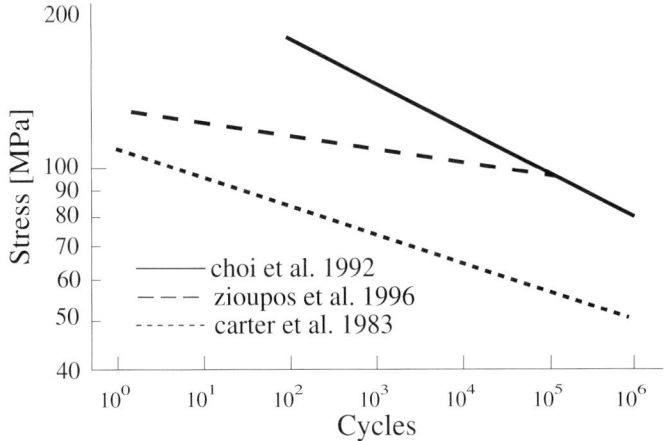

Figure 6.9. Fatigue behavior of cortical bone.

The osteoclast and osteoblast activity controls the extracellular matrix absorption or production, respectively. Therefore, it has been suggested that the Woehler S–N diagrams in Figure 6.9 have to be interpreted as a lower bound limit if the osteoblast activity is the dominant process, and an upper bound limit if the resorption process prevails [KEb]. In general, dynamic sinusoidal stress is applied at 2 Hz. Figure 6.9 reports the S–N diagrams of human cortical bone. Human cortical bone (average 69-year-old) cyclically stressed in tension at 37°C has provided lower fatigue strength values than bone (average 27-year-old) at 25°C [CAc, ZIb]. A different fatigue behavior is observed using the four-point bending method for bone tested at 25°C [CHc].

## 6.4 Dental Tissues

The dental tissues consist of enamel, dentine, cementum, and the pulp. The arrangement of the hard materials is depicted in Figure 6.10, where the junctions, CEJ (cementum–enamel junction) and DEJ (dentine–enamel junction), can be clearly distinguished.

Dentine forms the bulk of the tooth. Like compact bone, it comprises an inorganic mineral component, hydroxyapatite, and an organic matrix, which is mainly collagen. Again, as with bone, type I fibrillar collagen is the main constituent of the extracellular matrix of dentine. However, it has been established by chromatography that dentine collagen exhibits a cross-link distribution different from that of bone collagen [KUa]. Dentinal

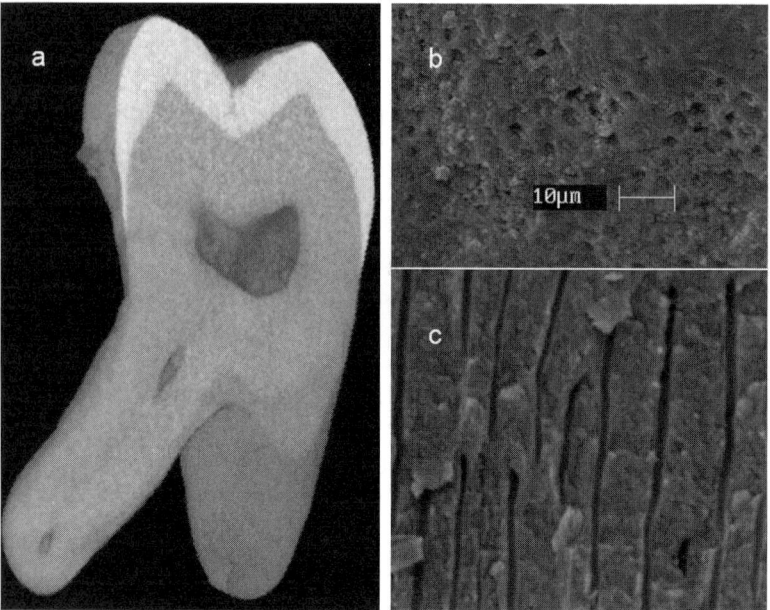

Figure 6.10. (a) 3-D reconstruction of a micro X-ray CT of a human third molar; (b) SEM image of a horizontal section from the dentine crown; (c) SEM image of a longitudinal section from the dentine crown (courtesy of Dr. D. Prisco, School of Dentistry, University of Naples Federico II, Naples, Italy.)

tubules across dentine [VEa] and the intertubular and peritubular dentine (ITD and PTD, respectively) may be distinctly identified. Dentinal tubules permeate through dentine, their pattern following an S-shaped curvature. Lateral branches of the tubules, with a smaller diameter, complete the network [CAa]. There is a different distribution of dentinal tubules and diameter in the pulp–DEJ direction [GAb]. The tubules are wider and more numerous near the pulp.

Enamel covers the crown of the tooth, and its structure consists of tightly packed hydroxyapatite crystals. These are organized in highly oriented patterns, forming enamel rods. These rods, which extend from the DEJ to the surface of the enamel, are arranged in circumferential rows around the main axis of the tooth.

Dentine and enamel are 65% and 95–98% mineralized, respectively. It should be noted that mature enamel is not a living tissue. Compared to dentine, enamel is harder but also more brittle. Moreover, thermal properties suggest that enamel has a higher coefficient of thermal conductivity than dentine. The overall natural organization of dentine and enamel is optimized in order to perform thermomechanical functions [CRb].

The pulp, which is richly innervated, is the soft tissue of the tooth. It occupies the central portion of the tooth represented by the coronal pulp chamber and the radicular root canal. Vessels enter into the pulp through the apical foramina. The principal cells of the pulp are the odontoblasts, whose processes extend into dentine.

Cementum covers the root of a tooth; it is a bonelike structure. This tissue is deposited as a thin layer from the CEJ to the apex of the tooth; this layer thickness is higher at the apex. Cementoblasts and cementocytes are the cells of cementum, however, unlike bone, cementum is avascular and incapable of remodeling. Cementum anchors the tooth to the surrounding alveolar bone through the periodontal ligament.

The periodontal ligament system comprises soft, fibrous connective tissue, which extends from the cement of the root to the alveolar bone, following an obliquely cervical direction. The periodontal ligament characterizes the joint between the tooth and the alveolar bone. Periodontal ligaments also act as a natural shock-absorbing system.

## 6.4.1 Elastic Properties

Dentine, like compact bone, is a heterogeneous multiphase material that exhibits a hierarchical composite structure. Therefore, like bone, mechanical properties depend on the scale level at which properties are measured. Also, experimental measurements are sensitive to the specimen source and testing conditions.

The Young's modulus of dental hard tissues, similarly to that of bone, is confined between upper and lower bounds related to the stiffness of hydroxyapatite and collagen, respectively. Given the geometrical constraints pertaining to dental hard tissues, it is very difficult to harvest specimens of appropriate dimensions in order to carry out classical mechanical testing. Therefore, the testing of dentine with dynamometers is generally performed using micromechanical methods. Specimen volumes are typically of the order of a few cubic millimeters, with cross-section areas often less than 1 mm$^2$ for tensile tests [SAa]. Data from measurements of the Young's modulus of dentine are presented in Table 6.11.

| Reference | CraigPey 58 | Tylesley 59 | BowenRod 62 | RensonBra 75 |
|---|---|---|---|---|
| Testing Mode | Compression | 4-Point Bending | Tension | Cantilever |
| Young's modulus [GPa] | 18.3 | 12.3 | 19.3 | 11.1–19.3 |

Table 6.11. Young's modulus of dentine.

| Temperature | 0°C | 23°C | 37°C | 50°C | 80°C |
|---|---|---|---|---|---|
| Young's Modulus [GPa] | 15.20 | 13.94 | 13.26 | 12.06 | 9.35 |

Table 6.12. Dependence of Young's modulus of dentine on temperature [WAb].

The mechanical properties of demineralized dentine are of great importance. Most of the bonding systems used in restorative dentistry are carried out through the etching of dentine, which has the effect of removing the mineral phase of the tissue. The Young's moduli in tension of mineralized and demineralized human dentine are 13.7 GPa and 0.25 GPa, respectively [SAb]. The temperature-dependence of the elastic properties of dentine are also notable. In contrast to bone tissue, which experiences a rather constant temperature, dental tissues are thermally stressed over a wider temperature range. Table 6.12 shows the dependence on temperature of Young's modulus in compression [WAb].

In each tooth, there is a variability in the mechanical properties of dentine. Anisotropy and inhomogeneity are induced by the tubules network, varying distributions of both intrafibrillar, and extrafibrillar minerals, and water content, inter alia. Variations are also observed among different teeth. Therefore, mechanical properties of dentine are site-dependent, and the elastic moduli measured through classic mechanical tests have to be interpreted as average values. The properties of caries-affected dentine are very important. This degenerated tissue is less stiff and contains more water than normal dentine [ITa].

Indentation tests are an alternative to classical micromechanical testing, permitting a detailed mapping of dentine and enamel elastic properties. The average values of Vickers' hardness of dentine and enamel are 57–60 kg/mm$^2$ and 294–408 kg/mm$^2$, respectively [WIb, FOb]. The averaged Young's moduli (derived from Vickers' hardness, assuming a Poisson ratio of 0.25) of enamel in the axial and transverse directions are 94 GPa and 80 GPa, and the modulus that has been derived for dentine is 20 GPa [XUa]. The Young's modulus of dentine derived from microindentation through a Knoop indenter decreases toward the DEJ. Using a proportional constant of 0.45 between Young's modulus and Knoop hardness, the elastic modulus decreases from 11.2 GPa at a distance of 2.5 mm to 8.7 GPa close to the DEJ [MEb]. Also, microindentation measurements suggest that the elastic modulus of dentine decreases toward the DEJ. Furthermore, gradients in the Young's modulus have been detected also in the facial–lingual direction [KIf].

It is important to notice that hardness measurements through indentation testing are strongly dependent on the size of the indenter. The impression size in dentine using traditional indenters (e.g. Vickers, Knoop, etc.) is of the same order as heterogeneities such as enamel rods or dentinal tubules, preventing accurate assessment of material properties. Using AFM, the elastic moduli derived from nanoindentation are 19.3 GPa and 90.6 GPa for dentine [VAb] and enamel [WIc], respectively. The hardness of hydrated PTD and ITD are 2.3 GPa, and 0.5 GPa, respectively [KIa]. A drawback in early AFM measurements was the inability to keep specimens wet during testing. Using modified AFM, it has been possible to apply loads in the range of 1 μN to 100 N through the tip of the indenter on fully hydrated specimens [BAa]. The Young's modulus in the axial direction increases monotonically from 15.3 GPa near the DEJ to 16.3 GPa at a distance of about 1 mm from the pulp. Also, a rapid decrease of the ITD Young's modulus is observed as dentine absorbs water from the dry state to the rehydrate state [KIc].

It is suggested that the dependence of mechanical properties of dentine from the site is related to mineral content and crystal thickness. Dentine exhibits minimum values of hardness and elastic modulus close to the DEJ. Towards the DEJ, the mineral content decreases and the crystal thickness increases [TEa].

Several AFM studies are reported regarding the gradient properties across the dentine–DEJ–enamel regions, the results showing general agreement. The Young's modulus gradually increases through the DEJ region (with a thickness of about 10 μm), increasing from about 20 GPa in the dentine region to about 80 GPa in the enamel region [MAb, FOa, MAc, BAb]. The anisotropy in the mechanical properties of dentine derives from the ultrastructural organization of collagen fibrils and mineral crystals. The contribution of collagen to the elastic modulus of enamel is negligible, and even viscoelastic properties do not change across the enamel region. Dentine strength, toughness, and bonding performances, on the other hand, are dependent on the collagenic network properties. Surprisingly, resonant ultrasound spectroscopy suggests that dentine is stiffer in the direction perpendicular to the tubules. The derived values for the Young's modulus in the axial and transverse directions are 23.2 GPa and 25.0 GPa, respectively, which would be consistent with cuffs of PTD and collagen fibrils that lie in planes transverse to each tubule axis [KId].

The size of human enamel specimens available for testing does not allow the measurements of elastic properties according to standard mechanical tests. Therefore, almost all of the literature concerning fracture properties of enamel refers to indentation techniques. AFM measurements through nanoindentation provide an excellent tool to distinguish the site-dependence properties and the anisotropy of human enamel. The Young's modulus, derived from nanohardness measurements assuming a

Poisson ratio of 0.28, is higher in the direction parallel to enamel rods (E = 87.5 GPa). In the direction parallel to enamel rods the derived Young's modulus is 72.7 GPa, therefore an anisotropy of about 30% is observed. Moreover, the elastic modulus is lower in both the tail and interrod enamel [HAb].

### 6.4.2  Ultimate Static Properties of Dentine

Concerning the harvesting of tissues for significant mechanical testing, third molars are the preferred specimen source, because dentine here is abundant in the crown region. Dentine specimens from the root of incisors and canines are also a popular choice. Hollow cylindrical specimens from the root site are likely to be axisymmetric, the tubules being almost perpendicular to the root axis. However, ultimate properties depend upon the organ, site, and orientation of tubules with respect to loading direction. The anisotropy and the site dependence of mechanical strength are shown in Table 6.13. It is interesting to observe that the tensile strength anisotropy, similarly to the elastic modulus, is such that strength is higher for specimens with tubules oriented perpendicular to the applied load; this is found for both root or crown dentine [LEb, MIa]. An increase of strength is measured toward the DEJ [STa, KOa]. Dentine close to the DEJ is less stiff but has a higher strength both in tension and in shear.

The anisotropy in tensile strength is also observed for demineralized dentine. Mid-coronal demineralized specimens contribute approximatively 30% to the strength of dentine with tubules oriented in the perpendicular direction [SAa]. The same result is found for root dentine [MIa]. Moreover, demineralized wet dentine shows a well-defined toe region and an ultimate strain that is one folder higher than mineralized dentine [SAb]. The yield strength in three-point bending of dentine from human molars is 75 MPa

| Organ | Region | Mode | Tubule Orientation | Strength [MPa] | Reference |
|---|---|---|---|---|---|
| Third molar | Mid-Coronal | T | Perpendicular | 105.5 | Sano et al. 94 |
| Incisor canine | Root | T | Parallel | 41.1 | Letchirakarn et al. 01 |
|  |  | T | Perpendicular | 59.6 |  |
| Third molar | Middle cervical | T | Perpendicular | 83.6 | Staninec et al. 02 |
|  | Middle occlusal | T | Perpendicular | 61.7 |  |
| Third molar | Outer coronal | S | Perpendicular | 76.7 | Konishi et al. 02 |
|  | Pulpal coronal | S | Perpendicular | 52.7 |  |
| Molar | Coronal and root | B | Perpendicular | 160 | Imbeni et al. 02 |
| Third molar | Coronal | T | Parallel | 73.1 | Miguez et al. 04 |
|  |  | T | Perpendicular | 140.4 |  |
|  | Root | T | Parallel | 63.2 |  |
|  |  | T | Perpendicular | 95.9 |  |

Table 6.13.  Mechanical strength of dentine. (T = tension, S = shear, B = bending.)

| Temperature | 0°C | 23°C | 37°C | 50°C | 80°C |
|---|---|---|---|---|---|
| Compressive strength [MPa] | 297 | 271 | 260 | 247 | 208 |

Table 6.14. Dependence of the compressive strength of dentine on temperature [WAb].

[IMa]. Table 6.14 shows the dependence on temperature of the compressive strength of dentine [WAb].

### 6.4.3 Viscoelastic Properties

The mechanical properties of dental tissues are time-dependent. For the case of the compressive elastic modulus, for instance, a power law is observed for hollow cylindrical specimens obtained from the root of incisors. The exponent (0.017) is towards the lower end of the range of values reported for bone. Moreover, the independence of the creep compliance from the applied stress suggests a linear viscoelastic behavior. The reduction in the relaxation modulus in water at 25°C after 1 hour is 6–17%, whereas at 37°C after six hours the reduction is 10–30% [DUa, JAa].

It must be borne in mind that viscoelastic testing, through creep or stress relaxation, generally implies long duration experiments. It is, therefore, important to consider the extent to which tissue degradation might occur over such periods. A 30% of reduction in the Young's modulus of dentine is reported for water storage, this reduction being ascribed to calcium phosphate dissolution. In order to collect reliable measurements over the course of long duration tests, the use of Hanks' Balanced Salt Solution (HBSS) is recommended [HAa, KId].

Viscoelastic measurements of demineralized dentine have been performed using both AFM and micromechanical axial tests [BAa, PAa]. The Young's modulus derived from AFM measurements of demineralized, dehydrated, and rehydrated dentine from the crowns of third molars are 149 kPa, 2.1 GPa, and 381 kPa, respectively. Viscoelastic behavior of fully demineralized dentine is observed, and it is suggested that collagen contribution to dentine modulus is negligible [BAa]. The low value reported for the modulus of demineralized dentine compared to other investigations [SAb] is ascribed to the testing method.

Viscoelastic behavior is also observed on demineralized dentine specimens from mid-coronal dentine, using tension and compression stress relaxation and creep tests in a wet environment. In particular, stress relaxation and creep tests in tension suggest a linear viscoelastic behavior for demineralized dentine, whereas a different behavior is observed in compression because creep results are dependent on the applied stress [PAa]. Using nano-DMA via AFM, with the indenter modulated at 200 Hz, a complete mapping of the storage and loss moduli of dentine and enamel have been obtained. Average storage moduli of 21 GPa and 63 GPa are measured for ITD and enamel, respectively. PTD displays a storage modulus higher than ITD. Moreover, the loss modulus of ITD is higher than those measured for PTD and enamel. This suggests a higher dissipation capability of ITD, which is consistent with the high concentration of collagen fibrils in the dentine matrix [BAb].

### 6.4.4 Fracture Properties

The fracture toughness of dentine is midway in the range observed for cortical bone (0.23–6.56 MN/m$^{-1.5}$), and is at least one order of magnitude greater than dentine restorative materials [TAa]. Using the three-point bending technique on dentine specimens with a chevron notch in the middle section, the anisotropy in the toughness of dentine and enamel can be examined. For dentine, the work of fracture over a plane oriented parallel to the tubules is higher than that perpendicular to the tubules. On the other hand, for enamel, the work of fracture over a plane oriented parallel to enamel rods is lower than that measured perpendicular to the rods [RAa].

Using the CT geometry loaded as mode I of coronal dentine specimens, with tubules oriented parallel to the surface of fracture, it is suggested that fracture toughness is invariant with temperature ($K_{Ic,37°C} = 3.1$ MNm$^{-1.5}$), whereas the strain energy release rate increases with temperature. Two toughening effects are suggested for the moderate level of fracture toughness shown by dentine for a crack propagating parallel to tubules. The first of these is the blunting of the tip of a propagating crack by each tubule that is crossed [ELa]. The second is the strengthening effect due to mineralized collagen fibrils, which form a planar, feltlike structure perpendicular to the tubules [NAb].

The dependence of fracture toughness upon the orientation of tubules has been observed also on dentine from human molars. Test specimens are shaped as notchless triangular prisms of dentine, inserted in a metal short rod CNT holder. Three tubule orientations have been tested: plane of crack propagation perpendicular, parallel, and parallel-transverse to the plane of dentinal tubules. Results indicate that the fracture toughness is 1.13 MNm$^{-1.5}$, 2.02 MNm$^{-1.5}$, and 1.97 MNm$^{-1.5}$, respectively [IWa]. The anisotropy in elastic, ultimate, and fracture properties of dentine would certainly indicate that dentine is stiffer, stronger, and tougher for specimens with tubules oriented perpendicular to the applied load. By using the three-point bending method on dentine specimens from the coronal and root regions of human molars, with tubules oriented perpendicularly to the fractured surface, the measured fracture toughness is 2.72 MNm$^{-1.5}$. However, it has been argued that these values of fracture toughness have to

be interpreted as an apparent value which overestimates by about 50% the
real fracture toughness of dentine. In fact, by using the same test method
on fatigue precracked samples, the real fracture toughness is found to be
1.79 MNm$^{-1.5}$ [IMa].

The size of human enamel specimen available for testing prevents the
measurements of fracture toughness according to standard fracture tests.
Therefore, almost all of the literature concerning fracture properties of
enamel refers to indentation-fracture techniques. By using the microindentation technique, through a Vickers' indenter, the fracture toughness of
human enamel differs for molar, incisor, and canine teeth, the lowest values
being measured for molar enamel. For each tooth, fracture toughness of
enamel increases in the incisal-cervical direction. The range of values observed is 0.7–1.27 MNm$^{-1.5}$ [HAd]. Similar values for the fracture toughness
of human enamel, spanning from 0.61 MNm$^{-1.5}$ to 0.84 MNm$^{-1.5}$, have
been measured using the indentation microfracture method, and higher
values are detected in the direction of enamel rods [OKa].

### 6.4.5  Fatigue Properties

Several investigations have been carried out on dentine using in vitro testing. Compared to cortical bone, dentine is biologically less active. Consequently, the fatigue behavior of dentine, in vitro, should reflect more the
effects of cyclic loading in vivo during mastication.

The fatigue behavior of dentine obtained from the crown and the root
of human molars, with the tubules oriented perpendicularly to the applied
stress and to the plane of the crack, has been investigated using cantilever
geometry. Fatigue testing has been carried out in an HBSS environment.
At room temperature, the endurance strength at 106–107 cycles, for a minimum to maximum stress ratio of 0.1, is 25 MPa and 45 MPa, for cyclic
frequencies of 2 Hz and 20 Hz, respectively. Therefore, the fatigue behavior of dentine is time-dependent [NAb]. Under the same testing conditions,
the endurance strength of dentine at body temperature cycling at a frequency of 10 Hz, reduces with increasing stress ratio. For a stress ratio
of −1, the endurance strength is about 30–50% of the single-cycle tensile
test, implying that dentine displays a metallike fatigue behavior [NAc].

## 6.5  References

[AKa]  Akkus, O., Jepsen, K.J., and Rimnac, C.M., Microstructural aspects
of the fracture process in human cortical bone, *J. Mat. Sci.*, **35**
(2000), 6065–6074.

[APa]  Apicella, A., Liguori, A., Masi, E., and Nicolais, L., Thick laminate composite modeling in total hip replacement, in **Experimental Techniques and Design in Composite Materials** Sheffield Academis Press, (1994), pp. 323–338.

[ASa]  Ascenzi, A., Baschieri, P., and Bonucci, E., The bending properties of single osteons. *J. Biomech.*, **23** (1990), 763–771.

[ASb]  Ascenzi, A. and Bonucci, E., The compressive properties of single osteons. *Anatom. Rec.*, **161** (1968), 377–391.

[ASc]  Ascenzi, A. and Bonucci, E., The tensile properties of single osteons. *Anatom. Rec.*, **158** (1967), 375–386.

[ASd]  Ascenzi, A., Benvenuti, A., Bigi, A., Foresti, E., Koch, M.H., Mango, F., Ripamonti, A., and Roveri, N., X-ray diffraction on Cyclically loaded osteons. *Calcif. Tissue Int.*, **62** (1998), 266–273.

[ASe]  Ashman, R.B., Cowin, S.C., Van Buskirk, W.C., and Rice, J.C., A continuous wave technique for the measurement of the elastic properties of cortical bone. *J. Biomech.*, **17** (1984), 349–61.

[AUa]  Augat, P., Link, T., Lang, T.F., Lin, J.C., Majumdar, S., and Genant, H.K., Anisotropy of the elastic modulus of trabecular bone specimens from different anatomical locations. *Med. Eng. Phys.*, **2** (1998), 124–31.

[BAa]  Balooch, M., Wu-Magidi, I.C., Balazs, A., Lundkvist, A.S., Marshall, S.J., Marshall, G.W., Siekhaus, W.J., and Kinney, J.H., Viscoelastic properties of demineralized human dentin measured in water with atomic force microscope (AFM)-based indentation. *J. Biomed. Mater. Res.*, **15** (1998), 539–44.

[BAb]  Balooch, G., Marshall, G.W., Marshall, S.J., Warren, O.L., Asif, S.A., and Balooch, M., Evaluation of a new modulus mapping technique to investigate microstructural features of human teeth. *J. Biomech.*, **37** (2004), 1223–32.

[BAc]  Bargren, J.H., Andrew, C., Bassett, L., and Gjelsvik, A., Mechanical properties of hydrated cortical bone. *J. Biomech.*, **7** (1974), 239–45.

[BAd]  Bayraktar, H.H. and Keaveny, T.M., Mechanisms of uniformity of yield strains for trabecular bone. *J. Biomech.*, **37** (2004), 1671–8.

[BEa]  Behiri, J.C. and Bonfield, W., Fracture mechanics of bone—the effects of density, specimen thickness and crack velocity on longitudinal fracture. *J. Biomech.*, **17** (1984), 25–34.

[BOa]  Bonfield, W. and Datta, P.K., Fracture toughness of compact bone. *J. Biomech.*, **9** (1976), 131–4.

[BOb]  Bonfield, W. and Grynpas, M.D., Anisotropy of the Young's modulus of bone. *Nature*, **270** (1977), 453–4.

[BOc] Boskey, A., Bone mineral crystal size. *Osteoporos. Int.*, **14** (2003), 16–21.

[BOd] Boyde, A. and Jones, S.J., Scanning electron microscopy of bone: Instrument, specimen, and issues. *Microscopy Res. Tech.*, **33** (1998), 92–120.

[BOe] Bowen, R.L. and Rodriguez, M.S., Tensile strength and modulus of elasticity of tooth structure and several restorative materials. *J. Am. Dent. Assoc.*, **64** (1962), 378–87.

[BRa] Bromage, T.G., Goldman, H.M., McFarlin, S.C., Warshaw, J., Boyde, A., and Riggs, C.M., Circularly polarized light standards for investigations of collagen fiber orientation in bone. *Anat. Rec. B. New Anat.*, **274** (2003), 157–68.

[BRb] Brown, S.J., Pollintine, P., Powell, D.E., Davie, M.W.J., and Sharp, C.A., Regional differences in mechanical and material properties of femoral head cancellous bone in health and osteoarthritis. *Calcif. Tissue Int.*, **71** (2002), 227–234.

[BUa] Bumrerraj, S. and Katz, J.L., Scanning acoustic microscopy study of human cortical and trabecular bone. *Ann. Biomed. Eng.*, **29** (2001), 1034–42.

[BUb] Burr, D.B., Schaffler, M.B., and Frederickson, R.G., Composition of the cement line and its possible mechanical role as a local interface in human compact bone. *J. Biomech.*, **21** (1988) 939–45.

[CAa] Cagidiaco, M.C. and Ferrari, M., Organization of the matrix. In: Bonding to dentin. Livorno: O. Debatte and F. Ed. (1995), 11–26.

[CAb] Carter, D.R. and Hayes, W.C., The compressive behavior of bone as two-phase porous structure. *J. Bone Joint Surg.*, **59** (1977), 954–962.

[CAc] Carter, D.R. and Caler, W.E., Cycle-dependent and time-dependent bone fracture with repeated loading. *J. Biomech. Eng.*, **105** (1983), 166–70.

[CAd] Cassidy, J.J., Hiltner, A., and Baer, E., Hierarchical structure of the intervertebral disc. *Connect Tissue Res.*, **1** (1989), 75–88.

[CAe] Causa, F., Manto, L., Borzacchiello, A., De Santis, R., Netti, P.A., Ambrosio, L., and Nicolais, L., Spatial and structural dependence of mechanical properties of porcine intervertebral disc. *J. Mater. Sci. Mater. Med.*, **12** (2002), 1277 80.

[CHa] Charras, G.T., Lehenkari, P.P., and Horton, M.A., Atomic force microscopy can be used to mechanically stimulate osteoblasts and evaluate cellular strain distributions. *Ultramicroscopy.*, **86** (2001), 85–95.

[CHb] Choi, K., Kuhn, J.L., Ciarelli, M.J., and Goldstein, S.A., The elastic moduli of human subchondral, trabecular, and cortical bone tissue and the size-dependency of cortical bone modulus. *J. Biomech.*, **11** (1990), 1103–1113.

[CHc] Choi, K. and Goldstein, S.A., A comparison of the fatigue behavior of human trabecular and cortical bone tissue. *J. Biomech.*, **25** (1992), 1371–81.

[CIa] Ciarelli, T.E., Fyhrie, D.P., Schaffler, M.B., and Goldstein, S.A., Variations in three-dimensional cancellous bone architecture of the proximal femur in female hip fractures and in controls. *J. Bone Miner. Res.*, **15** (2000), 32–40.

[CLa] Claes, L.E., Wilke, H.J., and Kiefer, H., Osteonal structure better predicts tensile strength of healing bone than volume fraction. *J. Biomech.*, **28** (1995), 1377–90.

[COa] Courtney, A.C., Hayes, W.C., and Gibson, L.J., Age-related differences in post-yield damage in human cortical bone. Experiment and model. *J. Biomech.*, **29** (1996), 1463–71.

[COb] Cowin, S.C. and Turner, C.H., On the relationship between the orthotropic Young's moduli and fabric. *J. Biomech.*, **25** (1992), 1493–1494.

[COc] Cowin, S.C., Bone poroelasticity. *J. Biomech.*, **32** (1999), 217–38.

[COd] Cowin, S.C., The relationship between the elasticity tensor and the fabric tensor. *Mech. Mater.*, **4** (1985), 134–147.

[COe] Cowin, S.C., Wolff's law of trabecular architecture at remodeling equilibrium. *J. Biomech. Eng.*, **1** (1986), 83–8.

[CRa] Craig, R.G. and Peyton, F.A., Elastic and mechanical properties of human dentin. *J. Dent. Res.*, **37** (1958), 710–8.

[CRb] Craig, R.G. and Peyton, F.A., Thermal conductivity of teeth structures, dentin cements, and amalgam. *J. Dent. Res.* **40** (1961), 411–418.

[CUa] Cuppone, M., Seedhom, B.B., Berry, E., Ostell, A.E., The longitudinal Young's modulus of cortical bone in the midshaft of human femur and its correlation with CT scanning data. *Calcif. Tissue Int.*, **74** (2004), 302–9.

[CUb] Currey, J.D., Stress concentrations in bone. *Q J. Microsc. Sci.*, **103** (1962), 111–133.

[CUc] Currey, J., Sacrificial bonds heal bone. *Nature*, **414** (2001), 699.

[CUd] Currey, J.D. and Butler, G., The mechanical properties of bone tissue in children. *J. Bone Joint Surg. Am.*, **57** (1975), 810–4.

[DAa] Dabestani, M. and Bonfield, W., Elastic and anelastic microstrain measurement in human cortical bone. In: **Implant Materials in Biofunction**. de Putter, C., de Lange, G.L., de Groot, K., Lee, A.J.C. Eds. Elsevier Science, Amsterdam (1988), 435–440.

[DEa] De Santis, R., Mollica, F., Prisco, D., Rengo, S., Ambrosio, L., and Nicolais, L., A 3-D analysis of mechanically stressed dentin-adhesive-composite interfaces using X-ray micro-CT. *Biomaterials*, **26** (2005), 257–270.

[DEb] De Santis, R., Anderson, P., Tanner, K.E., Ambrosio, L., Nicolais L., Bonfield, W., and Davis, G.R., Bone fracture analysis on the short rod chevron-notch specimens using the X-ray computer micro-tomography. *J. Mat. Sci. Mat. Med.*, **11** (2000), 629–636.

[DEc] Deligiann, D.D., Missirlis, Y.F., Tanner, K.E., and Bonfield, W., Mechanical behavior of trabecular bone of the human femoral head in females. *J. Mater. Sci. Mat. Med.*, **2** (1991), 168–175.

[DEd] Deligiann, D.D., Maris, A., and Missirlis, Y.F., Stress relaxation behavior of trabecular bone specimens. *J. Biomech.*, **27** (1994), 1469–1476.

[DIa] Ding, M., Dalstra, M., Danielsen, C.C., Kabel, J., Hvid, I., and Linde, F., Age variations in the properties of human tibial trabecular bone. *J. Bone Joint Surg. Br.*, **6** (1997), 995–1002.

[DOa] Dowling, N.E., **Mechanical Behavior of Materials**. Prentice-Hall, Englewood Cliffs, NJ (1996).

[DUa] Duncanson, M.G., Jr. Korostoff, E., Compressive viscoelastic properties of human dentin: I. Stress-relaxation behavior. *J. Dent. Res.*, **54** (1975), 1207–12.

[ELa] El Mowafy, O.M. and Watts, D.C., Fracture toughness of human dentin. *J. Dent. Res.*, **65** (1986), 677–81.

[ERa] Erdemt, U., Instrument science and technology: Force and weight measurement, *J. Phys E: Sci. Instrum.*, **15** (1982), 857–872.

[EVa] Evans, F.G. and Vincentelli, R., Relation of collagen fibers orientation to some mechanical properties of human cortical bone. *J. Biomech.*, **2** (1969), 63–71.

[FAa] Fan, Z. and Rho, J.Y., Effects of viscoelasticity and time-dependent plasticity on nanoindentation measurements of human cortical bone. *J. Biomed. Mater. Res. A*, **67** (2003), 208–14.

[FAb] Fantner, G.E., Hassenkam, T., Kindt, J.H., Weaver, J.C., Birkedal, H., Pechenik, L., Cutroni, J.A., Cidade, G.A.G., Stucky, G.D., Morse, D.E., and Hansma, P.K., Sacrificial bonds and hidden length dissipate energy as mineralized fibrils separate during bone fracture. *Nature*, **4** (2005), 612–616.

[FOa] Fong, H., Sarikaya, M., White, S.N., and Snead, M.L., Nano-mechanical properties profiles across dentin-enamel junction of human incisor teeth. *Mater. Sci. Eng. C*, **7** (2000), 119–128.

[FOb] Forss, H., Seppa, L., and Lappalainen, R., In vitro abrasion resistance and hardness of glass-ionomer cements. *Dent. Mater.*, **7** (1991), 36–9.

[GAa] Galante, J., Rostoker, W., and Ray, R.D., Physical properties of trabecular bone. *Calcif Tissue Res.*, **5** (1970), 236–46.

[GAb] Garberoglio, R. and Brannstrom, M., Scanning electron microscopic investigation of human dentinal tubules. *Arch. Oral Biol.*, **21** (1976), 355–62.

[GIa] Giesen, E.B., Ding, M., and Dalstra, M., and van Eijden, T.M., Mechanical properties of cancellous bone in the human mandibular condyle are anisotropic. *J. Biomech.*, **34** (2001), 799–803.

[GIb] Gilmore, R.S. and Katz, J.L., Elastic properties of apatites, *J. Mat. Sci.*, **17** (1982), 1131–1141.

[GOa] Goldstein, S.A., Wilson, D.L., Sonstegard, D.A., and Matthews, L.S., The mechanical properties of human tibial trabecular bone as a function of metaphyseal location. *J. Biomech.*, **12** (1983), 965–969.

[GUa] Guo, X.E., Liang, L.C., and Goldstein, S.A., Micromechanics of osteonal cortical bone fracture. *J. Biomech. Eng.*, **120** (1998), 112–7.

[HAa] Habelitz, S., Marshall, G.W., Jr., Balooch, M., and Marshall, S.J., Nanoindentation and storage of teeth. *J. Biomech.*, **35** (2002), 995–8.

[HAb] Habelitz, S., Marshall, S.J., Marshall, G.W., Jr., and Balooch, M., Mechanical properties of human dental enamel on the nanometre scale. *Arch. Oral Biol.*, **46** (2001), 173–83.

[HAc] Hara, T., Takizawa, M., Sato, T., and Ide, Y., Mechanical properties of buccal compact bone of the mandibular ramus in human adults and children: Relationship of the elastic modulus to the direction of the osteon and the porosity rati. *Bull. Tokyo Dent. Coll.*, **39** (1998), 47–55.

[HAd] Hassan, R., Caputo, A.A., and Bunshah, R.F., Fracture toughness of human enamel. *J. Dent. Res.*, **60** (1981), 820–7.

[HEa] Hengsberger, S., Kulik, A., and Zysset, P., Nanoindentation discriminates the elastic properties of individual human bone lamellae under dry and physiological conditions. *Bone*, **30** (2002), 178–84.

[HOa]   Hogan, H.A., Micromechanics modeling of Haversian cortical bone
        properties. *J. Biomech.*, **25** (1992), 549–56.

[HOb]   Homminga, J., McCreadie, B.R., Ciarelli, T.E., Weinans, H.,
        Goldstein, S.A., and Huiskes, R., Cancellous bone mechanical prop-
        erties from normals and patients with hip fractures differ on the
        structure level, not on the bone hard tissue level. *Bone*, **30** (2002),
        759–64.

[HVa]   Hvid, I. and Hansen, S.L., Trabecular bone strength patterns at the
        proximal tibial epiphysis. *J. Orthop. Res.*, **4** (1985), 464–472.

[HUa]   Huiskes, R. Ruimerman, R., van Lenthe, G.H., and Janssen, J.D.,
        Effects of mechanical forces on maintenance and adaptation of form
        in trabecular bone. *Nature*, **405** (2000), 704–706.

[IMa]   Imbeni, V., Nalla, R.K., Bosi, C., Kinney, J.H., and Ritchie, R.O., In
        vitro fracture toughness of human dentin. *J. Biomed. Mater. Res. A*,
        **66** (2003), 1–9.

[ITa]   Ito, S., Saito, T., Tay, F.R., Carvalho, R.M., Yoshiyama, M., and
        Pashley, D.H., Water content and apparent stiffness of non-caries
        versus caries-affected human dentin. *J. Biomed. Mater. Res. B Appl.
        Biomater.*, **72** (2005), 109–16.

[IWa]   Iwamoto, N. and Ruse, N.D., Fracture toughness of human dentin.
        *J. Biomed. Mater. Res. A*, **66** (2003), 507–12.

[JAa]   Jantarat, J., Palamara, J.E., Lindner, C., and Messer, H.H., Time-
        dependent properties of human root dentin. *Dent. Mater.*, **18** (2002),
        486–93.

[JAb]   Jasiuk, I. and Ostoja-Starzewski, M., Modeling of bone at a single
        lamella level. *Biomech. Model Mechanobiol.*, **3** (2004), 67–74.

[JEa]   Jepsen, K.J. and Davy, D.T., Comparison of damage accumulation
        measures in human cortical bone. *J. Biomech.*, **30** (1997), 891–4.

[KAa]   Kabel, J., van Rietbergen, B., Odgaard, A., and Huiskes, R., Con-
        stitutive relationships of fabric, density, and elastic properties in
        cancellous bone architecture. *Bone*, **25** (1999), 481–486.

[KAa]   Kafka, V. and Jirova, J., A structural mathematical model for the
        viscoelastic anisotropic behavior of trabecular bone. *Biorheology*,
        **20** (1983), 795–805.

[KAb]   Katz, J.L. and Meunier, A., Material properties of single osteons
        and osteonic lamellae using high frequency scanning acoustic mi-
        croscopy. In: **Bone Structure and Remodeling**. Odgaard, A.,
        Weinas, H., Eds. Word Scientific, Amsterdam (1994), 157–165.

[KAc]   Katz, J.L., Anisotropy of Young's modulus of bone. *Nature*, **283**
        (1980), 106–7.

[KEa]  Keaveny, T.M. and Hayes, W.C., A 20-year perspective on the mechanical properties of trabecular bone. *J. Biomech. Eng.* **115** (1993), 534–42.

[KEb]  Keaveny, T.M., Morgan, E.F., Niebur, G.L., and Yeh, O.C., Biomechanics of trabecular bone. *Annu. Rev. Biomed. Eng.*, **3** (2001), 307–333.

[KIa]  Kinney, J.H., Balooch, M., Marshall, S.J., Marshall, G.W., Jr., Weihs, T.P., Hardness and Young's modulus of human peritubular and intertubular dentine. *Arch. Oral Biol.*, **41** (1996), 9–13.

[KIb]  Kinney, J.H., Balooch, M., Marshall, S.J., Marshall, G.W., Jr., and Weihs, T.P., Atomic force microscope measurements of the hardness and elasticity of peritubular and intertubular human dentin. *J. Biomech. Eng.*, **118** (1996), 133–5.

[KIc]  Kinney, J.H., Balooch, M., Marshall, G.W., and Marshall, S.J., A micromechanics model of the elastic properties of human dentine. *Arch. Oral Biol.*, **44** (1999), 813–22.

[KId]  Kinney, J.H., Marshall, S.J., and Marshall, G.W., The mechanical properties of human dentin: A critical review and re-evaluation of the dental literature. *Crit. Rev. Oral Biol. Med.*, **14** (2003), 13–29.

[KIe]  Kinney, J.H., Gladden, J.R., Marshall, G.W., Marshall, S.J., So, J.H., and Maynard, J.D., Resonant ultrasound spectroscopy measurements of the elastic constants of human dentin. *J. Biomech.*, **37** (2004), 437–41.

[KIf]  Kishen, A., Ramamurty, U., and Asundi, A., Experimental studies on the nature of property gradients in the human dentine. *J. Biomed. Mater. Res.*, **51** (2000), 650–9.

[KOa]  Konishi, N., Watanabe, L.G., Hilton, J.F., Marshall, G.W., Marshall, S.J., and Staninec, M., Dentin shear strength: Effect of distance from the pulp. *Dent. Mater.*, **18** (2002), 516–20.

[KOb]  Kopperdahl, D.L. and Keaveny, T.M., Yield strain behavior of trabecular bone. *J. Biomech.*, **31** (1998), 601–608.

[KUa]  Kuboky, Y. and Mechanic, G.L., Comparative molecular distribution of cross-link in bone and dentine collagen. Structure-function relationship. *Calcif. Tissue Int.*, **34** (1982), 306–308.

[LAa]  Lakes, R.S., Nakamura, S., Behiri, J.C., and Bonfield, W., Fracture mechanics of bone with short cracks. *J. Biomech.*, **23** (1990), 967–75.

[LAb]  Lang, H.P., Hegner, M., and Gerber, C., Cantilever array sensors. *Mater. Today*, **8** (2005), 30–36.

[LEa] Lenz, C. and Nackenhorst, U., A numerical approach to mechanosensation of bone tissue based on a micromechanical analysis of a single osteon. *PAMM*, **4** (2004), 342–343.

[LEb] Lertchirakarn, V., Palamara, J.E., and Messer, H.H., Anisotropy of tensile strength of root dentin. *J. Dent. Res.*, **80** (2001), 453–6.

[LIa] Liebschner, M.A. and Keller, T.S., Hydraulic strengthening affects the stiffness and strength of cortical bone. *Ann. Biomed. Eng.* **33** (2005), 26–38.

[LIb] Linde, F., Norgaard, P., Hvid, I., Odgaard, A., and Soballe, K., Mechanical properties of trabecular bone. Dependency on strain rate. *J. Biomech.*, **24** (1991), 803–809.

[LIc] Linde, F. and Sorensen, H.C., The effect of different storage methods on the mechanical properties of trabecular bone. *J. Biomech.*, **26** (1993), 1249–52.

[MAa] Majumdar, S., Kothari, M., Augat, P., Newitt, D.C., Link, T.M., Lin, J.C., Lang, T., Lu, Y., and Genant, H.K., High-resolution magnetic resonance imaging: Three-dimensional trabecular bone architecture and biomechanical properties. *Bone*, **22** (1998), 445–54.

[MAb] Marshall, G.W., Jr., Balooch, M., Gallagher, R.R., Gansky, S.A., and Marshall, S.J., Mechanical properties of the dentinoenamel junction: AFM studies of nanohardness, elastic modulus, and fracture. *J. Biomed. Mater. Res.*, **54** (2001), 87–95.

[MAc] Marshall, S.J., Balooch, M., Habelitz, S., Balooch, G., Gallagher, R., and Marshall, G.W., The dentin-enamel junction-a natural, multi-level interface. *J. Europ. Ceramic Soc.*, **23** (2003), 2897–2904

[MAd] Martin, R.B. and Burr, D.B., A hypothetical mechanism for the stimulation of osteonal remodeling by fatigue damage. *J. Biomech.*, **15** (1982), 137–9.

[MAe] Martin, B., Aging and strength of bone as a structural material. *Calcif. Tissue Int.*, **53 Suppl 1** (1993), S34–39.

[MCa] McElhaney, J.H., Dynamic response of bone and muscle tissue. *J. Appl. Physiol.*, **21** (1966), 1231–36.

[MEa] Mente, P.L. and Lewis, J.L., Experimental method for the measurement of the elastic modulus of trabecular bone tissue. *J. Orthop. Res.*, **7** (1989), 456–61.

[MEb] Meredith, N., Sherriff, M., Setchell, D.J., and Swanson, S.A., Measurement of the microhardness and Young's modulus of human enamel and dentine using an indentation technique. *Arch. Oral Biol.* **41** (1996), 539–45.

[MIa]  Miguez, P.A., Pereira, P.N., Atsawasuwan, P., and Yamauchi, M., Collagen cross-linking and ultimate tensile strength in dentin. *J. Dent Res.*, **83** (2004), 807–10.

[MIb]  Misch, C.E., Qu, Z., and Bidez, M.W., Mechanical properties of trabecular bone in the human mandible: implications for dental implant treatment planning and surgical placements. *J. Oral Maxillofac. Surg.*, **57** (1999), 700–706.

[MOa]  Morgan, E.F. and Keaveny, T.M., Dependence of yield strain of human trabecular bone on anatomic site. *J. Biomech.*, **5** (2001), 569–77.

[MOb]  Mosekilde, L., Mosekilde, L., and Danielsen, C.C., Biomechanical competence of vertebral trabecular bone in relation to ash density and age in normal individuals. *Bone*, **2** (1987), 79–85.

[NAa]  Nalla, R.K., Kinney, J.H., and Ritchie, R.O., On the fracture of human dentin: Is it stress- or strain-controlled? *J. Biomed. Mater. Res.A*, **67** (2003), 484–95.

[NAb]  Nalla, R.K., Imbeni, V., Kinney, J.H., Staninec, M., Marshall, S.J., and Ritchie, R.O., In vitro fatigue behavior of human dentin with implications for life prediction. *J. Biomed. Mater. Res.A* **66** (2003), 10–20.

[NAc]  Nalla, R.K., Kinney, J.H., Marshall, S.J., and Ritchie, R.O., On the in vitro fatigue behavior of human dentin: Effect of mean stress. *J. Dent. Res.*, **83** (2004), 211–5.

[NAd]  Natali, A.N., and Hart, R.T., Mechanics of hard tissues, in Integrated Biomaterials Science, R. Barbucci, Ed. Kluwer Academic and Plenum, New York (2002) 459–489.

[NIa]  Nicholson, P.H., Muller, R., Lowet, G., Cheng, X.G., Hildebrand, T., Ruegsegger, P., van der Perre, G., Dequeker, J., and Boonen, S., Do quantitative ultrasound measurements reflect structure independently of density in human vertebral cancellous bone? *Bone*, **23** (1998), 425–31.

[NIb]  Nicholson, P.H.F., Cheng, X.G., Lowet, G., Boonen, S., Davie, M.W.J., Dequeker, J., and Van der Perre, G., Structural and material mechanical properties of human vertebral cancellous bone. *Med. Eng. Phys.*, **19** (1997), 729–737.

[NIc]  Nicolais, L., Mechanics of composites (particulate and fiber polymeric laminate properties). *Polym. Eng. Sci.*, **15** (1975), 137–149.

[NOa]  Norman, T.L., Vashishth, D., and Burr, D.B., Mode I fracture toughness of human bone. In: **Advances in Bioengineering**, Vanderby, R., Ed. ASME, New York (1991), pp. 361–364.

[NOb] Norman, T.L., Vashishth, D., and Burr, D.B., Fracture toughness of human bone under tension. *J. Biomech.*, **28** (1995), 309–20.

[NOc] Norman, T.L., Nivargikar, S.V., and Burr, D.B., Resistance to crack growth in human cortical bone is greater in shear than in tension. *J. Biomech.*, **29** (1996), 1023–31.

[NOd] Norman, T.L. and Wang, Z., Microdamage of human cortical bone: Incidence and morphology in long bones. *Bone*, **20** (1997), 375–9.

[OKa] Okazaki, K., Nishimura, F., and Nomoto, S., Fracture toughness of human enamel. *Shika Zairyo Kikai*, **8** (1989), 382–7.

[OMa] O'Mahony, A.M., Williams, J.L., Katz, J.O., and Spencer, P., Anisotropic elastic properties of cancellous bone from human edentulous mandible. *Clin. Oral Impl. Res.*, **11** (2000), 415–421.

[OUa] Ouyang, J., Yang, G.T., Wu, W.Z., Zhu, Q.A., and Zhong, S.Z., Biomechanical characteristics of human trabecular bone. *Clin. Biomech.* **12** (1997), 522–524.

[PAa] Pashley, D.H., Agee, K.A., Wataha, J.C., Rueggeberg, F., Ceballos, L., Itou, K., Yoshiyama, M., Carvalho, R.M., and Tay, F.R., Viscoelastic properties of demineralized dentin matrix. *Dent. Mater.*, **19** (2003), 700–6.

[RAa] Rasmussen, S.T., Patchin, R.E., Scott, D.B., and Heuer, A.H., Fracture properties of human enamel and dentin. *J. Dent. Res.*, **55** (1976), 154–64.

[RAb] Raum, K., Jenderka, K.V., Klemenz, A., and Brandt, J., Multilayer analysis: Quantitative scanning acoustic microscopy for tissue characterization at a microscopic scale. *IEEE Trans. Ultras Ferroelectr. Freq. Contr.*, **50** (2003), 507–516.

[REa] Reilly, D.T., Burstein, A.H., and Frankel, V.H., The elastic modulus for bone. *J. Biomech.*, **7** (1974), 271–5.

[REb] Reilly, D.T. and Burstein, A.H., The elastic and ultimate properties of compact bone tissue. *J. Biomech.*, **8** (1975), 393–405.

[REc] Renson, C.E. and Braden, M., Experimental determination of the rigidity modulus, Poisson's ratio and elastic limit in shear of human dentine. *Arch. Oral Biol.*, **20** (1975), 43–7.

[RHa] Rho, J.Y., Ashman, R.B., and Turner, C.H., Young's modulus of trabecular and cortical bone material: Ultrasonic and microtensile measurements. *J. Biomech.*, **2** (1993), 111–119.

[RHb] Rho, J.Y., Kuhn-Spearing, L., and Zioupos, P., Mechanical properties and the hierarchical structure of bone. *Med. Eng. Phys.*, **2** (1998), 92–102.

[RHc]  Rho, J.Y., Roy, M.E., 2nd, Tsui, T.Y., and Pharr, G.M., Elastic properties of microstructural components of human bone tissue as measured by nanoindentation. *J. Biomed. Mater. Res.*, **45** (1999), 48–54.

[RHd]  Rho, J.Y., An ultrasonic method for measuring the elastic properties of human tibial cortical and cancellous bone. *Ultrasonics*, **8** (1996), 777–783.

[ROa]  Rohl, L., Larsen, E., Linde, F., Odgaard, A., and Jorgensen, J., Tensile and compressive properties of cancellous bone. *J. Biomech.* **12** (1991), 1143–9.

[ROb]  Roy, M.E., Rho, J.Y., Tsui, T.Y., Evans, N.D., and Pharr, G.M., Mechanical and morphological variation of the human lumbar vertebral cortical and trabecular bone. *J. Biomed. Mater. Res.*, **44** (1999), 191–7.

[RUa]  Runkle, J.C. and Pugh, J., The micro-mechanics of cancellous bone. II. Determination of the elastic modulus of individual trabeculae by a buckling analysis. *Bull. Hosp. Joint Dis.*, **36** (1975), 2–10.

[SAa]  Sano, H., Shono, T., Sonoda, H., Takatsu, T., Ciucchi, B., Carvalho, R., and Pashley, D.H., Relationship between surface area for adhesion and tensile bond strength-evaluation of a micro-tensile bond test. *Dent. Mater*, **10** (1994), 236–40.

[SAb]  Sano, H., Ciucchi, B., Matthews, W.G., and Pashley, D.H., Tensile properties of mineralized and demineralized human and bovine dentin. *J. Dent. Res.*, **73** (1994), 1205–11.

[SCa]  Schaffler, M.B., Choi, K., and Milgrom, C., Aging and matrix microdamage accumulation in human compact bone. *Bone*, **17** (1995), 521–25.

[SCb]  Schoenfeld, C.M., Lautenschlager, E.P., and Meyer, P.R., Mechanical properties of human cancellous bone in the femoral head. *Med. Biol. Engng.*, **12** (1974), 313–317.

[SCc]  Schwartz-Dabney, C.L., Dechow, P.C., Schwartz-Dabney, C.L., and Dechow, P.C., Variations in cortical material properties throughout the human dentate mandible. *Am. J. Phys. Anthropol.*, **120** (2003), 252–77.

[SEa]  Sevostianov, I. and Kachanov, M., Impact of the porous microstructure on the overall elastic properties of the osteonal cortical bone. *J. Biomech.*, **33** (2000), 881–8.

[SIa]  Silver, F.H., Seehra, G.P., Freeman, J.W., and DeVore, D., Viscoelastic properties of young and old human dermis: A proposed molecular mechanism for elastic energy storage in collagen and elastin. *J. Appl. Pol. Sci.*, **79** (2001), 134–142.

[SIb]  Simkin, A. and Robin, G., Fracture formation in differing collagen fiber pattern of compact bone. *J. Biomech.*, **7** (1974), 183–8.

[SKa]  Skovoroda, A.R., Emelianov, S.Y., and O'Donnell, M., Tissue elasticity reconstruction based on ultrasonic displacement and strain images, *IEEE Trans. Ultrason. Ferroelectr. Freq. Contr.*, **42** (1995), 747–745.

[STa]  Staninec, M., Marshall, G.W., Hilton, J.F., Pashley, D.H., Gansky, S.A., Marshall, S.J., and Kinney, J.H., Ultimate tensile strength of dentin: Evidence for a damage mechanics approach to dentin failure. *J. Biomed. Mater. Res.*, **63** (2002), 342–5.

[TAa]  Tam, L.E. and Yim, D., Effect of dentine depth on the fracture toughness of dentine-composite adhesive interfaces. *J. Dent.*, **25** (1997), 339–46.

[TEa]  Tesch, W., Eidelman, N., Roschger, P., Goldenberg, F., Klaushofer, K., and Fratzl, P., Graded microstructure and mechanical properties of human crown dentin. *Calcif Tissue Int.*, **69** (2001), 147–57.

[THa]  Thompson, J.B., Kindt, J.H., Drake, B., Hansma, H.G., Morse, D.E., and Hansma, P.K., Bone indentation recovery time correlates with bond reforming time. *Nature*, **414** (2001), 773–776.

[TUa]  Turner, C.H., Rho, J., Takano, Y., Tsui, T.Y., and Pharr, G.M., The elastic properties of trabecular and cortical bone tissues are similar: results from two microscopic measurement techniques. *J. Biomech.*, **32** (1999), 437–41.

[TYa]  Tyldesley, W.R., The mechanical properties of human enamel and dentine. *Briti. Dent. J.*, **106** (1959), 269–278.

[ULa]  Ulrich, D., van Rietbergen, B., Weinans, H., and Ruegsegger, P., Finite element analysis of trabecular bone structure: A comparison of image-based meshing techniques. *J. Biomech.*, **31** (1998), 1187–92.

[VAa]  Van Lenthe, G.H. and Huiskes, R., How morphology predicts mechanical properties of trabecular structures depends on intra-specimen trabecular thickness variations. *J. Biomech.*, **9** (2002), 1191–1197.

[VAb]  Van Meerbeek, B., Willems, G., Celis, J.P., Roos, J.R., Braem, M., Lambrechts, P., and Vanherle, G., Assessment by nano-indentation of the hardness and elasticity of the resin-dentin bonding area. *J. Dent. Res.*, **72** (1993), 1434–42.

[VAc]  Van Rietbergen, B., Huiskes, R., Eckstein, F., and Ruegsegger, P., Trabecular bone tissue strains in the healthy and osteoporotic human femur. *J. Bone Miner. Res.*, **18** (2003), 1781–8.

[VAd] Vashishth, D., Behiri, J.C., and Bonfield, W., Crack growth resistance in cortical bone: Concept of microcrack toughening. *J. Biomech.*, **30** (1997), 763–9.

[VEa] Veis, A., Dentin. In: **Extracellular Matrix. Tissue Function**, Vol. 1. Comper WD, The Netherlands (1996), 41–75.

[WAa] Watanabe, L.G., Marshall, G.W., Jr., and Marshall, S.J., Dentin shear strength: effects of tubule orientation and intratooth location. *Dent. Mater.*, **12** (1996), 109–15.

[WAb] Watts, D.C., el Mowafy, O.M., and Grant, A.A., Temperature-dependence of compressive properties of human dentin. *J. Dent. Res.* **66** (1987), 29–32.

[WEa] Weiner, S. and Wagner, H.D., The material bone: Structure-mechanical function relations. *Ann. Rev. Mater. Sci.* **28** (1998), 271–298.

[WEb] Weyland, M. and Midgley, P.A., Electron tomography. *Mater. Today*, **7** (2004), 32–40.

[WIb] Willems, G., Lambrechts, P., Braem, M., Celis, J.P., and Vanherle, G., A classification of dental composites according to their morphological and mechanical characteristics. *Dent. Mater.*, **8** (1992), 310–9.

[WIc] Willems, G., Celis, J.P., Lambrechts, P., Braem, M., and Vanherle, G., Hardness and Young's modulus determined by nanoindentation technique of filler particles of dental restorative materials compared with human enamel. *J. Biomed. Mater. Res.*, **27** (1993), 747–55.

[WOa] Wolff, J., **Das Gesetz der Transformation der Knochen**. Published with support from the Royal Academy of Sciences in Berlin. A. Hirschwald, ed. Berlin, 1892. English trans. by P. Maquet and R. Furlong. **The Law of Bone Remodeling**. Belin, Springer-Verlag, Heidelberg (1986).

[XUa] Xu, H.H., Smith, D.T., Jahanmir, S., Romberg, E., Kelly, J.R., Thompson, V.P., and Rekow, E.D., Indentation damage and mechanical properties of human enamel and dentin. *J. Dent. Res.*, **77** (1998), 472–80.

[XUb] Xu, J., Rho, J.Y., Mishra, S.R., and Fan, Z., Atomic force microscopy and nanoindentation characterization of human lamellar bone prepared by microtome sectioning and mechanical polishing technique. *J. Biomed. Mater. Res. A*, **67** (2003), 719–26.

[YAa] Yamamoto, E., Crawford, P.R., Chan, D.D., and Keaveny, T.M., Development of residual strains in human vertebral trabecular bone after prolonged static and cyclic loading at low load levels. *J. Biomech.* **39**(10) (2006), 1812–8.

[YAb] Yamashita, J., Furman, B.R., Rawls, H.R., Wang, X., and Agrawal, C.M., The use of dynamic mechanical analysis to assess the viscoelastic properties of human cortical bone. *J. Biomed. Mater. Res.* **58** (2001), 47–53.

[YOa] You, L.D., Weinbaum, S., Cowin, S.C., and Schaffler, M.B., Ultrastructure of the osteocyte process and its pericellular matrix. *Anat. Rec. A Discov. Mol. Cell Evol. Biol.*, **278** (2004), 505–13.

[YUa] Yu, Z. and Boseck, S., Scanning acoustic microscopy and its applications to material characterization. *Rev. Modern Phys.*, **67** (1995), 863–891.

[ZHa] Zhang, N. and Grimm, M.J., Measurement of elastic moduli of individual trabeculae of vertebtrae using scanning acoustic microscopy. *2001 Bioengineering Conference ASME.* Snowbird, Utah. **50** (2001), 283–284.

[ZIa] Zilch, H., Rohlmann, A., Bergmann, G., and Kolbel, R., Material properties of femoral cancellous bone in axial loading. Part II: Time dependent properties. *Arch. Orthop. Trauma Surg.*, **97** (1980), 257–62.

[ZIb] Zioupos, P., Wang, X.T., and Currey, J.D., Experimental and theoretical quantification of the development of damage in fatigue tests of bone and antler. *J. Biomech.*, **29** (1996), 989–1002.

[ZIc] Zioupos, P., X, T.W., and Currey, J.D., The accumulation of fatigue microdamage in human cortical bone of two different ages in vitro. *Clin. Biomech.*, **11** (1996), 365–375.

[ZId] Zioupos, P., Currey, J.D., and Hamer, A.J., The role of collagen in the declining mechanical properties of aging human cortical bone. *J. Biomed. Mater. Res.*, **45** (1999), 108–16.

[ZYa] Zysset, P.K., Guo, X.E., Hoffler, C.E., Moore, K.E., and Goldstein, S.A., Elastic modulus and hardness of cortical and trabecular bone lamellae measured by nanoindentation in the human femur. *J. Biomech.*, **32** (1999), 1005–12.

# 7

## Mechanics in Tumor Growth

L. Graziano and L. Preziosi

*Department of Mathematics*
*Polytechnic of Turin*
*I-10127 Torino, Italy*

Abstract. This chapter focuses on the mechanical aspects of tumor growth. After describing some of the main features of tumor growth and in particular the phenomena involving stress and deformation, the chapter deals with the multiphase framework recently developed to describe tumor growth and shows how the concept of evolving natural configurations can be applied to the specific problem. Some examples are then described according to the type of constitutive equation used, specifically focusing on contact inhibition of growth, nutrient-limited avascular growth, and interaction with the environment.

## 7.1 Introduction

The attempt to give a unified description of what a tumor is, unfortunately is still hopeless, both because there are several tumors with different origin and characteristics and because there are several concurrent causes of tumor development. Using probably a naive description, one can say that the cells forming a compact tumor, like other cells in the body, live in a watery environment full of proteins. These include all sorts of nutrients the cells need to survive and duplicate, and chemical factors, in particular growth promoting factors, growth inhibitory factors, and chemo-tactic factor, which trigger subcellular chemical pathways determining the behavior of the cell. The extracellular space is also filled with a network of cross-linked proteins (e.g. elastin, collagen, proteoglycans) collectively known as the extracellular matrix (ECM), which forms the structure of the tissue.

Both in a physiological situation and in a pathological one, the interactions that a cell has with its neighbors and with the extracellular matrix is very complex. In particular, focusing on the physiological behavior, cells pull on the extracellular matrix to move and want to be attached to it to duplicate. They like the growth factors and proteins embedded in the extracellular matrix and continuously remodel it by digesting part of the ECM or cleaving some of the constituents by the continuous production of matrix-degrading enzymes, for example, matrix metallo-proteinases (MMP). At the same time, some cells (in particular fibroblasts) rebuild the extracellular matrix. As described in the following, this process is affected by the stress applied to the tissue, as it can be easily understood by recalling bad experiences (one hopes not personal) such as the therapeutic action of braces and the traction applied to heal a fractured bone, or just the fact that exercise and physical training have a good effect on our body whereas prolonged rest is detrimental for both bones and muscles.

Cells also prefer to feel the presence of other cells of the same type, either by the transduction of specific chemical signals or by cell contact. If they feel lonely they commit suicide by a process called anoikis.

On the other hand, cells replicate if they sense that there is sufficient space for doing it or if they are chemically stimulated. Conversely, if they sense that there are a sufficient number of cells around, they can alter their activity and enter a quiescent state ready to reactivate their replication program if, for instance, a neighboring cell dies.

Most of the complex processes briefly sketched above are influenced by the production and reception of chemical signals. In most cases the behavior of a cell depends on the balance between two (or more) contradictory signals. For instance, mitosis can be stimulated by the overexpression of a growth-promoting signal or by the underexpression of a growth-inhibitory signal. An excessive presence of extracellular matrix can be caused by the excessive production of ECM, by a decreased production of matrix degrading enzymes (MDEs), or by an increased production of tissue inhibitors of metallo-proteinases (TIMPs), that is, the molecules that make MDEs ineffective, so that even if the production of MDEs is normal their effect is damped by TIMP upregulation.

The behavior of a cell also depends on the balance between chemical and mechanical inputs, so that a cell might give up duplication for the presence of growth-inhibitory factors in spite of the fact that there is space around, or vice versa might duplicate, stimulated by growth-promoting factors, in spite of the lack of space.

This is not a pathological situation. For instance, in wound healing the endothelial cells covering the wall of a capillary duplicate in response to the stimulus of vascular endothelial growth factors (VEGF) produced by hypoxic cells in spite of the absence of any mechanical stimulus. Cells then move toward the injury forming new capillaries to bring the materials necessary for healing the cut.

In this scenario, how then can a normal tissue generate a hyperplasia and then a tumor? The reasons can be disparate but, generally speaking, have to do with the occurrence of some failure in the complex mechanisms controlling the "circle of life," including the following.

- The cell becomes insensitive to growth inhibitory signals, for instance losing retinoblastoma suppressor.

- The cell produces growth-promoting signals, for instance activating H-Ras oncogene.

- The cell delays apoptosis (i.e. natural death), for instance producing IGF survival factors.

- The cell completely loses its program-to-death, becoming immortal.

- The cell acquires a limitless replicative potential by turning on telomerase.

- The cell becomes insensitive to mechanical cues, the so-called cadherin switch.

- The cell does not need to be properly attached to the ECM, the so-called integrin switch.

- The cell does not need to feel the presence of similar cells to survive.

The last two characteristics are linked to the diffusion of metastasis and the formation of secondary tumors.

The problem is then very complex. The majority of the models present in the literature focus on the chemical aspects of tumor growth, which is, however, fundamental, and are based on reaction–diffusion equations and mass balance equations with suitable closure for the velocity field. For a more detailed discussion on these aspects, the interested reader is referred to the books [ADa, CHd, PRa] and the special issues [BEb, CHb, CHc] specifically devoted to tumor modeling, and to the recent review articles [ARa, MAb] where even more references can be found. In fact, in this chapter we only focus on the mechanical aspects of tumor growth, which include the effect of stress on cell growth and apoptosis, the involvement of stress on the surrounding tissue, or simply the link between stress and deformation in deducing tumor growth models. Therefore, the effect of chemical factors plays a secondary role here, although we are well aware of their importance. Actually, even when focusing on the mechanical aspects of tumor growth, it is clear that the duplication or death of cells is chemically regulated inside the cell. So, at some stage, the mechanical signal has to be translated into a chemical message that goes to the nucleus and determines the behavior of the cell.

The main advantage of the introduction of such a mechanical framework is in the ability to deal with stress, with its influence on the evolution of the tissue itself, and with the mechanical interaction with other surrounding tissues. The main limitation is due to the fact that data on the response of multicell aggregates to traction and compression are not available yet for tumors, although similar studies have been done for other tissues (mainly bones and cartilage, but also brain, lungs, heart, skin, and so on). Furthermore, in order to use models with many constituents it is necessary to discriminate the stress contribution due to the different constituents. However, in our opinion, once the experiments are done the modeling framework has a great potential.

## 7.2   Mechanics and Mechanotransduction in Tumor Growth

### 7.2.1   Cadherin Switch

In normal tissue the rate of proliferation decreases when cells come in contact, a phenomenon often called contact inhibition of growth [DEb, DIa, KAa, NEa, POa, STa]. A quantification of this phenomenon is represented in Figure 7.1 which reports some experimental results by Tzukatani et al. [TZa] on human breast epithelial cells grown in vitro over a suitable substratum and by Orford et al. [ORa] on canine kidney-derived nontransformed epithelial cells.

It can be seen that after an initial exponential growth cell density saturates forming a monolayer of cells. On the other hand, "tumor" cells continue proliferating forming a multilayer leading to the conjecture that they need to feel more contact or larger pressure to stop their proliferation program. The starters of this growth control mechanism are the cadherins, the transmembrane receptors involved in homophilic cell–cell interactions, because of their crucial role in cell–cell adhesion and in mechanotrasduction [LEa, NEa, STa, STc, TAa, TZa, UGb] (see Figure 7.2).

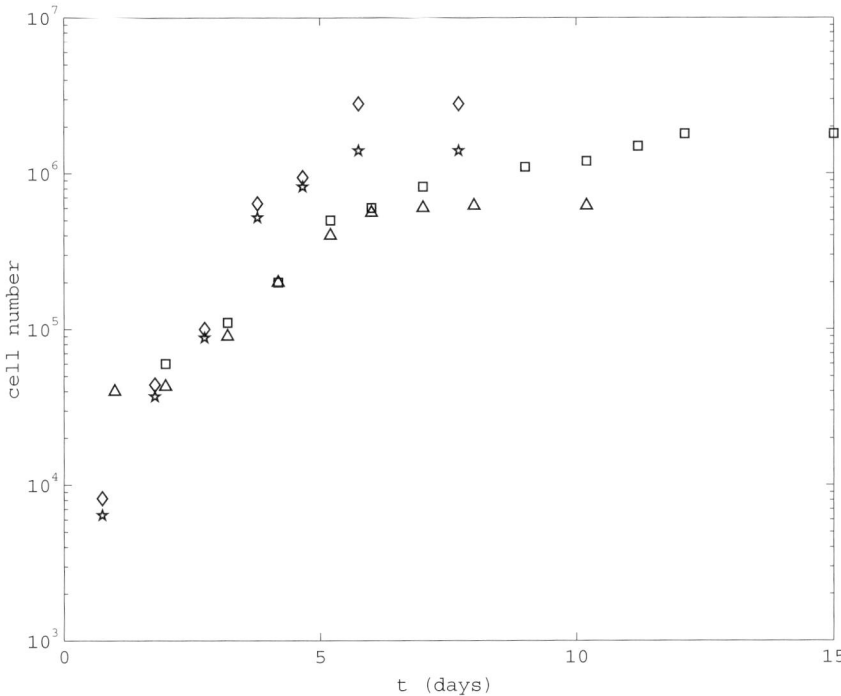

Figure 7.1. Examples of contact inhibition of growth reported in the experiments by Tzukatani et al. [TZa] using human breast epithelial cells (Squares = late passage and triangles = early passage) and by Orford et al. [ORa] using canine kidney-derived nontransformed epithelial cells (stars = wild type and diamonds = S37A mutant).

Their involvement has been checked in several ways. For instance, Warchol [WAa] spread synthetic beads coated with N-cadherin ligands over a substratum and seeded some cells on it. They then found that due to the interaction with the beads cells stopped duplicating. Similarly, Caveda et al. [CAe] found that coating the underlying substratum with the extracellular domain of recombinant VE-cadherin suppressed cell proliferation. Conversely, Castilla et al. [CAc] found that the disruption of the intercellular cadherin junctions triggers the production of growth factors that contribute to induce proliferation.

On the basis of these observations, it is clear that if a cell is not so sensitive to the control mechanisms above it is subject to deregulated growth, a phenomenon that is considered such an important milestone in the development of tumors to deserve to be named "cadherin switch" in analogy with the "angiogenic switch" leading to the vascularization of tumors that is briefly described in Section 7.2.4.

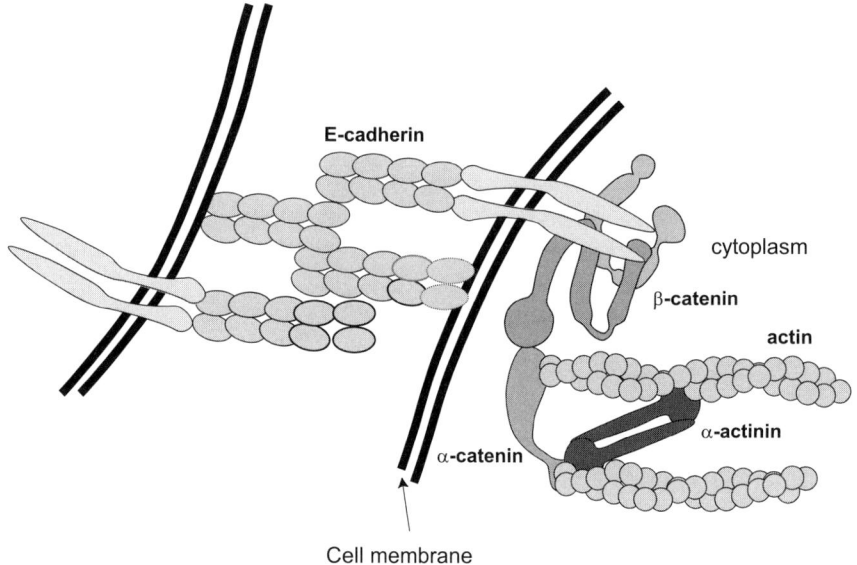

Figure 7.2. Cadherin-cadherin junction.

In fact, it is known that loss of contact responsiveness is commonly associated with the formation of hyperplasia and malignant transformation such as gastric carcinoma [BEa, ODa], adenocarcinoma [TZb], epithelial tumors [CAd, CHg], colon polyps and carcinoma [GOa], gynecological cancers [RIa], and intimal thickening [UGa] (see also the review by Harja and Fearon [HAa]).

However, cadherins only represent the tip of the iceberg. They are more visible than other hidden players for their transmembrane location, but there are many other candidates that can be responsible for a possible incorrect mechanotransduction. The second family of suspects is the catenins, the proteins cadherin link to for a functional cell-to-cell adhesion (see Figure 7.2). In fact, Stockinger et al. [STc] showed that epithelial cells exhibited a strong $\beta$-catenin activity at low densities ($\leq 40\%$ confluency), which was five- to sevenfold reduced when cells reached a confluency $> 80\%$. In fact, it is taught that in physiological conditions upon reaching confluency the expressed cadherins sequester catenins, downregulating their activity. Because it is known that the upregulation of catenins is necessary for cell duplication, as we show in the following and is sketched in Figure 7.3, the final result is that cell adhesion negatively affects cell proliferation.

To test the link between cadherins and catenins Caveda et al. [CAe] transfected Chinese hampster ovary cells with a cytoplasmic truncated

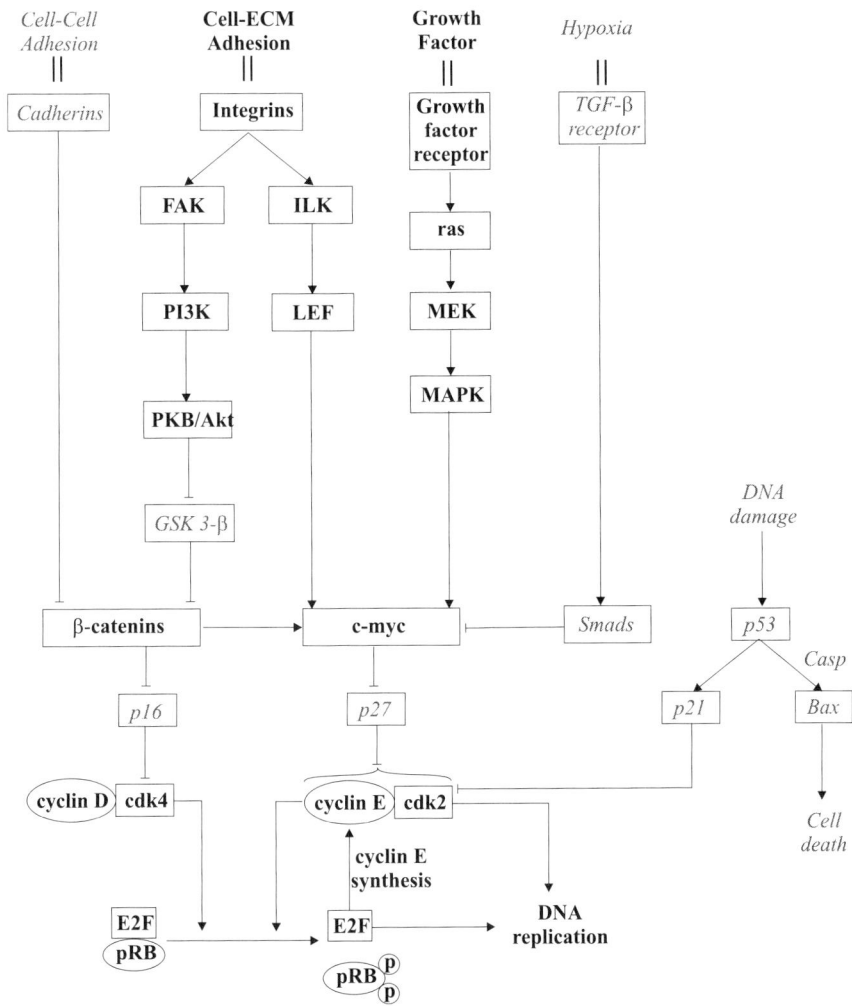

Figure 7.3. Sketch of some protein cascades involved in the cell cycle and in particular of those involving cadherins, integrins, and catenins. In order to put in evidence the on–off mechanisms, arrows, corresponding to stimulatory activities, connect proteins with the same fonts, and blockades, corresponding to inhibitory activities, connect proteins written using different fonts. At the end, forgetting the details of the cascades, one can, for instance, extract that cell–cell adhesion inhibits proliferation. cdk stands for cyclin-dependent kinase, pRB for hypophosphorilated retinoblastoma, and the added **p**s indicate its phosphorylation. The curly bracket indicates that p27 binds to the cdk2/cyclin complex.

mutant of VE-cadherin. They found that the deletion of the cytoplasmic tail of VE-cadherin abolishes its growth inhibitory activity without
affecting its adhesive properties.

More in detail, Dietrick et al. [DEb] explain the mechanism of contact
inhibition of growth as follows.

- Tissue compression and overexpression of cadherins cause the underexpression of catenins.

- The underexpression of catenins determines the accumulation of the
  cyclin-dependent kinase (cdk) inhibitors p16, p21, and p27.

- Their overexpression inhibits the entry in the S phase causing cell
  cycle arrest in the G1 phase [COa, KAa, POa]. More in detail, referring to Figure 7.3,
  - p16 blocks the activity of cdk4 by dissociating cyclin D from
    cdk4 and binding to cdk4;
  - p27 inhibits cdk2-cyclin E activity directly by binding to the
    complex.

In Figure 7.3, the process above is schematized by reading the bold words
as underexpressed proteins. Conversely, reading them as overexpressed
quantities one has, for instance, that upregulation of catenins leads to the
expression of cyclin-dependent kinase and then to DNA replication and
mitosis. This procedure is particularly useful for linear cascades, although
it may fail in the presence of feedback loops.

In order to fully exploit the protein cascade in tumor modeling, one
should have all the affinity constants and reaction rates, which at present
is not the case. In addition, the spatial localization of the proteins
involved in the cascade should also be taken into account. So, at present,
the way generally used in the literature is to proceed, whenever possible, with a spatially homogeneous Boolean reasoning, which means using on/off relationships. For this reason in Figure 7.3 we tried to highlight the overexpressed and underexpressed proteins by using different
fonts.

### 7.2.2  Interaction with the Extracellular Matrix and Integrin Switch

Another main component of both normal and tumor tissues is the extracellular matrix (ECM), a fibrous structure composed of many constituents
produced by a variety of stromal cells, mainly fibroblasts. The ECM is
constantly renewed through the concomitant production of matrix metalloproteinases (MMP) and new ECM components.

In stationary conditions the remodeling of ECM is a slow process. For instance, in the human lung the physiological turnover of ECM is 10–15% per day [J0a], which leads to an estimated complete turnover in a period of nearly a week. However, when a new tissue has to be formed (e.g. to repair a wound), then the rate of production is one or two orders of magnitude faster [CHf, DEa]. Hence, it seems that the production of ECM constituents is also affected by the pressure felt by the cells. However, this relation is rather complicated. It is, for instance, well known that for bones, teeth, and muscles [KIa, KJa, MAc] the remodeling process is strongly affected by the stresses and strains to which the tissue is subject. This is a physiologically functional process because it allows keeping the stroma young and reactive. In fact, prolonged rest or space flight are detrimental to bones and muscles, whereas exercise and physical training have an opposite effect.

The percentage of ECM content changes considerably from tissue to tissue (see Table 7.1), from normal to tumor tissues, and also within the same tumor with tumor progression (see [ZHa]). For instance, Takeuchi et al. [TAb] found that breast tumors presented a denser and more fibrous stroma with several differences in the chemical composition. On the other hand, it is well known that the first hints of the possible presence of breast nodules are obtained by palpating the breast and feeling stiffer regions

| Tissue [Source] | Water (%) | Collagen (%) | Proteo-glycans (%) | Elastin (%) |
|---|---|---|---|---|
| Intervertebral disk | | | | |
| Nucleus pulposus | 80 | 5 | 13 | |
| Annulus fibrosus [EYa] | 65 | 23 | 7 | |
| Aorta [MOa] | 65 | 9 | 1 | 16 |
| Articular cartilage (femoral) [MOa] | 80 | 13 | 5 | Trace |
| Corneal stroma [MAe, TSa] | 77 | 16 | 1 | |
| Ligament (cruciate) [WOa] | 68 | 25 | 1 | <1.6 |
| Meniscus [MOe] | 74 | 23 | 1 | Trace |
| Skeletal muscle [LE] | | 14 | | |
| Tendon [K0a, WOa] | 55–70 | 25 | 0.5 | <1 |
| Skin [WOa] | 60 | 26 | 1 | 2–4 |
| Subcutaneous tissue [LEb] | | 21 | | |
| Prostate cancer [ZHa] | | 7–26 | | |

Table 7.1. Constituents of several tissues. Notice the strong variability in the collagen content of prostate cancer which depends on the grade of the tumor.

| Tissue | Elastic Modulus (Pa) |
|---|---|
| Normal mammary gland | $167 \pm 31$ |
| Average breast tumor | $4049 \pm 938$ |
| Stroma attached to tumor | $916 \pm 269$ |
| Reconstituted basement membrane | $175 \pm 37$ |
| Collagen (2.0 mg/ml) | $328 \pm 87$ |
| Collagen (4.0 mg/ml) | $1589 \pm 380$ |

Table 7.2. Examples of elastic moduli of normal and abnormal breast tissue and stroma (data from [PAc]).

(see Table 7.2). Increased presence of ECM was also observed in other pathologies such as cardiac hyperthrophy, intima hyperplasia, cardiac fibrosis, liver fibrosis, pulmonary fibrosis, asthma, glomerulonephritis, and colon cancer [BRc, J0a, MAa, PUa].

The alteration in the ECM composition can be due to several probably concurring reasons.

- Increased synthesis of ECM proteins

- Decreased activity of matrix degrading enzymes (MDEs)

- Upregulation of tissue-specific inhibitors of metalloproteinases (TIMPs)

On the other hand, excessive degradation of ECM due to excessive production of MMP-13 characterizes chronic inflammatory diseases such as osteoarthritic cartilage, rheumatoid synovium, chronic ulcer, intestinal ulcerations, periodontitis, and many malignant tumors [YAa].

The interaction between ECM and cells is very important because cells need to properly adhere in order to survive. They only duplicate if they are anchored to the ECM. As shown in Figure 7.3, the mechanotransduction cascade is mainly activated by integrins.

On the other hand, in the process of invasion and formation of metastases tumor cells detach from the original site, invade the surrounding tissue, intravasate entering the blood or lymphatic system, and extravasate to reach a secondary site. It is then clear that the formation and diffusion of mestatases require that cells acquire the ability of surviving without interacting with the ECM. In fact, as for cadherins, it is found that tumors have altered integrins. This, in turn, alters the downstream integrin signaling pathway, so that one could argue that there is an integrin switch in addition to the mentioned cadherin and angiogenic switches.

Actually, Paszek et al. [PAc] prove that through the integrin signaling pathway the stiffness of the ECM promote malignant behavior consisting in growth enhancement and loss of tissue polarity which, for instance, leads to the absence of lumen formation in ductal carcinoma and the formation of hyperplasia, the first step toward tumorigenesis.

### 7.2.3 Nutrient-Limited Growth and Tumor Structure

In order to give a more complete picture of the dynamics of tumor growth we need to mention some important effects that are related to the contribution of nutrients and vascularization in tumor growth. In particular, the former has to do with the existence of a nutrient-limited dimension of the tumor. We do not enter into detail because nutrient-limited growth has little to do with mechanics, which is the focus of this chapter and refer the reader to [ARa] for a recent descriptive and detailed review.

Generally speaking, tissues receive vital nutrients and oxygen perfusing through the vessel wall and diffusing in the extracellular space. When tumor cells cluster in a tissue forming a multicellular spheroid they receive their nutrients through the boundary of the tumor. Nutrients then diffuse toward the center of the tumor. When the tumor is small, all cells are well-nourished and proliferate rapidly. As the colony increases in size, due to the strong metabolic activity characterizing tumor cells, the cells toward the center are progressively starved of oxygen and nutrients and, as a consequence, their proliferation rate decreases. If the oxygen concentration falls below a critical threshold value then the cells are unable to survive and undergo cell death generating a necrotic core.

Eventually avascular tumors will reach an equilibrium size ($\sim$2 mm in diameter [FOb]), at which the rates of cell proliferation and apoptosis, averaged over the tumor volume, balance. At this stage the tumor typically comprises an outer rim of proliferating cells, a central core of necrotic debris, and an intermediate region of quiescent cells which are alive, but do not proliferate due to nutrient deprivation [SUa].

### 7.2.4 Angiogenic Switch

The switch from the slow and relatively harmless avascular growth phase described above to the rapid and life-threatening vascular growth phase occurs during a process termed angiogenesis [CAb, CrRa, FOb]. We briefly describe them in the following, although the interested reader can find more information on the process in the recent reviews by Bussolino et al. [BUa] and by Mantzaris et al. [MAb].

It is possible to divide the angiogenic process into the following well-differentiated stages which sometimes partially overlap.

1. Due to the lack of oxygen and nutrients certain tumor cells secrete a range of diffusible proteins and chemicals that are known collectively

as tumor angiogenic factors (TAFs), in particular, vascular endothelial growth factors (VEGF).

2. The reception of TAFs causes a loss of interconnection between the endothelial cells that line the blood vessels, and a reduction of vascular tonus. This, in particular, induces an increase in the vessel permeability.

3. TAFs also stimulate the endothelial cells to release some proteolitic enzymes (serine-proteins, iron-proteins) that degrade the basement membrane surrounding the capillary facilitating cell movement, to proliferate, and to migrate chemotactically (i.e. up the TAF gradient), toward the source of angiogenic stimulus, that is, the tumor.

4. Capillary sprouts then form by the accumulation of endothelial cells. The stage of differentiation is characterized by the exit of the endothelial cells from the cell cycle and by their capacity of surviving in suboptimal conditions and of building themselves primitive capillary structures, not yet physiologically active.

5. When capillary tips come into close proximity, they fuse together by a process called anastomosis, forming closed loops through which blood may circulate. Secondary sprouts emanate from the new loops and so the process continues, with increasing numbers of capillary tips being formed, until the new vessels penetrate the tumor.

6. In the stage of maturation, the newborn vessel is completed by the formation of new extracellular matrix and by the arrival of other cells named pericytes and sometimes of flat muscle cells. During this phase a major role is played by some molecules called angiopoietins leading to the development of the simple endothelial tubes into a more elaborate vascular tree composed of several cell types. In fact, they contribute to the maintenance of vessel integrity through the establishment of appropriate cell–cell and cell–matrix connections.

7. After the formation of the vascular network, a remodeling process starts. This involves the loss of some physiologically useless capillaries and the remodeling of the extracellular matrix. Shear stress and pressure inside the vessel are the most important drivers of the remodelling process.

At this point the tumor receives much more nutrient and the tumor cells become very aggressive with their mitotic rate increasing considerably, leading to much faster growth.

It has to be mentioned that mechanics has an important role not only in the formation and regression of the blood vessel, but also on the interaction with the outer environment. For instance, growing around the vessel which is usually immature (Step 5 above), the tumor compresses it and might cause its collapse. In turn this makes the surrounding tissue hypoxic and leads to new stimulation of the angiogenic pathway above, so that the formation of new vessels and tumor growth is subject to cyclic behavior.

## 7.3 Multiphase Models

Referring to Araujo and McElwain [ARa] for a recent review, we here recall that the first models dealing with avascular tumor growth worked under the hypothesis that the tumor is made by only one type of cells occupying a constant volume ratio $\bar{\phi}_T$; for example, they fill the space as a bunch of rigid spheres in a close-packed configuration, so that the mass balance equation

$$\rho \left[ \frac{\partial \phi_T}{\partial t} + \nabla \cdot (\phi_T \mathbf{v}_T) \right] = \rho \Gamma_T , \qquad (3.1)$$

where $\rho$ can be taken as the constant density of water, and $\phi_T$ is the volume ratio occupied by tumor cells, can be written as

$$\bar{\phi}_T \nabla \cdot \mathbf{v}_T = \Gamma_T, \qquad (3.2)$$

where $\Gamma_T$ is a function of the concentration of nutrients and of a plethora of important chemical factors that are diluted in the extracellular liquid surrounding the cells and influence all vital functions of the tumor.

Enforcing a symmetry condition, usually spherical symmetry, allows the reduction of the number of space variables to one and the velocity vector to a scalar, so that one can directly integrate (3.2) to have the velocity in any point

$$v_T(r,t) = \frac{1}{r^2 \bar{\phi}_T} \int_0^r \Gamma_T(x,t) x^2 \, dx, \qquad (3.3)$$

and, in particular, the evolution of the free border of the tumor

$$\frac{dR}{dt}(t) = v_T(R(t),t) = \frac{1}{R^2(t)\bar{\phi}_T} \int_0^{R(t)} \Gamma_T(x,t) x^2 \, dx. \qquad (3.4)$$

The core of these types of models consisted then in describing how the growth term $\Gamma_T$ depends on the chemical factors and nutrients diffusing in the environment. The evolution of the tumor border then is a byproduct of the geometrical reasoning. In fact, Eq. (3.4) corresponds to a global mass balance on the tumor mass that determines how the tumor grows, without resorting to any force balance. In particular, the nutrient-limited radius can possibly be obtained by solving

$$\int_0^{R(t)} \Gamma_T(x,t)x^2\, dx = 0\,, \tag{3.5}$$

which is related to a global balance between proliferation in the outer rim and death in the core. A similar consideration holds when including the existence of quiescent and necrotic regions in the tumor.

At this point, if the tumor is immersed in a homogeneous environment with known mechanical properties, then by knowing the (radial) displacement of the tumor border, one could compute the stress in the surrounding tissue. As examined in [AMd], there were some difficulties in generalizing this method to three-dimensional problems and to problems involving more populations. For this reasons, some years ago several authors (see, for instance, [BRa, BRb, BYb, FRa, FRb, FRc, JAa]) started linking, in the simplest possible way, motion to stress describing the tumor as a deformable porous material. Recently, the multiphase approach was described by Araujo and McElwain [ARb] for a general mixture of $n$ constituents. In the following we focus on specific applications referring to [AMd] and [ARb] for the general case.

### 7.3.1  A Basic Triphasic Model: ECM, Tumor Cells, and Extracellular Liquid

As discussed above a tumor is made of at least three main constituents occupying a relevant percentage of space: tumor cells, extracellular matrix (ECM), and extracellular liquid. In addition one should consider the nutrients and chemical factors diffusing in the liquid and absorbed/produced by the cells. However, in this section we do not focus on them but on the constituents filling the available space and present a very general triphasic model. After that, we consider some special cases that are applied in the following section to describe specific phenomena.

The starting point is to write the mass balance equations for the constituents

$$\frac{\partial \phi_0}{\partial t} + \nabla \cdot (\phi_0 \mathbf{v}_0) = \Gamma_0\,, \tag{3.6}$$

$$\frac{\partial \phi_T}{\partial t} + \nabla \cdot (\phi_T \mathbf{v}_T) = \Gamma_T\,, \tag{3.7}$$

$$\frac{\partial \phi_\ell}{\partial t} + \nabla \cdot (\phi_\ell \mathbf{v}_\ell) = \Gamma_\ell\,, \tag{3.8}$$

where $\phi_0$, $\phi_T$, and $\phi_\ell$ are the volume ratios occupied by ECM, tumor cells, and extracellular liquid, respectively, and $\mathbf{v}_0$, $\mathbf{v}_T$, and $\mathbf{v}_\ell$ are the relative velocities.

The saturation assumption implies that

$$\phi_0 + \phi_T + \phi_\ell = 1 \,, \tag{3.9}$$

and if the mixture is closed, then the growth terms satisfy

$$\Gamma_0 + \Gamma_T + \Gamma_\ell = 0 \,. \tag{3.10}$$

The momentum balance equations per constituent are written

$$\rho\phi_0 \left( \frac{\partial \mathbf{v}_0}{\partial t} + \mathbf{v}_0 \cdot \nabla \mathbf{v}_0 \right) = \nabla \cdot \mathbf{T}_0 + \mathbf{b}_0 + \mathbf{m}_0^\sigma \,, \tag{3.11}$$

$$\rho\phi_T \left( \frac{\partial \mathbf{v}_T}{\partial t} + \mathbf{v}_T \cdot \nabla \mathbf{v}_T \right) = \nabla \cdot \mathbf{T}_T + \mathbf{b}_T + \mathbf{m}_T^\sigma \,, \tag{3.12}$$

$$\rho\phi_\ell \left( \frac{\partial \mathbf{v}_\ell}{\partial t} + \mathbf{v}_\ell \cdot \nabla \mathbf{v}_\ell \right) = \nabla \cdot \mathbf{T}_\ell + \mathbf{b}_\ell + \mathbf{m}_\ell^\sigma \,, \tag{3.13}$$

where $\mathbf{T}_i$ is the partial stress tensor, $\mathbf{b}_i$ is the body force, and $\mathbf{m}_i^\sigma$ is the interaction force acting on the $i$th constituent due to its interaction with the other constituents.

Before proceeding we recall that in a Lagrangian frame of reference (related to the constituent) the mass balance equations for the tumor cells can be written as

$$\frac{d}{dt}(\phi_T J_T) = \Gamma_T J_T \,, \tag{3.14}$$

where $J_T = \det \mathbf{F}_T$ and $\mathbf{F}_T$ is the deformation gradient relative to the tumor constituent. Similar relations hold for the other constituents. The meaning of Lagrangian is discussed in the following.

The main contribution to the interaction forces can be assumed to be proportional to the velocity difference between the constituents. Compatibly with thermodynamics the saturation assumption implies the existence of a Lagrangian multiplier $P$ which is then identified with the extracellular liquid pressure in the constitutive equations, so that one can write

$$\mathbf{m}_\ell^\sigma = P\nabla\phi_\ell - \mathbf{M}_{\ell T}(\mathbf{v}_\ell - \mathbf{v}_T) - \mathbf{M}_{\ell 0}(\mathbf{v}_\ell - \mathbf{v}_0)$$
$$- \frac{\Gamma_\ell}{2}\mathbf{v}_\ell + \frac{\Gamma_\ell - \Gamma_T}{6}\mathbf{v}_T + \frac{\Gamma_\ell - \Gamma_0}{6}\mathbf{v}_0 \,,$$

$$\mathbf{m}_T^\sigma = P\nabla\phi_T - \mathbf{M}_{\ell T}(\mathbf{v}_T - \mathbf{v}_\ell) - \mathbf{M}_{T0}(\mathbf{v}_T - \mathbf{v}_0)$$
$$- \frac{\Gamma_T}{2}\mathbf{v}_T + \frac{\Gamma_T - \Gamma_\ell}{6}\mathbf{v}_\ell + \frac{\Gamma_T - \Gamma_0}{6}\mathbf{v}_0 \,, \tag{3.15}$$

$$\mathbf{m}_0^\sigma - P\nabla\phi_0 - \mathbf{M}_{T0}(\mathbf{v}_0 - \mathbf{v}_T) - \mathbf{M}_{\ell 0}(\mathbf{v}_0 - \mathbf{v}_\ell)$$
$$- \frac{\Gamma_0}{2}\mathbf{v}_0 + \frac{\Gamma_0 - \Gamma_\ell}{6}\mathbf{v}_\ell + \frac{\Gamma_0 - \Gamma_T}{6}\mathbf{v}_T \,,$$

where $\mathbf{M_{ij}}$ refers to the interaction between the $i$th and the $j$th constituent and

$$\mathbf{T}_\ell = -(\mathrm{P}\phi_\ell)\mathbf{I} + \hat{\mathbf{T}}_\ell \,,$$

$$\mathbf{T}_T = -(\mathrm{P}\phi_T)\mathbf{I} + \hat{\mathbf{T}}_T \,, \tag{3.16}$$

$$\mathbf{T}_0 = -(\mathrm{P}\phi_0)\mathbf{I} + \hat{\mathbf{T}}_0 \,,$$

where $\hat{\mathbf{T}}_i$ is named excess stresses. The terms in (3.15) proportional to the mass production rates $\Gamma_i$ are, however, negligible, as discussed in [PRb].

In biological phenomena inertia can be neglected and also the interaction force between the extracellular matrix and the liquid is negligible (because in most cases the ECM has a fibrous structure filling a moderate amount of space) with respect to the interaction force between cell and liquid and above all cell and ECM. However, this last assumption is not essential and can be dropped.

In addition, as a first approximation we consider the ECM as rigid. Under these assumptions one can simplify (3.6)–(3.8) and (3.11)–(3.13) writing

$$\begin{cases} \dfrac{\partial \phi_0}{\partial t} = \Gamma_0 \,, \\[2mm] \dfrac{\partial \phi_T}{\partial t} + \nabla \cdot (\phi_T \mathbf{v}_T) = \Gamma_T \,, \\[2mm] \dfrac{\partial \phi_\ell}{\partial t} + \nabla \cdot (\phi_\ell \mathbf{v}_\ell) = \Gamma_\ell \,, \\[2mm] \mathbf{0} = -\phi_T \nabla P + \nabla \cdot \hat{\mathbf{T}}_T + \mathbf{b}_T + \mathbf{M}_{\ell T}(\mathbf{v}_\ell - \mathbf{v}_T) - \mathbf{M}_{T0}\mathbf{v}_T \,, \\[2mm] \mathbf{0} = -\phi_\ell \nabla P + \nabla \cdot \hat{\mathbf{T}}_\ell - \mathbf{M}_{\ell T}(\mathbf{v}_\ell - \mathbf{v}_T) \,, \end{cases} \tag{3.17}$$

because the rigidity assumption implies that the stress tensor $\mathbf{T}_0$ simply reacts to the forces applied to the ECM. The body force $\mathbf{b}_T$ relates, for instance, to chemotactic or haptotactic action on the tumor cells, and $\mathbf{b}_\ell$ is assumed to vanish.

If as usual in the porous media model it is assumed that $\hat{\mathbf{T}}_\ell = \mathbf{0}$, the last equation gives rise to Darcy's law and can then more familiarly be written as

$$\mathbf{v}_\ell - \mathbf{v}_T = -\mathbf{K}\nabla P \,, \tag{3.18}$$

where $\mathbf{K}$ is related to the permeability and is a function of the liquid volume ratio.

In order to eliminate the interaction force between the liquid and the cells, it might be convenient to add the two momentum equations to obtain

$$-(1 - \phi_0)\nabla P + \nabla \cdot \hat{\mathbf{T}}_T - \mathbf{K}_0^{-1}\mathbf{v}_T + \mathbf{b}_T = \mathbf{0}, \qquad (3.19)$$

where $\mathbf{K}_0 = \mathbf{M}_{T0}^{-1}$ is related to the permeability of the sticky granular flow in the porous structure constituted by the ECM network. This equation can then take the place of the first momentum equation in (3.17).

Hence the basic model can be written as

$$\begin{cases} \dfrac{\partial \phi_0}{\partial t} = \Gamma_0, \\[2mm] \dfrac{\partial \phi_T}{\partial t} + \nabla \cdot (\phi_T \mathbf{v}_T) = \Gamma_T, \\[2mm] \nabla \cdot (\phi_T \mathbf{v}_T + \phi_\ell \mathbf{v}_\ell) = 0, \\[2mm] \mathbf{v}_\ell - \mathbf{v}_T = -\mathbf{K}\nabla P, \\[2mm] \mathbf{v}_T = \mathbf{K}_0 \left[ -(1 - \phi_0)\nabla P + \nabla \cdot \hat{\mathbf{T}}_T + \mathbf{b}_T \right], \end{cases} \qquad (3.20)$$

where $\phi_\ell = 1 - \phi_0 - \phi_T$ and (3.20)$_3$ was obtained by summing the mass balance equations and using (3.9) and (3.10).

## Limit Case: Neglecting the Mechanical Interaction with the Extracellular Liquid

Consider the case in which the permeability tensor is isotropic. If, for the sake of simplicity, $K_0 \gg K$, as is plausible, by substituting the gradient of pressure from Darcy's law (3.18) to (3.19), it can be readily realized that as a first approximation

$$\mathbf{v}_T = K_0 \left( \nabla \cdot \hat{\mathbf{T}}_T + \mathbf{b}_T \right). \qquad (3.21)$$

We notice that if $\mathbf{b}_T$ is proportional to the gradient of some chemical concentration

$$\mathbf{b}_T = \chi \nabla c, \qquad (3.22)$$

and the partial stress tensor is neglected, Eq.(3.21) implies the usual chemotactic closure

$$\mathbf{v}_T = w \nabla c, \qquad (3.23)$$

where $w = K_0 \chi$.

In particular, one has the classical chemotactic models

$$\frac{\partial \phi_T}{\partial t} + \nabla \cdot (w \phi_T \nabla c) = \Gamma_T. \qquad (3.24)$$

Chemotaxis can then be conceived as a force balanced by the drag force exerted by the substratum and not as a convenient closure of the mass balance equation.

It is well known that Eq. (3.24) with the concentration either given or evolving according to a typical reaction–diffusion equation may be characterized by a solution that blows up in finite time. On the other hand, Kowalczyk [KOa] showed that if mechanics is properly accounted for, that is, if (3.21) is used with a suitable constitutive equation for the stress, the blowup of the solution is prevented. For instance, it is enough to assume that the ensemble of cells behaves as an elastic fluid with a convex pressure–volume ratio dependence.

In the case of more chemotactic/haptotactic effects, one has the model

$$
\begin{cases}
\dfrac{\partial \phi_0}{\partial t} = \Gamma_0 \,, \\[2mm]
\dfrac{\partial \phi_T}{\partial t} + \nabla \cdot \left[ K_0 \phi_T \left( \nabla \cdot \hat{\mathbf{T}}_T + \sum_i \chi_i \nabla c_i \right) \right] = \Gamma_T \,,
\end{cases}
\tag{3.25}
$$

which, of course, has to be associated with suitable reaction–diffusion equations for the chemical factors $c_i$.

The first equation describes the possible deposition or degradation of the extracellular matrix. If the stress tensor is isotropic, $\hat{\mathbf{T}}_T = -\Sigma(\phi_T)\mathbf{I}$, then $(3.25)_2$ may be rewritten as

$$
\frac{\partial \phi_T}{\partial t} + \nabla \cdot \left( \sum_i w_i \phi_T \nabla c_i \right) = \nabla \cdot (K_0 \phi_T \Sigma'(\phi_T) \nabla \phi_T) + \Gamma_T \,,
\tag{3.26}
$$

where $\Sigma'$ is the derivative of $\Sigma$ with respect to the volume ratio $\phi_T$ and $w_i = K_0 \chi_i$.

## Limit Case: Constant ECM

If we assume now that the amount $\phi_0$ of ECM is maintained constant in the system, the first equation in (3.20) can be dropped. Substituting then $\mathbf{v}_\ell$ from Darcy's law in the third equation one has that

$$
\nabla \cdot ((1 - \phi_0)\mathbf{v}_T - \phi_\ell \mathbf{K} \nabla \mathbf{P}) = 0 \,,
\tag{3.27}
$$

$$
\nabla \cdot ((1 - \phi_0)\mathbf{v}_\ell + \phi_T \mathbf{K} \nabla \mathbf{P}) = 0 \,.
\tag{3.28}
$$

In one-dimensional problems, this implies that cells move up the pressure gradient, and the extracellular liquid moves in the opposite direction, which is in agreement with the experimental results by Dorie et al. [DOa, DOb] on the internalization of cells.

As the interstitial pressure is higher inside the tumor than at its outer boundary, cells move toward the center of the tumor and the extracellular liquid flows toward the boundary. A recirculation flow then forms: tumor cell near the center die due to nutrient deprivation and generate reusable extracellular fluid. This liquid flows to the boundary where it is taken up by proliferating cells, that are then internalized in the tumor.

In particular, in the limit of a negligible amount of ECM, then the interaction term with the ECM drops and we can simplify the last equation in (3.20) as

$$-\nabla P + \nabla \cdot \hat{\mathbf{T}}_T + \mathbf{b}_T = \mathbf{0}, \qquad (3.29)$$

which for particular constitutive equations will lead to the model proposed in Section 7.4.2.

## 7.4 Constitutive Equations

As usual, the modeling procedure above needs the specification of the constitutive equations describing the mechanical response to strain. This is not a standard step in this case because tumor cells are generated and die during the evolution. There is then a difficulty in defining a reference configuration and in using a Lagrangian coordinate system. In particular, the meaning of deformation also loses the immediate meaning it had in classical continuum mechanics when dealing with inert matter. In fact, when dealing with a growing tumor, with respect to what should we measure deformations? The material is always changing. That is why the concept of evolving natural configuration described several times in this book becomes very helpful.

Of course, the problem is circumvented if one can model the tumor as a fluid, because in this case it is possible to use an Eulerian approach. This is what was done in the first models developed in a multiphase framework. An issue to keep in mind when formulating constitutive equations for living tissues and in particular tumors is what can be actually measured by biologists. In fact, testing the mechanical behavior of living tissues is much more difficult than for inert matter, and has not been done yet for tumors. In the following we present some examples of models of tumor growth using different constitutive equations.

### 7.4.1 Elastic Fluid: An Example Describing Contact Inhibition of Growth

Referring to Section 7.2.1, in this section we focus on the fact that when cells are in a crowded environment they sense the presence of other cells and their behavior then crucially depends on how they can stand the pressure (see Figure 7.1). We then focus on how this can affect both mitosis and production of extracellular matrix and matrix-degrading enzymes and in particular how a misperception of the compression state can generate hyperplasia, fibrosis, and tumor lesions.

Of course, we are well aware that cellular mechanotrasduction is not the only cause of formation of hyperplasia and tumors and that chemical factors will operate to regulate the reproduction rates. However, we recall that the aim of this chapter is to describe the mechanical aspects of tumor growth and therefore we focus on what happens when the only thing that transforms a normal cell into an abnormal cell is how it senses and responds to the stress exerted on it.

Here the stress and therefore its influence on the evolution of the cell population occurs through three contributions: cell replication, the production of extracellular matrix, and the release of matrix-degrading enzymes.

The main aim is to show how an underestimation of the compression state of the local tissue and then of the subsequent stress which is exerted on a cell, can generate by itself a clonal advantage on the surrounding cells leading to the replacement and the invasion of the healthy tissue. In order to do that the model in [CHe] focuses on the evolution of

- The volume ratio $\phi_n$ occupied by normal cells

- The volume ratio $\phi_T$ occupied by tumor cells or more precisely by abnormal cells that give rise to hyperplasia and dysplasia

- The volume ratio $\phi_m$ occupied by host extracellular matrix

- The volume ratio $\phi_M$ occupied by extracellular matrix produced by tumor cells, which is known to be structurally and chemically different from that produced by normal cells

- The concentration $c$ of matrix-degrading enzymes, for example, the plasminogen activators or matrix metalloproteinases

In the following an important role is played by the overall volume ratio occupied by cells and extracellular matrix

$$\psi = \phi_n + \phi_T + \phi_m + \phi_M \tag{4.1}$$

and by its relation with the stress.

We assume that what makes the difference between a normal and a tumor cell stays in the growth term and in its dependence on the stress level. Generalizing Eq. (3.25) to the populations above, one can then write

$$
\begin{cases}
\dfrac{\partial \phi_m}{\partial t} = \Gamma_m, \\[2mm]
\dfrac{\partial \phi_M}{\partial t} = \Gamma_M, \\[2mm]
\dfrac{\partial \phi_n}{\partial t} + \nabla \cdot \left[ K_{0n}\phi_n \left( \nabla \cdot \hat{\mathbf{T}} + \chi_{nm}\nabla\phi_m + \chi_{nM}\nabla\phi_M \right) \right] = \Gamma_n, \\[2mm]
\dfrac{\partial \phi_T}{\partial t} + \nabla \cdot \left[ K_{0T}\phi_T \left( \nabla \cdot \hat{\mathbf{T}} + \chi_{Tm}\nabla\phi_m + \chi_{TM}\nabla\phi_M \right) \right] = \Gamma_T,
\end{cases}
\tag{4.2}
$$

In [CHe] the simplest constitutive equation is used and the ensemble of cell is described as an elastic fluid

$$
\hat{\mathbf{T}} = -\Sigma \mathbf{I}, \tag{4.3}
$$

where $\Sigma$ vanishes below a value $\psi_0$, is increasing for $\psi > \psi_0$, and tends to infinity at $\psi = 1$. Of course, treating the multicellular spheroid as a viscous fluid as shown in the following section would be easy to do, make the model closer to reality, and confer more stability to the solution.

Haptotactic effects are expressed in the last two term on the left-hand side of Eqs. (4.2)$_3$ and (4.2)$_4$, which, with respect to the existing literature, take into account the different behavior of cells in the presence of the two kinds of ECM. In fact, cells preferentially adhere to normal ECM, whereas the ECM produced by the tumor cells favors their motility (and also their proliferation, although here this aspect is neglected as explained above). Hence, the coefficients $\chi_{ij}$ and $K_{0j}$ are allowed to be different and to depend on the overall volume ratio $\psi$ defined in (4.1) and on the percentage of extracellular matrix $\psi_m = \phi_m + \phi_M$. There is, in fact, an optimal concentration of ECM. Motility decreases both at smaller contents of ECM because of the lack of substratum to move on and at larger contents of ECM because of the increase of adhesive sites (see [HIa, PAa]). In addition, it vanishes for high values of the overall volume ratio $\psi$ because of the occupation of space by the cells and by the ECM.

Contact inhibition of growth is modeled by [CHe] assuming that mitosis stops when the volume ratio (or the compression) overcomes a given threshold. Their assumption is that the threshold value for a tumor is slightly larger than the physiological one. Actually, it may even tend to infinity, meaning that the cells are completely insensitive to compression and continue replicating independently of the compression level.

Of course, the growth terms depend on other quantities, such as the amount of nutrient and growth factors, but, as already stated, in this section we only focus on the possible role of stress in tumor invasion. In doing this we tacitly assume that all the constituents necessary to grow and undergo mitosis can be abundantly found in the extracellular liquid that is a passive constituent in the global mass balance equation.

We then consider the following growth terms

$$\Gamma_i = [\gamma_i H_\sigma(\psi - \psi_i) - \delta_i(\psi)]\phi_i\,, \qquad i = n, T\,. \tag{4.4}$$

In fact, not only growth but also apoptosis may be influenced by the compression level, as shown in [HEb].

In (4.4) $H_\sigma(\psi - \psi_i)$ is a mollifier of the step function, which is at least continuous, is constantly equal to 1 for $\psi$ smaller than the threshold value $\psi_i$, and vanishes for $\psi > \psi_i + \sigma$. According to the discussion above the threshold values $\psi_n$ and $\psi_T$ are such that $\psi_n < \psi_T$.

Coming back to the evolution of the extracellular matrix, it contains many macromolecules, including fibronectin, laminin, and collagen, which are produced in a stress-dependent way by the cells and are degraded by MDEs [CHa, MAd, PAb, STb]. Hence the remodeling process is obtained assuming the following growth terms in (4.2)$_1$ and (4.2)$_2$,

$$\Gamma_m = \mu_n(\Sigma)\phi_n - \nu c\phi_m\,,$$

$$\Gamma_M = \mu_T(\Sigma)\phi_T - \nu c\phi_M\,. \tag{4.5}$$

Although it is known that the degradation by MDEs does not distinguish between the two types of ECM, in (4.5) the production coefficients of ECM by normal and tumor cells might be different, as occurs in the case of fibrosis and of many tumors.

Finally, active MDEs are produced (or activated) by the cells, diffuse throughout the tissue, and undergo some form of decay (either passive or active). So one has to introduce the following reaction–diffusion equation governing the evolution of MDE concentration

$$\frac{\partial c}{\partial t} = \kappa \nabla^2 c + \pi_n(\Sigma)\phi_n + \pi_T(\Sigma)\phi_T - \frac{c}{\tau}\,. \tag{4.6}$$

The functions $\pi_n$ and $\pi_T$ model the production of active MDEs by normal and tumor cells, respectively, which might be different and certainly depend on the compression level.

Actually, in the following we focus on the effect of stress on growth. So, neglecting the effect of ECM and haptotaxis on cell motion, the complete system of equations can be summarized as

$$
\begin{cases}
\dfrac{\partial \phi_n}{\partial t} = \nabla \cdot [\phi_n K_0(\psi, \psi_m)\Sigma'(\psi)\nabla\psi] + \gamma_n H_\sigma(\psi - \psi_n)\phi_n - \delta_n(\psi)\phi_n \,, \\[2ex]
\dfrac{\partial \phi_T}{\partial t} = \nabla \cdot [\phi_T K_0(\psi, \psi_m)\phi_T\Sigma'(\psi)\nabla\psi] + \gamma_T H_\sigma(\psi - \psi_T)\phi_T - \delta_T(\psi)\phi_T \,, \\[2ex]
\dfrac{\partial \phi_m}{\partial t} = \mu_n(\Sigma)\phi_n - \nu c\phi_m \,, \\[2ex]
\dfrac{\partial \phi_M}{\partial t} = \mu_T(\Sigma)\phi_T - \nu c\phi_M \,, \\[2ex]
\dfrac{\partial c}{\partial t} = \kappa\nabla^2 c + \pi_n(\Sigma)\phi_n + \pi_T(\Sigma)\phi_T - \dfrac{c}{\tau} \,.
\end{cases}
\tag{4.7}
$$

We mention that if initially the tumor fills a compact region of space $\Omega$, one then has a free boundary problem related to (4.7) with the interface between tumor and normal tissue moving with the common velocity of the cells

$$
\mathbf{n} \cdot \frac{d\mathbf{x}_T}{dt} = \mathbf{n} \cdot \mathbf{v} = -K_0(\psi, \phi_m + \phi_M)\Sigma'(\psi)\mathbf{n} \cdot \nabla\psi \,,
\tag{4.8}
$$

and with $(4.7)_2$ valid inside $\Omega$ and $(4.7)_1$ outside it.

In addition, continuity of stress and velocity enforce the following interface conditions

$$
\phi_n = \phi_T \,, \qquad \mathbf{n} \cdot \nabla\phi_n = \mathbf{n} \cdot \nabla\phi_T \,.
\tag{4.9}
$$

Still referring to [CHe] for more details, in the case of constant production of ECM and MDEs an homogeneous stationary solution is given by

$$
\begin{aligned}
&\phi_n = \hat{\phi}_n = \psi_n + H_\sigma^{-1}\left(\frac{\delta_n}{\gamma_n}\right) - M_n \,, &&\phi_m = M_n \,, \\
&\phi_T = 0 \,, &&\phi_M = 0 \,, \\
&c = \pi_n\tau\left[\psi_n - M_n + H_\sigma^{-1}\left(\frac{\delta_n}{\gamma_n}\right)\right],
\end{aligned}
\tag{4.10}
$$

where

$$
M_n = \frac{\mu_n}{\nu\pi_n\tau} \,,
\tag{4.11}
$$

and

$$
H_\sigma^{-1}\left(\frac{\delta_n}{\gamma_n}\right) \in (0, \sigma) \,.
\tag{4.12}
$$

Symmetrically, another stationary solution is

$$
\begin{aligned}
&\phi_n = 0 \,, &&\phi_m = 0 \,, \\
&\phi_T = \hat{\phi}_T = \psi_T - M_T + H_\sigma^{-1}\left(\frac{\delta_T}{\gamma_T}\right), &&\phi_M = M_T \,, \\
&c = \pi_T\tau\left[\psi_T - M_T + H_\sigma^{-1}\left(\frac{\delta_T}{\gamma_T}\right)\right],
\end{aligned}
\tag{4.13}
$$

where

$$M_T = \frac{\mu_T}{\nu \pi_T \tau} . \tag{4.14}$$

By a simple linear stability analysis with respect to homogeneous perturbations it can be proved that the former configuration is unstable with an exponential growth rate for the tumor population equal to

$$\gamma_T H_\sigma \left( \psi_n - \psi_T + H_\sigma^{-1} \left( \frac{\delta_n}{\gamma_n} \right) \right) - \delta_T . \tag{4.15}$$

Because tumor cells are characterized by a smaller sensitivity to compression (i.e. $\psi_n < \psi_T$, $\delta_n = \delta_T$, $\gamma_n = \gamma_T$), or a smaller aptoptotic rate (i.e. $\delta_T < \delta_T$, $\psi_n = \psi_T$, $\gamma_n = \gamma_T$), or a larger growth rate (i.e. $\gamma_T > \gamma_n$, $\psi_n = \psi_T$, $\delta_n = \delta_T$), then

$$\psi_n + H_\sigma^{-1} \left( \frac{\delta_n}{\gamma_n} \right) < \psi_T + H_\sigma^{-1} \left( \frac{\delta_T}{\gamma_T} \right) . \tag{4.16}$$

In order to describe in more detail what happens in the early stages, consider the case in which all parameters for normal and tumor cells are equal but for $\psi_T > \psi_n + H_\sigma^{-1}(\delta/\gamma)$ (we dropped the indices to stress the equality).

Still referring to [CHe] for more detail, if we assume that at a certain instant, considered as the initial time, some normal cells undergo some genetic mutation that makes them less sensitive to the compression level, so that, for instance, $\phi_T(t = 0, \mathbf{x}) = a_0(\mathbf{x})$, then at early times one has the solution

$$\phi_T(t, \mathbf{x}) = a_0(\mathbf{x}) e^{(\gamma - \delta)t} \tag{4.17}$$

$$\phi_M(t, \mathbf{x}) = \frac{a_0(\mathbf{x})}{\frac{\gamma - \delta}{\mu} + \frac{\psi_n}{M} - 1} \left[ e^{(\gamma - \delta)t} - e^{-\nu c_0 t} \right] , \tag{4.18}$$

and

$$\phi_n(t, \mathbf{x}) = \psi_n - M + H_\sigma^{-1} \left( \frac{\delta}{\gamma} \right) - \phi_T(t, \mathbf{x}) , \tag{4.19}$$

$$\phi_m(t, \mathbf{x}) = M - \phi_M(t, \mathbf{x}) , \tag{4.20}$$

$$c(t, \mathbf{x}) = \pi \tau \left[ \psi_T - M + H_\sigma^{-1} \left( \frac{\delta}{\gamma} \right) \right] , \tag{4.21}$$

with $M = M_n = M_T$.

In particular, we stress that $\phi_n(t, \mathbf{x}) + \phi_T(t, \mathbf{x})$, $\phi_m(t, \mathbf{x}) + \phi_M(t, \mathbf{x})$, and $c(t, \mathbf{x})$ remain constant, which implies $\Sigma'(\psi) = 0$ and absence of motion. For this reason, in [CHe] this phase is called the relaxed replacement phase. In fact, tumor cells simply substitute normal cells without causing any compression of the tissue, as shown by the early development

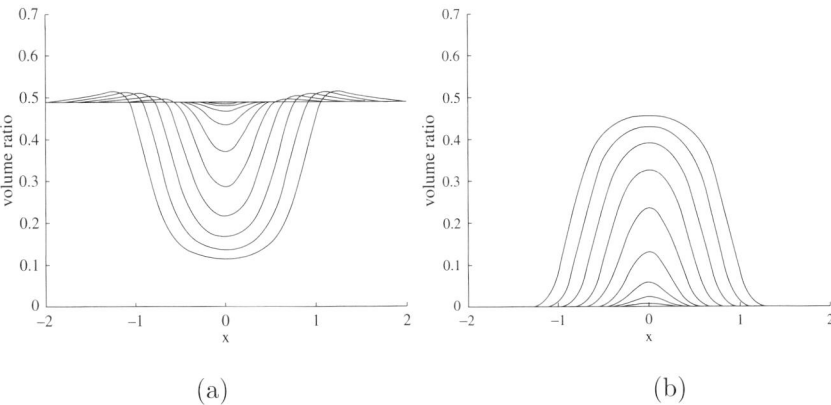

(a)  (b)

Figure 7.4. Early development of a tumor. Replacement of normal cells (left) by tumor cells (right) at times $\tilde{t} = \gamma t = 1, \ldots, 10$ without compression until $\tilde{t} \approx 5$ and progressive compression of the surrounding tissue for larger times. But for $\psi_n = 0.6$ and $\psi_T = 0.7$, the same parameters are used for both tumor and normal tissue $\delta/\gamma = 0.1$, $\mu = 0.1$ days$^{-1}$, $\pi/c_0 = 400$ days$^{-1}$, $\nu c_0 = 0.25$ days$^{-1}$, $\gamma\tau = 0.005$, $\pi/c_0 = 400$ days, $M_0 = 0.2$, $\sigma = 0.1$, and $a_0(x) = 0.001 \exp\{-30\tilde{x}^2\}$ where distances are scaled with $\sqrt{\gamma/KE}$.

in Figure 7.4a for $\gamma t > 5$ and Figure 7.5a. A similar behavior can be shown if the production rates are stress-dependent.

We observe that if $a_0(\mathbf{x})$ has a compact support, as it should be because the source of the tumor is localized, then the solution for $\phi_T$ will always have a compact support, because $(4.7)_2$ is parabolic degenerate.

After a time that can be estimated by

$$t \approx \frac{1}{\gamma - \delta} \log \left[ \frac{\delta}{\gamma} \frac{\psi_n - M + H_\sigma^{-1}\left(\frac{\delta}{\gamma}\right)}{\max \ a_0(\mathbf{x})} \right], \tag{4.22}$$

the amount of tumor cells produced is larger than the amount of normal cells that would normally die. The tumor then starts compressing the tissue, as observed experimentally, and the growth of the hyperplasia is accompained by a compression of the normal tissue near the interface separating the two tissues shown by the maxima in Figures 7.4a and 7.5a. Cells start moving away from the compressed regions. At the same time the ECM is completely replaced by that produced by the tumor.

From Figure 7.5 it is evident that the tumor front travels with a constant velocity. Of course, in the model the influence of nutrients is neglected, so from the biological point of view this makes sense until one can assume that the nutrients are abundantly supplied to the entire tumor. Including

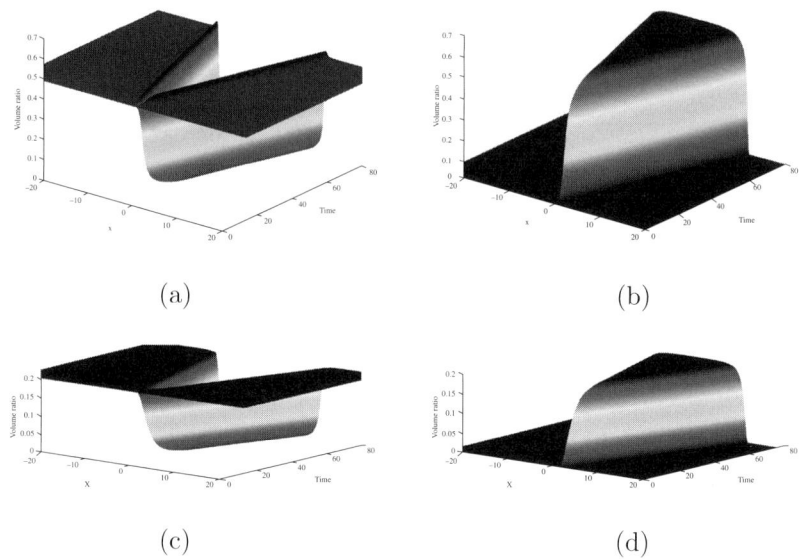

(a)                                                        (b)

(c)                                                        (d)

Figure 7.5. Tissue invasion for longer times and for the same parameters used in Figure 7.4. The traveling wave characteristic and the transition layer are evident. The compression of the normal tissue due to the expansion of the hyperplasia is also put in evidence by the peaks in (a) corresponding to the normal tissue; (b) refers to the tumor tissue; (c) to the ECM produced by the normal tissue; and (d) to that produced by the tumor.

nutrients and starvation in the model would lead to nutrient-limited growth as we show in the following section.

In [CHe] it is shown that the speed of the traveling wave solution can be evaluated as

$$v \approx \sqrt{2K_0\Sigma'(\psi_n)\delta\left(1 - \frac{\delta}{\gamma}\right)(\psi_T - \psi_n)} \tag{4.23}$$

(if $\psi_T = (1 + \epsilon)\psi_n$ with $\epsilon \ll 1$) and there is a transition layer between the normal and abnormal tissue having width

$$z_2 - z_1 \approx \sqrt{\frac{2K_0\Sigma'(\psi_n)(\psi_T - \psi_n)}{\delta\left(1 - \frac{\delta}{\gamma}\right)}}. \tag{4.24}$$

In fact, to second order one has that

$$\phi_T \approx \hat{\phi}_T - \frac{\gamma - \delta}{2KE}(z - z_1)^2 \quad \text{for } z \in [z_1, 0],$$

$$\phi_n \approx \hat{\phi}_n + \frac{\delta}{2KE}(z - z_2)^2 \quad \text{for } z \in [0, z_2], \tag{4.25}$$

where $\hat{\phi}_n$ and $\hat{\phi}_T$ are respectively defined in (4.10) and (4.13) and

$$z_1 \approx -\frac{v}{\gamma - \delta}, \qquad z_2 \approx \frac{v}{\delta}. \tag{4.26}$$

Figure 7.6a shows a comparison between the traveling wave solution (4.25) and the one obtained numerically, whereas Figures 7.6b,c compare the theoretical values of $v$, $z_1$, $z_2$, and then of the transition layer thickness $z_1 + z_2$ with those obtained from the simulations.

As mentioned in Section 7.2.2 (see also Table 7.1), often tumors are characterized by a considerable change in the content of ECM. In fact, for instance, self-palpation is encouraged in order to identify possible breast tumors by sensing a stiffer nodule with respect to the surrounding tissue. Figure 7.7 reports what happens if $\mu_T > \mu_n$, corresponding to the generation of fibrotic tissue, with a smaller amount of cells and a compressed tissue (see Figure 7.7e). In this case the normal ECM is produced by the cells at a rate larger than physiological, so that at the end the hyperplasic tissue replacing the normal one is also characterized by a larger amount of ECM. In particular, doubling the rate of ECM production leads to a fibrotic tissue with a ratio of cells versus ECM content nearly equal to 1.02, compared with 2.45 in the physiological situation. A similar thing is obtained halving the rate of production of MDEs.

On the other hand, a hypoproduction of ECM (or a hyperproduction of MDEs) leads to a tissue characterized by a ratio of cells versus ECM content nearly equal to 6.86 as shown in Figure 7.8 with a larger amount of cells and a smaller amount of ECM (see Figure 7.1a).

Comparing Figures 7.7e and 7.8e, one can notice that in the two situations the overall volume ratio $\psi$ is very similar. However, the composition of the tissue is dramatically different with the obvious changes in the mechanical properties of the tissue. This is due to the fact that the overall volume ratio is mainly influenced by the value at which growth stops whereas the tissue composition is influenced by the other parameters. A similar thing would occur by a pathological production of MDEs.

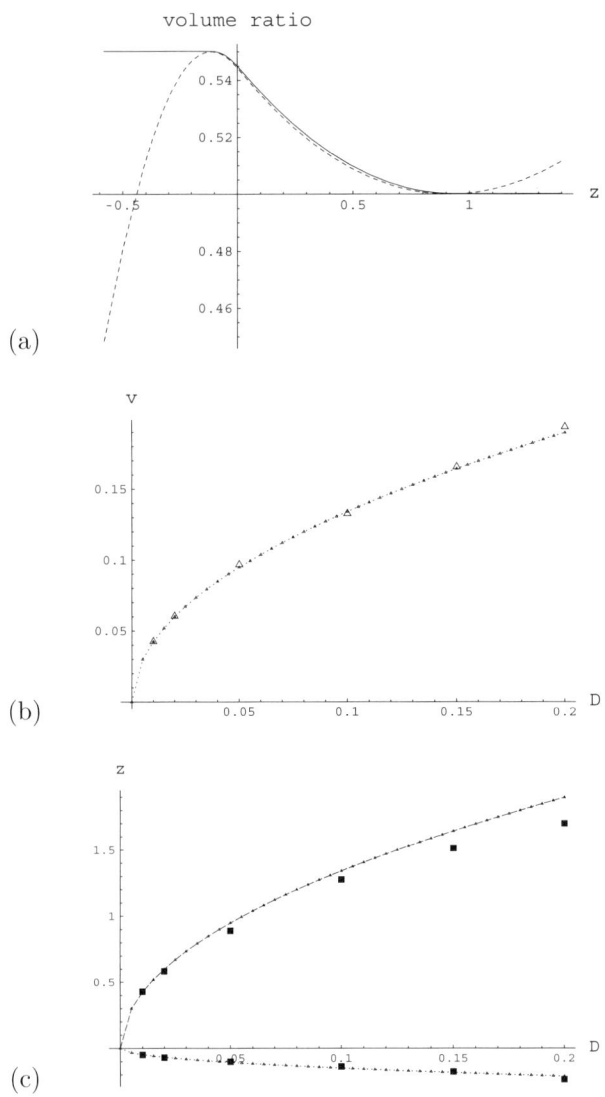

Figure 7.6. Traveling wave solution for $\delta/\gamma = 0.1$. (a) Comparison between analytical approximation (dotted line) and numerical solution (full line) for $\psi_n = 0.5$ and $\psi_T = 0.55$. Negative $z$s correspond to the tumor, positive $z$s to the host tissue. Distances are scaled with $z = \sqrt{\gamma/KE}x$. (b) Velocity of propagation as a function of $D = \psi_T - \psi_n$. Analytical estimates are given by the curves and numerical results by the squares. (c) The lower curves refer to $z_1$ and the upper curves to $z_2$. The thickness of the transition layer is then given by the distance between the two curves.

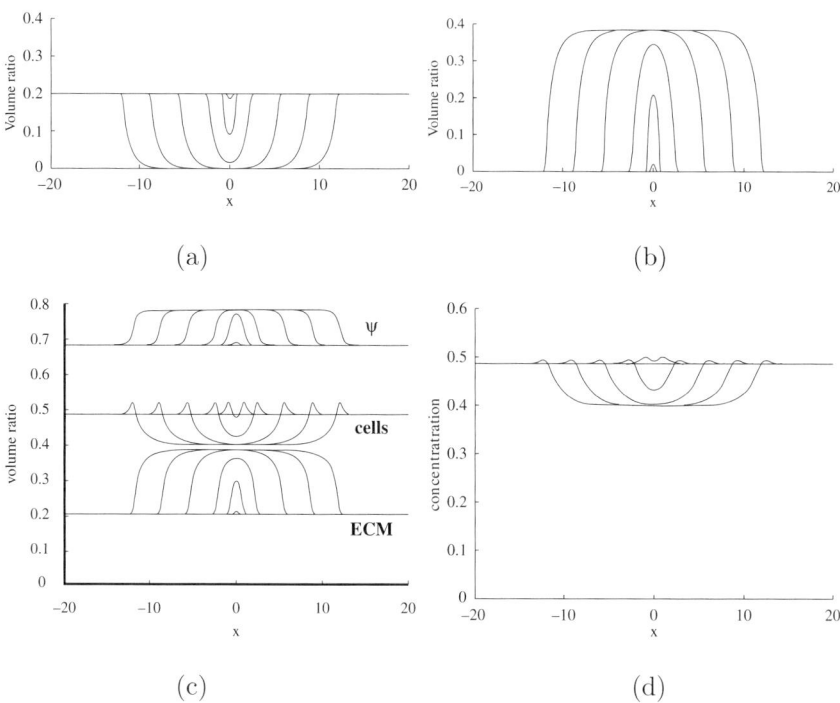

Figure 7.7. Formation of fibrosis for $\mu_T = 0.2$ days$^{-1}$, with all other parameters as in Figure 7.4 and at times $\tilde{t} = \gamma t = 0, 2.5, 5, 10, 20, 40, 60, 80$. Volume ratio of normal cells and tumor cells (a) and ECM produced by normal cells and by tumor cells (b). (c) The lower set of curves refers to the volume ratio of ECM $\psi_m = \phi_m + \phi_M$, the central ones to the volume ratio of cells $\phi_n + \phi_T$, and the upper ones to the overall volume ratio $\psi$, (i.e. the sums of the two above). (d) Concentration of MDEs.

## 7.4.2 Viscous Fluid: An Example Showing Nutrient-Limited Growth

In this section, following [BYb], we assume that the solid constituent of the mixture can be modeled as an ensemble of sticky cells floating in a liquid environment and neglecting the presence of ECM.

In [BYb] the ensemble of cells is described as a "viscous growing fluid," so that also in this case one does not need to consider the deformations of the material with respect to some reference configuration, but only to deal with their rates. In this respect, it is possible to use an Eulerian framework

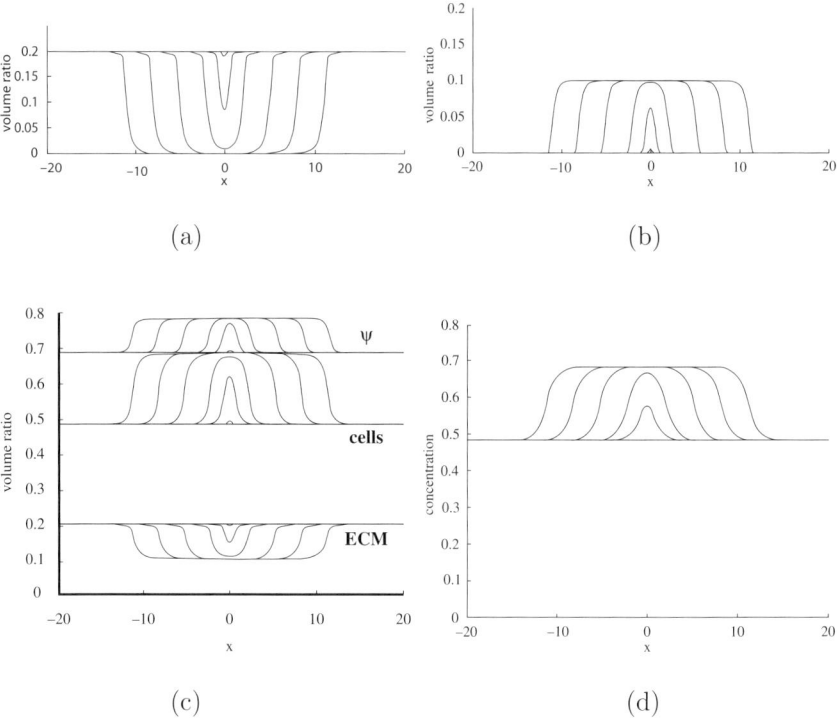

Figure 7.8. Hypoproduction of ECM for $\mu_T = 0.05$ days$^{-1}$, with all other parameters as in Figure 7.4 and at times $\tilde{t} = \gamma t = 0, 2.5, 5, 10, 20, 40, 60, 80$. Volume ratio of normal cells and tumor cells (a) and ECM produced by normal cells and by tumor cells (b). (c) The lower set of curves refers to the volume ratio of ECM $\psi_m = \phi_m + \phi_M$, the central ones to the volume ratio of cells $\phi_n + \phi_T$, and the upper ones to the overall volume ratio $\psi$ (i.e. the sums of the two above). (d) Concentration of MDEs.

and the mathematical description of the "growing fluid" just involves an additional source of mass.

To complete the picture we might also consider $N$ chemical factors and nutrients diffusing in the extracellular liquid

$$\frac{\partial c_i}{\partial t} + \nabla \cdot (c_i \mathbf{v}_\ell) = \nabla \cdot (k_i \nabla c_i) + \gamma_i \phi_T - \delta_i \phi_T c_i, \qquad i = 1, \ldots, N, \quad (4.27)$$

where the last two terms refer, respectively, to the possible production and absorption by tumor cells and $\gamma_i$ and $\delta_i$ might not be constant.

Actually, in [BYb] only a general nutrient $n$ perfusing through the vessels far away from the tumor and absorbed by the cells is considered, so that just the equation

$$\frac{\partial n}{\partial t} + \nabla \cdot (n\mathbf{v}_\ell) = \nabla \cdot (k_n \nabla n) - \delta_n \phi_T n , \qquad (4.28)$$

need be added to (3.20) with $\mathbf{b}_T = \mathbf{0}$ and

$$\hat{\mathbf{T}}_T = (-\Sigma + \lambda_T \nabla \cdot \mathbf{v}_T)\mathbf{I} + 2\mu_T \mathbf{D}_T , \qquad (4.29)$$

where $\mathbf{D}_T = (\nabla \mathbf{v}_T + \nabla \mathbf{v}_T^T)/2$ is the rate of strain tensor. It must be stressed that neither $\lambda_T$ nor $\mu_T$ will be constant as viscous forces among cells increase, at least linearly, with their volume ratio. Of course, as in the previous section, $\Sigma$ is also a function of the volume ratio measuring the response to compression and is taken positive in compression.

The growth term $\Gamma_T$ is constructed in [BYb] on the basis of the following phenomenological observations.

- Proliferation occurs if the nutrient concentration exceeds the threshold value $\hat{n}$. Where $n$ $(>\hat{n})$ is close to $\hat{n}$ the proliferation rate is proportional to $n - \hat{n}$; as $n$ increases, the proliferation rate eventually saturates.

- Cell proliferation is strongly affected by the presence of other cells that exert stress on the membrane of the replicating cell. In particular, the proliferation rate approaches zero as the volume ratio approaches one.

- Apoptosis is proportional to the volume ratio of cells.

A suitable function $\Gamma_T$, which combines these features and is continuous across $n = \hat{n}$, is given by

$$\Gamma_T = \frac{\gamma \phi_T}{1 + \sigma \Sigma(\phi_T)} \frac{(n - \hat{n})_+}{1 + \nu n} - \delta \phi_T , \qquad (4.30)$$

where $(f)_+$ is the positive part of $f$. Hence, when

$$\frac{(n - \hat{n})_+}{1 + \nu n} < \frac{\delta}{\gamma}[1 + \sigma \Sigma(\phi_T)] , \qquad (4.31)$$

there is a net loss of cells which at the end will lead to a limit radius related to the amount of nutrient available.

The stress–volume ratio relation is obtained in [BYb] under the following considerations.

- Two cells that are far apart ignore each other.

- If the distance between two cells falls below a threshold value then they attract each other.

- When cells in contact are pulled apart, an adhesive force competes with cell separation.

- If two cells are too close together, they experience a repulsive force.

- The repulsive force becomes infinite in the limit as the cells are packed so densely that they fill the whole control volume.

In the one-dimensional case, the previous description can be reformulated as follows.

- Cell in regions where $\phi_T < \hat{\phi}$ experience neither attractive nor repulsive forces.

- The attractive force attains a maximum value ($\Sigma = \Sigma_1$) when $\phi_T = \phi_1 > \hat{\phi}$.

- The attractive and repulsive forces balance when $\phi_T = \phi_2 > \phi_1$.

- The repulsive force becomes infinite as $\phi_T$ tends to one.

It has to be mentioned that a continuous function satifying the properties above would be decreasing in the interval $(\hat{\phi}, \phi_1)$, giving rise to a problem that might become ill-posed (see Eq. (4.35) below) if the solution achieves values in the interval above, giving rise to dramatic instability problems. However, Witelski [WIa] showed that a shock layer forms corresponding to "a quick jump over the bad section ... where the diffusion coefficient is negative" (see also [ELa]). Hence if one starts from initial conditions away from $(\hat{\phi}, \phi_1)$ the solution never achieves values in that interval, but the solution might lose regularity by forming a sharp front more or less as in phase transition problems. However, having in mind the experiments discussed in Section 7.4.4 and sketched in Figure 7.13, one can also assume that the multicell spheroid fractures for $\phi_T < \phi_1$ under the action of tensile stresses, keeping the validity of the model for $\phi_T > \phi_1$ where $\Sigma(\phi_T)$ is increasing.

We finally mention that the Young's modulus for a tumor is of the order of 1 kPa (see Table 7.2), whereas, as discussed in Section 7.4.4, the maximum tension is of the order of 0.1 kPa [BAa].

Summarizing, recalling (3.20), one has

$$
\begin{cases}
\dfrac{\partial \phi_T}{\partial t} + \nabla \cdot (\phi_T \mathbf{v}_T) = \dfrac{\gamma \phi_T}{1 + \sigma \Sigma(\phi_T)} \dfrac{(n - \hat{n})_+}{1 + \nu n} - \delta \phi_T \,, \\[2ex]
\nabla \cdot (\phi_T \mathbf{v}_T + \phi_\ell \mathbf{v}_\ell) = 0 \,, \\[2ex]
\nabla P = -\Sigma' \nabla \phi_T + \nabla(\lambda_T \nabla \cdot \mathbf{v}_T) + \nabla \cdot [\mu_T(\nabla \mathbf{v}_T + (\nabla \mathbf{v}_T)^T)] \,, \\[2ex]
\dfrac{\partial n}{\partial t} + \nabla \cdot (n \mathbf{v}_\ell) = \nabla \cdot (k_n \nabla n) - \delta_n \phi_T n \,,
\end{cases}
\tag{4.32}
$$

where $\mathbf{v}_\ell$ is given by Darcy's law

$$\mathbf{v}_\ell = \mathbf{v}_T - K\nabla\mathrm{P}\,. \tag{4.33}$$

The growth problem is a free-boundary problem with a material interface fixed on the tumor cells. This interface moves with the cell velocity

$$\mathbf{n}\cdot\frac{d\mathbf{x}_T}{dt} = \mathbf{n}\cdot\mathbf{v}_T\,. \tag{4.34}$$

An interesting simplification occurs in one-dimensional problems with viscous contributions neglected. In this case, the system reduces to

$$\frac{\partial\phi_T}{\partial t} = \frac{\partial}{\partial x}\left(K\Sigma'\phi_T\frac{\partial\phi_T}{\partial x}\right) + \Gamma_T\,, \tag{4.35}$$

$$\frac{\partial n}{\partial t} + \frac{\partial}{\partial x}\left(\frac{\hat{K}\Sigma'\phi_T}{1-\phi_T}\frac{\partial\phi_T}{\partial x}n\right) = k_n\frac{\partial^2 n}{\partial x^2} - \delta_n\phi_T n\,, \tag{4.36}$$

with

$$v_T = -\phi_\ell K\Sigma'\frac{\partial\phi_T}{\partial x}\,, \tag{4.37}$$

$$v_\ell = \phi_T K\Sigma'\frac{\partial\phi_T}{\partial x}\,, \tag{4.38}$$

and $P + \Sigma = $ constant. In particular, Eq. (4.35) is similar to the equation encountered in many one-dimensional poroelastic problems.

Figure 7.9a describes the trend toward the stationary state and then how nutrient limits the growth of avascular tumors. Initially a stress-free tumor of size $L = 0.1$ is implanted. We observe that at $\tilde{t} = 500$ the tumor is still so small that all cells have sufficient nutrient to replicate (i.e. $n > \hat{n}$ everywhere). The maximum cell compaction occurs at the tumor center and, due to the repulsive forces they experience, cells move toward the border, causing the tumor to increase in size. At $\tilde{t} = \delta_n t = 1000$ the nutrient concentration near the center of the tumor falls below $n_0$ and cells there start dying. The location of the maximum cell volume fraction moves toward the tumor boundary, and in the center a (local) minimum appears. Cells that are located in the central region of the tumor, between the two symmetric maxima, move toward the center, whereas those in the outer regions, between the maxima and the tumor boundary, move toward the boundary. Of course, the tumor still grows, but at a reduced rate. In the stationary configuration (the lowest curve) the maximum cell volume fraction occurs on the tumor boundary. For the choice of parameters in the figure, at equilibrium the nutrient concentration at the tumor center only just exceeds that which triggers central necrosis and the formation of a sharp front dividing the compact tumor from the necrotic core.

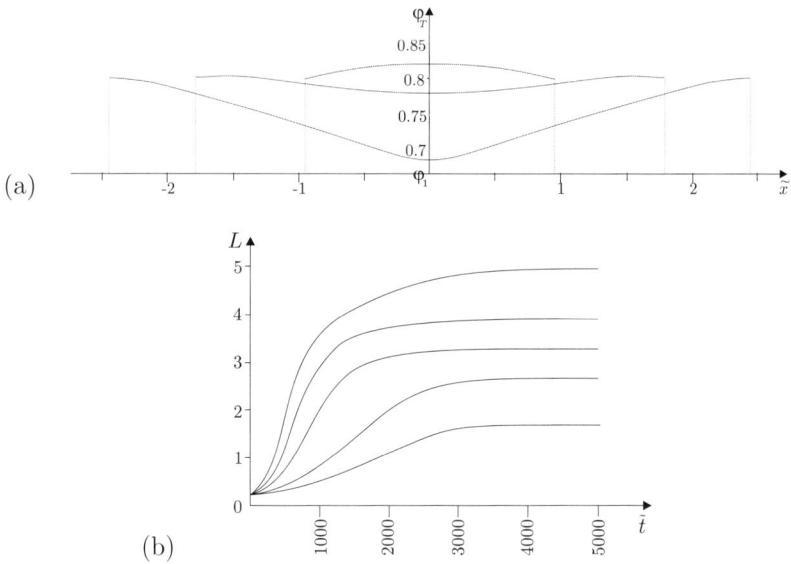

Figure 7.9. Evolution toward the steady state (thicker line) for $\delta/\delta_n = 0.001$, and $\sigma = 0$. (a) Volume ratio plotted versus space ($\tilde{x} = \sqrt{\delta_n/\kappa_n}x$) for $\tilde{\gamma} = \gamma n_{\text{ext}}/\delta_n = 0.0125$. Transient times are $\tilde{t} = \delta_n t = 500, 1000$. (b) Temporal evolution of the tumor size for different values of $\tilde{\gamma} = 0.0025, 0.005, 0.0075, 0.01, 0.0125$ (from lower to upper curve).

Figure 7.9b shows the temporal evolution of the tumor size for different values of growth rate $\tilde{\gamma} = \gamma n_{\text{ext}}/\delta_n$, where $n_{\text{ext}}$ is the amount of nutrient at the tumor border.

Another result of the model is that as the cell proliferation rate decreases more rapidly with increasing cellular stress, the equilibrium tumor size becomes smaller and the cells more uniformly distributed across the tumor. Actually, in [BYb] it is shown that if this influence if sufficiently large then no nontrivial equilibrium solutions exist and the tumor is eliminated. From the application viewpoint, this suggests that if there were a method to make tumor cells more sensible to mechanical compression (e.g. making their mitotic or apoptotic rate depend on the stress), this could be used to control the size of the tumor.

The following viscous-type constitutive equation

$$\mathbf{T} = -P\mathbf{I} + 2\mu_T \left[ \mathbf{D}_T - \frac{1}{3}(\nabla \cdot \mathbf{v}_T)\mathbf{I} \right], \tag{4.39}$$

has been also used by Franks and coauthors [FRa, FRb, FRc] under the additional strong hypothesis that all constituents filling the space move with the same velocity (live tumor cells and what they call surrounding

material in [FRc], with the addition of death tumor cells in [FRa, FRb]).
In fact, the stress in (4.39) probably refers to the mixture. For instance,
the model in [FRb], which describes tumor growth in a breast duct and
also focuses on the mechanical interaction with the duct walls, is written
in the notation of this chapter as

$$\begin{cases} \dfrac{\partial \phi_T}{\partial t} + \nabla \cdot (\phi_T \mathbf{v}) = [\mathrm{An} - \mathrm{B}(1 - \delta \mathrm{n})]\phi_T \,, \\[2ex] \dfrac{\partial \phi_D}{\partial t} + \nabla \cdot (\phi_D \mathbf{v}) = \mathrm{B}(1 - \delta \mathrm{n})\phi_T \,, \\[2ex] \dfrac{\partial \phi_\ell}{\partial t} + \nabla \cdot (\phi_\ell \mathbf{v}) = 0 \,, \\[2ex] \dfrac{\partial n}{\partial t} + \nabla \cdot (n\mathbf{v}) = D\nabla^2 n - \gamma An\phi_T \,, \\[2ex] \nabla P = \mu \left[ \nabla^2 \mathbf{v} + \tfrac{1}{3}\nabla(\nabla \cdot \mathbf{v}) \right] \,, \end{cases} \qquad (4.40)$$

where $\phi_D$ is the volume ratio of dead cells, which is joined with the constitutive equation (4.39) and the saturation assumption $\phi_T + \phi_D + \phi_\ell = 1$.
The mixture is not closed, so that the global volume of the tumor is increased by the source term $An$. For the sake of completeness, we mention
that in [FRa] and [FRc] the authors also add diffusion terms to the first
three equations in (4.40).

### 7.4.3 Evolving Natural Configurations in Tumor Growth

In the previous sections we have described some models that use fluid-
like constitutive equations. However, this is only a rough approximation,
because tumors as most tissue show solidlike characteristics. As already
mentioned several times in this book, in this case in order to define, for
instance, an elastic mechanical response, one needs to measure the deformation with respect to some reference state. However, the basic question
is: "Deformation with respect to what, if the tissue is always changing?" In
order to address this problem Ambrosi and Mollica [AMb, AMc] used the
theory of evolving natural configurations splitting the evolution in growth
and elastic deformation. In their model the interaction with the ECM and
with the extracellular liquid are neglected and the tumor is described as a
one-constituent compressible elastic body.

The theory for materials with evolving natural configurations is an ideal
setting to investigate tumor growth as the growth of other tissues. In fact,
the essential difficulty in formalizing the dynamics of growth is to model

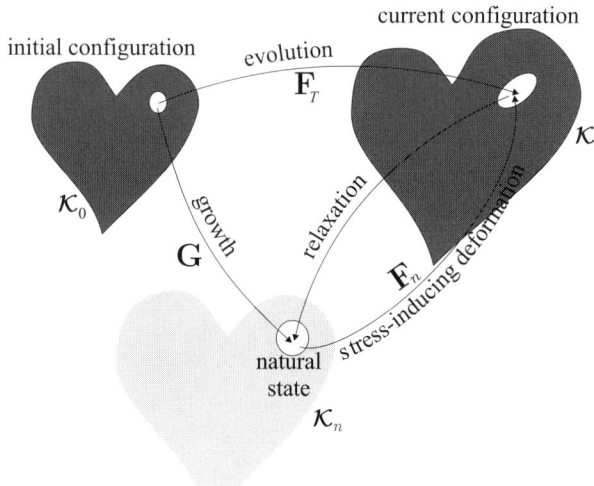

Figure 7.10. Evolving natural configuration.

simultaneously the change in mass, and the stresses that accompany it, possibly caused by growth itself or by the application of external loads.

With the theory for materials with evolving natural configurations one is able to separate such contributions and to model each of them individually. Following the notation in Figure 7.10, the aim is then to distinguish in the evolution of the tumor given through the deformation gradient $\mathbf{F}_T$ the contribution of pure growth from the stress-inducing deformation. In particular, it is natural to work so that no growth occurs during stress-inducing deformation. By the way, from the biological point of view, the two contributions could be easily testable in principle as growth occurs on a much longer time scale (hours up to a day) than deformation.

The deformation gradient $\mathbf{F}_T$ is a mapping from a tangent space onto another tangent space, and therefore it indicates how the body is deforming locally in going from $\mathcal{K}_0$ to $\mathcal{K}_t$. Working in the tangent space, take a neighborhood of a point and assume relieving its state of stress keeping its mass constant, so that it is allowed to relax to a stress-free configurations. The atlas of these configuration forms a natural configuration relative to $\mathcal{K}_t$ which we denote by $\mathcal{K}_n$. Of course, this natural configuration depends on time. We identify this deformation without growth with the tensor $\mathbf{F}_N$, which then describes how the body is deforming locally in going from the natural configuration $\mathcal{K}_n$ to $\mathcal{K}_t$. The tensor

$$\mathbf{G} \equiv \mathbf{F}_N^{-1} \mathbf{F}_T \,, \tag{4.41}$$

tells how the body is growing locally.

Hence, the following decomposition holds,

$$\mathbf{F}_T = \mathbf{F}_N \mathbf{G}.  \tag{4.42}$$

The tensor $\mathbf{F}_N$ is then connected to the stress response of the tumor to deformations and the tensor $\mathbf{G}$ is the one that is directly connected to growth and is therefore named the growth tensor.

There are then two things to determine constitutively: how the natural configurations evolve: that is, characterizing the growth tensor $\mathbf{G}$, and how the material behaves from each natural configuration. Being the density of a single cell is equal to the density of water, we assume that for any given "particle" the volume ratio in the natural configuration and in the original reference configuration are the same; that is, $\phi_T(t = 0) = \phi_N$. Denoting by $dV$, $dV_N$, and $dv$ the volume elements in the reference, natural, and current configuration, respectively, the related masses are then $dM = \rho\phi_N dV$, $dm = \rho\phi_N dV_N$, and $dm = \rho\phi_T dv$. Because mass is preserved between $\mathcal{K}_n$ and $\mathcal{K}_t$, one then has that

$$J_g = \det \mathbf{G} = \frac{dV_N}{dV} = \frac{dm}{dM},  \tag{4.43}$$

$$J_T = \det \mathbf{F}_T = \frac{dv}{dV} = \frac{\phi_N}{\phi_T}\frac{dm}{dM},  \tag{4.44}$$

and, in particular, because of (4.42),

$$J_N = \det \mathbf{F}_N = \frac{\phi_N}{\phi_T}.  \tag{4.45}$$

It can be readily realized looking at (4.43) that net growth corresponds to $J_g > 1$ and net death to $J_g < 1$. Of course, $J_g$ never vanishes, otherwise $\mathbf{F}_T$ would be singular. Applying the polar decomposition theorem to the growth tensor $\mathbf{G}$ we are sure that there exist a unique rotation $\mathbf{R}_g$ and a unique symmetric tensor $\mathbf{U}_g$ such that $\mathbf{G} = \mathbf{R}_g \mathbf{U}_g$. However, for the arbitrariness of the choice of the natural configuration with respect to rotations, we can certainly choose it so that $\mathbf{R}_g = \mathbf{I}$ and $\mathbf{G} = \mathbf{U}_g$.

Differentiating (4.45) and then using (3.14) rewritten here for sake of clarity

$$\frac{d}{dt}(\phi_T J_T) = \Gamma_T J_T,  \tag{4.46}$$

one has

$$\dot{J}_N = -\frac{\phi_N}{\phi_T^2}\dot{\phi}_T = -\frac{\phi_N}{\phi_T}\left(\frac{\Gamma_T}{\phi_T} - \frac{\dot{J}_T}{J_T}\right),  \tag{4.47}$$

or

$$\frac{\dot{J}_T}{J_T} - \frac{\dot{J}_N}{J_N} = \frac{\Gamma_T}{\phi_T}.  \tag{4.48}$$

Recalling the splitting (4.42), one finally has

$$\frac{\dot{J}_g}{J_g} = \frac{\Gamma_T}{\phi_T}, \tag{4.49}$$

or defining the rate of growth tensor

$$\mathbf{D}_g = \mathrm{sym}(\dot{\mathbf{G}}\mathbf{G}^{-1}), \tag{4.50}$$

one has from standard tensor calculus that

$$\dot{J}_g = J_g \mathrm{tr}\, \mathbf{D}_g, \tag{4.51}$$

and therefore

$$\mathrm{tr}\, \mathbf{D}_g = \frac{\Gamma_T}{\phi_T}. \tag{4.52}$$

Equation (4.52) reveals then that the first principal invariant of $\mathbf{D}_g$ is the right-hand side of the Eulerian mass balance equation (3.1) and the rate of mass production can be easily identified from one of the constitutive parameters of the material. In particular, for isotropic growth, $\mathbf{G} = g\mathbf{I}$ where $g$ is a scalar, (4.49) is rewritten

$$\frac{3\dot{g}}{g} = \frac{\Gamma_T}{\phi_T}, \tag{4.53}$$

which for known $\Gamma_T$ completely determines $\mathbf{G}$.

In general, however, growth requires giving constitutively a suitable evolution equation for the growth tensor, which may depend on a variety of quantities, for example,

$$\dot{\mathbf{G}} = \mathcal{L}_g(\mathbf{X}, t, \mathbf{S}, \mathbf{G}, \mathbf{c}), \tag{4.54}$$

where $\mathbf{c}$ is the set of nutrients and growth factors involved in growth and $\mathbf{S}$ is a suitable invariant measure of stress that might contain information on the direction of principal stresses. The involvement of the quantities above implies a strong coupling between the growth tensor and mechanical and chemical terms, so that, in general, one cannot look at growth in time as being separated from the overall mechanical response and the chemical background. This means that Eq. (4.54) has to be solved simultaneously with the other evolution equations.

Finally, we observe that one can replace the mass balance equation (4.46) with

$$\frac{d}{dt}(\phi_T J_N) = 0, \tag{4.55}$$

an equation not involving growth, which is instead given constitutively and is related to $\Gamma_T$ through (4.54) or (4.53) for isotropic growth. Equation (4.55) resembles the usual Lagrangian version of conservation of mass in the absence of mass sources but is related to a fictitious deformation from the natural configuration which is never achieved by the growing body.

The other constitutive relation to be specified regards the stress tensor. In their paper Ambrosi and Mollica [AMc] assumed that at any time the mechanical response of the tumor from the natural configuration is hyperelastic. In particular, they used a Blatz–Ko constitutive relation [BLa], a classical nonlinear elastic model, which can be seen as a generalization of the classical Mooney–Rivlin model for rubber. The Blatz–Ko material is the simplest hyperelastic compressible material and has been successfully applied to model polymeric foams, a system that shows some analogies with the mechanical behavior of cell aggregates [SEa].

The Cauchy stress tensor takes then the form

$$\mathbf{T}_T = \frac{\mu}{J_N}[-(J_N)^{-q}\mathbf{I} + \mathbf{B}_N], \qquad (4.56)$$

where $\mathbf{B}_N =: \mathbf{F}_n\mathbf{F}_N^T$ is the left Cauchy–Green stretch tensor, and $\mu$ and $q$ are positive material constants. Further comments on other possible constitutive choices are contained in the following section. Their growth model inside the tumor can then be summarized as

$$\begin{cases} \phi_T J_N = \phi_N\,, \\[2mm] \nabla \cdot \mathbf{T}_T = \mathbf{0}\,, \\[2mm] \dot{g} = \dfrac{g}{3}\dfrac{\Gamma_T}{\phi_T}\,, \\[2mm] \nabla \cdot (D\nabla n) - \delta\phi_T n = 0\,, \end{cases} \qquad (4.57)$$

where the stress tensor is given by (4.56) and

$$\Gamma_T = \gamma\frac{n - n_0}{N - n_0}e^{-(s/s_0)^2}\phi_T, \qquad (4.58)$$

where $s$ is the trace of the first Piola–Kirchoff stress tensor $\mathbf{P} = J_T\mathbf{T}_T\mathbf{F}_T^{-T}$. This very simple equation assumes that stress always inhibits growth whereas net growth occurs if a sufficient minimum quantity of nutrient is available, otherwise one has that the body resorbs mass.

The aim of the paper by Ambrosi and Mollica [AMb] was to compare the model with the experiments performed by Helmlinger et al. [HEa] who study the influence of external loading on tumor growth by letting a tumor spheroid grow in a gel. In the experimental setup agarose, a polysaccharide extracted from seaweed, is dissolved into boiling water. When the medium starts cooling down, the polysaccharide chains cross-link with each other, causing the solution to gel into a semi-solid matrix. The more agarose is dissolved in the boiling water, the firmer the gel is. While the solution is still fluid, the tumor cells are plugged into the polymerizing medium. After cooling, the tumor cells are trapped in the agarose gel that has known mechanical properties depending on the solid-phase concentration. The nutrient rapidly diffuses in the liquid phase of the gel, thus providing a constant concentration at the boundary of the spheroid. As the spheroid grows, it displaces the surrounding gel, which in turn exerts a uniform compression on the tumor spheroid. By varying the volume fraction of the solid component during the preparation of the gel, they are able to modulate its stiffness and hence to apply different stress fields on the tumor.

The main result obtained in [HEa] is that the stress field definitely reduces the final dimensions of the spheroids. At a cellular level, though, spheroids cultured in gels of increasing stiffness are characterized by a decreased apoptosis rate with no significant change in proliferation rate and hence increased cellular packing. Moreover, inner regions of free-suspension spheroids often exhibit large voids that were rarely seen in gel-cultured spheroids. The agarose gel was modeled as a poroelastic material with the constitutive equation given by

$$\mathbf{T} = \tilde{\gamma} \frac{\rho}{\rho_0} \frac{e^{\beta(I-3)}}{(I\!I\!I - \phi_0^2)^\alpha} \left[ \beta \mathbf{B} - \alpha \frac{I\!I\!I}{I\!I\!I - \phi_0^2} \mathbf{I} \right], \tag{4.59}$$

where $I$ and $I\!I\!I$ are the first and third invariants of the right Cauchy–Green deformation tensor $\mathbf{C} = \mathbf{F}^\mathsf{T} \mathbf{F}$ of the surrounding gel and $\alpha = \beta(1 - \phi_0^2)$. The free boundary problem for the growth of an elastic tumor in a gel is completed by interfacing (4.57) and

$$\nabla \cdot \mathbf{T} = \mathbf{0}, \tag{4.60}$$

where $\mathbf{T}$ is given in (4.59) with continuity of displacement and of the normal component of the stress thus providing the following interface conditions

$$[\![ \mathbf{u} \cdot \mathbf{n} ]\!] = 0, \qquad [\![ \mathbf{Tn} ]\!] = \mathbf{0}, \tag{4.61}$$

where the jump is evaluated across the the interface and $\mathbf{n}$ is its unit normal vector.

In agreement with the experiments, Ambrosi and Mollica [AMc] assumed spherical symmetry. Normalizing space length with the outer radius of the gel the tumor spheroid is assumed to be initially located in $R \in [0, \bar{R}]$, and the gel fills the space $R \in [\bar{R}, 1]$, where $\bar{R}$ is chosen sufficiently smaller than one so that the effect of the external constraints on the growth of the spheroid can be neglected.

Figure 7.11 shows the evolution of a freely growing tumor, that is, not in the gel, and then without external loads. After an initial exponential

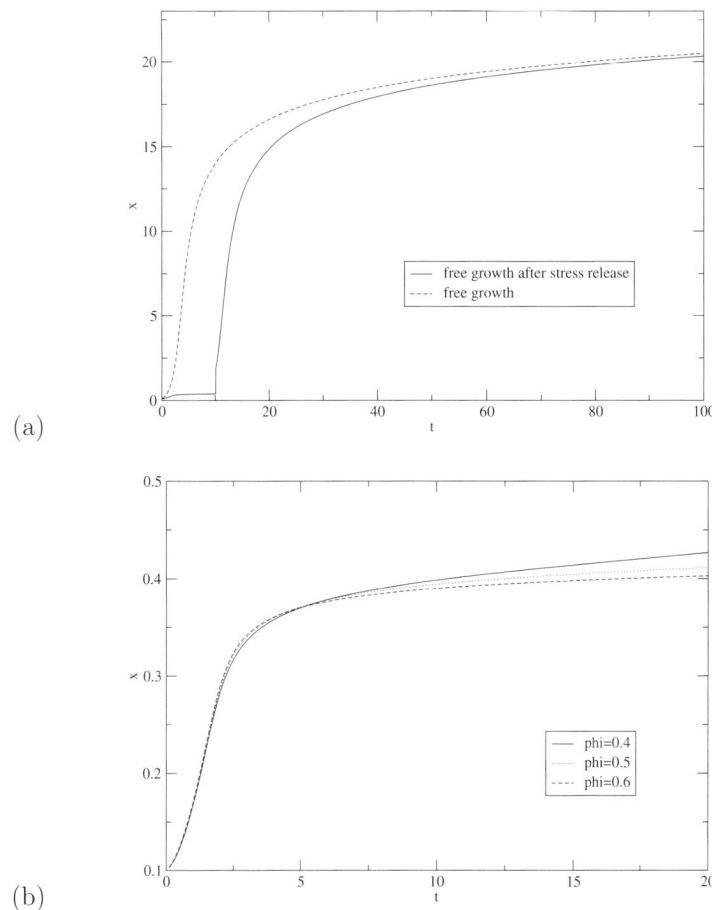

(a)

(b)

Figure 7.11. Comparison between free growth of a tumor spheroid (dashed line in (a)) and constrained growth ((b) and full line in (a)). At $t = 10$ the spheroid is ideally extracted; being unloaded, it retains residual stress only. The plot of the subsequent increase in size shows that the past history tends to be asymptotically forgotten: the radius approaches the free one and residual stress vanishes. In (b) the evolution of the position of the interface between tumor and gel is given for different gel concentrations.

growth and a transition period growth becomes linear in time. In the absence of external loads the change in growth rate is essentially due to the reduced availability of nutrient that occurs when the diameter of the spheroid overcomes the diffusion length of the nutrient in the spheroid. This effect is confirmed by the plots of concentration depicted in Figure 7.12a at different times: at $t = 20$ the nutrient has a nonnegligible concentration just

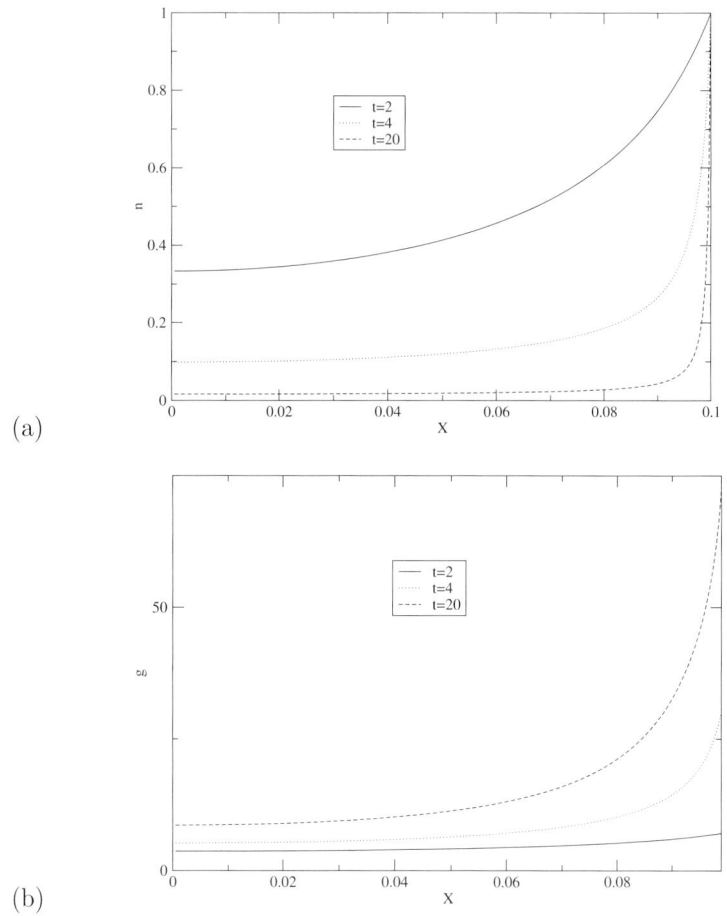

(a)

(b)

Figure 7.12. (a) Available nutrient concentration and (b) growth function in a freely growing tumor plotted versus undeformed radius at different times.

in a thin layer around the border (the proliferating rim) so that the growth is essentially on the surface. Note that the existence of a proliferating rim, as described in the experimental literature, here arises without any ad hoc introduction. In principle, residual stresses due to nonhomogeneous growth could inhibit proliferation too. However, the stress field generated in free growth is small and does not affect the size of the spheroid: in this case the key mechanism is the decreasing amount of available nutrient that influences the growth function $g$ through the relationship (4.58), yielding the behavior of Figure 7.12b.

The position of the interface between a spheroid embedded in a poroelastic medium versus time is tracked in Figure 7.11b for different concentrations of the solid component. The growth of the size of the spheroid is linear from the very beginning, until it becomes almost constant when the external stress starts inhibiting growth. Inasmuch as the nutrient has a nonconstant spatial distribution in the tumor, the growth is not homogeneous and some residual stress is generated. The stress in the tumor is always compressive, the largest value occurring at the interface between spheroid and gel.

Looking carefully at the final size of the spheroid, one understands that the diameter is still much smaller than the freely floating one, whereas experimentally the final size of the embedded spheroid is some tenths of the freely growing one. This observation suggests that, although growth inhibition by stress works fine as described in the present work, some mechanism of stress release must be included in the model in order to obtain a final size that is also in quantitative agreement with experiments.

After extracting from the gel a spheroid that was in its plateau phase, cells restart duplication, yielding the results shown in Figure 7.11. When comparing the dashed line indicating the diameter of the spheroid growing after stress release with a free-growth one, one finds that the former tends just asymptotically to reach the latter so that, in some sense, the inhomogeneous original growth never completely vanishes. The slope discontinuity of the full line in Figure 7.11a occurring at $t = 10$ corresponds to gel extraction.

### 7.4.4 Viscoelasticity and Pseudo-Plasticity in Tumor Growth

The models presented in Sections 7.4.1 to 7.4.3 can be certainly improved by taking into account the viscoelastic behavior that characterizes most biological materials. However, the characteristic times of the rate-dependent response of the materials involved are of the order of tens of seconds and therefore much less than the characteristic times of cell duplication (a day), so in our opinion viscoelasticity only plays a secondary role in problems coupled with growth. Of course, it can have important effects in mechanical problems characterized by times of the order of the relaxation and retardation times, but in describing them growth can be in our opinion neglected, so that the two descriptions are somewhat complementary. On the other hand, when treating the tumor as a solid there is an important effect that should not be neglected which has to do with the pseudo-plastic behavior of multicellular spheroids.

The cellular scale motivation of the macroscopic plastic behavior is the following. As shown in Figure 7.2, cells adhere to each other via cadherin junctions and to the extracellular matrix via integrin junctions. These bonds have a limited strength as measured, for instance, by Baumgartner et al. [BAa] and Canetta et al. [CAa]. In fact, the adhesive strength

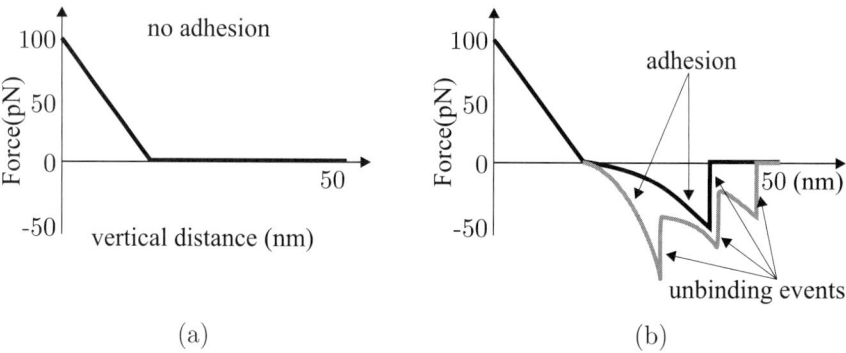

Figure 7.13. Adhesive force measurement. The positive branch refers to the compression of the cell by the bead. At larger distances the adhesive cell acts as an elastic nonlinear spring until single or multiple unbinding occurs (redrawn from [BAa]).

of a single bond was found to be in the range of 35–55 pN. Because the density of VE-cadherin on a cell surface is about 400–800 molecules/$\mu$m$^2$ of the surface one can estimate the resistance to pulling to be of the order of 0.1 kPa.

Typical experiments to test the adhesive strength of a cell consist of glueing a functionalized microsphere at the tip of an AFM cantilever (atomic force microscopy). After putting the microsphere in contact with the cell, the cantilever is pulled away at a constant speed (in the range 0.2–4 $\mu$m/sec). If there is no adhesion between the bid and the cell, the force measured has the behavior shown in Figure 7.13a. This is experimentally obtained, for instance, by the addition of an antibody of the VE-cadherin external domain. On the other hand, adhesion gives rise to the measurement of a stretching force and a characteristic jump indicating the rupture of an adhesive bond, as shown in Figure 7.13b. Actually, because a sphere binds to many receptors, it is common to experience multiple unbinding events occurring at different instants during the single experiment, as shown by the grey curve in in Figure 7.13b.

Transferring this concept to tumor mechanics it is clear that if an ensemble of cells is subject to a sufficiently high tension or shear, then some bonds break and some others form, leading to the necessity to introduce plasticity in the description. This in particular occurs during growth when the duplicating cell needs to displace its neighbors to make room for its sister cells as sketched in Figure 7.14.

Generalizing the concepts introduced in the previous section, what is left when relieving the state of stress of a particle in the configuration $\mathcal{K}_t$ keeping its mass constant, includes both growth and plastic deformation

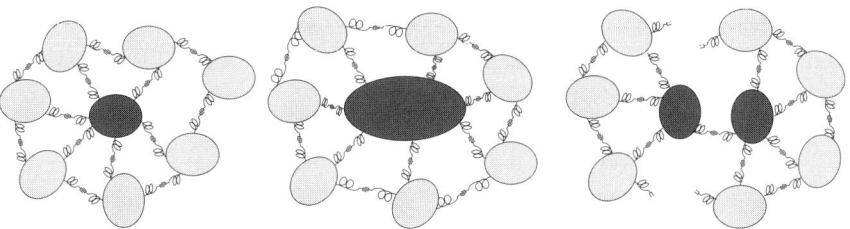

Figure 7.14. Sketch of pseudo-plastic behavior in tumor growth.

due to unbinding events. This means that focusing on the population of tumor cells, one need to generalize Figure 7.8 to a three-step process which includes plastic deformations; that is,

$$\mathbf{F}_T = \mathbf{F}_N \mathbf{F}_p \mathbf{G}_T . \tag{4.62}$$

From the biological point of view, it is not difficult to imagine the splitting, because as already stated the characteristic cell cycle time is much longer than the times involved both in plastic and elastic phenomena.

Denoting by $\mathcal{K}_p$ the intermediate configuration for the tumor between $\mathcal{K}_0$ and $\mathcal{K}_n$, we assume that for any given "particle" the volume ratio in $\mathcal{K}_p$ is the same as in the natural configuration and in the original reference configuration; that is, $\phi_p = \phi_T(t = 0) = \phi_N$. The generalization of (4.43)–(4.45) gives

$$J_T = \frac{\phi_N}{\phi_T} \frac{dm}{dM} , \tag{4.63}$$

$$J_g = \frac{\phi_N}{\phi_p} \frac{dm}{dM} = \frac{dm}{dM} , \tag{4.64}$$

$$J_p = \frac{\phi_p}{\phi_N} = 1 , \tag{4.65}$$

$$J_N = \frac{\phi_N}{\phi_T} . \tag{4.66}$$

Notice that differentiating (4.66) one has

$$\frac{d}{dt} \log(\phi_T J_N) = 0 , \tag{4.67}$$

an equation similar to (4.55), where the time derivative is computed along any constituent.

In order to be more specific, we consider the tumor as a triphasic mixture made of ECM, extracellular liquid, and cells, as in Section 7.4.1. However, contrary to Section 7.4.1, in this section we neglect ECM remodeling. The extracellular matrix can then be viewed as a substrate on which cells move and duplicate and from the mechanical point of view taking it into account results very useful because they can represent a suitable framework to properly define a Lagrangian coordinate system. Equation (3.8) with $\Gamma_0 = 0$ and $\Gamma_\ell = -\Gamma_T$ can be written in the Lagrangian form just defined as

$$\frac{d_0}{dt}(\phi_0 J_0) = 0\,, \tag{4.68}$$

$$\frac{d_0}{dt}(\phi_T J_0) + \text{Div}_0[\phi_T J_0 \mathbf{F}_0^{-1}(\mathbf{v}_T - \mathbf{v}_0)] = \Gamma_T J_0\,, \tag{4.69}$$

$$\frac{d_0}{dt}(\phi_\ell J_0) + \text{Div}_0[\phi_\ell J_0 \mathbf{F}_0^{-1}(\mathbf{v}_\ell - \mathbf{v}_0)] = -\Gamma_T J_0\,, \tag{4.70}$$

where $d_0/dt$ is the time derivative following the ECM, $\text{Div}_0$ is the divergence operator with respect to the ECM, and $\mathbf{F}_0$ is the deformation gradient of the ECM.

Note that by summing the mass balance equations, thanks to the saturation assumption and the fact that the mixture is closed, one has

$$\frac{d_0 J_0}{dt} + \text{Div}_0[J_0 \mathbf{F}_0^{-1}(\mathbf{v}_c - \mathbf{v}_0)] = 0\,, \tag{4.71}$$

where $\mathbf{v}_c = \phi_T \mathbf{v}_T + \phi_\ell \mathbf{v}_\ell + \phi_0 \mathbf{v}_0$.

Following a procedure similar to that used to obtain Eqs. (4.48)–(4.52) one can write

$$\frac{1}{J_0}\frac{d_0 J_0}{dt} - \frac{1}{J_N}\frac{d_0 J_N}{dt} = \frac{\Gamma_T}{\phi_T} - \frac{1}{\phi_T J_0}\text{Div}_0[\phi_T J_0 \mathbf{F}_0^{-1}(\mathbf{v}_T - \mathbf{v}_0)]\,, \tag{4.72}$$

and

$$\frac{1}{J_g}\frac{d_0 J_g}{dt} = -\frac{1}{J_p}\frac{d_0 J_p}{dt} - \frac{1}{J_0}\frac{d_0 J_0}{dt} + \frac{1}{J_T}\frac{d_0 J_T}{dt} \tag{4.73}$$

$$+\frac{\Gamma_T}{\phi_T} - \frac{1}{\phi_T J_0}\text{Div}_0[\phi_T J_0 \mathbf{F}_0^{-1}(\mathbf{v}_T - \mathbf{v}_0)]\,, \tag{4.74}$$

where the first term on the r.h.s. vanishes because of (4.65)–(4.52). This relation is not as good looking as (4.51) because we are using a Lagrangian framework based on a constituent different from the one that is duplicating. Equation (4.74) can be rewritten in a slightly more compact form as

$$\frac{d_0}{dt}\log\frac{J_g J_0}{J_T} = \frac{\Gamma_T}{\phi_T} - \frac{1}{\phi_T J_0}\text{Div}_0[\phi_T J_0 \mathbf{F}_0^{-1}(\mathbf{v}_T - \mathbf{v}_0)]\,, \tag{4.75}$$

and, recalling (4.69), Eq. (4.72) can be simplified to

$$\frac{d_0}{dt}\log(\phi_T J_N) = 0\,, \tag{4.76}$$

an equation similar to (4.55).

If tumor cells are assumed to grow isotropically, so that $\mathbf{G}_T = g_T\mathbf{I}$, one then has that $\mathbf{F}_N = g_T^{-1}\mathbf{F}_T\mathbf{F}_p^{-1}$ and

$$\frac{3}{g_T}\frac{d_0 g_T}{dt} = \frac{1}{J_0}\frac{d_0 J_0}{dt} + \frac{1}{J_T}\frac{d_0 J_T}{dt} + \frac{\Gamma_T}{\phi_T} - \frac{1}{\phi_T J_0}\mathrm{Div}_0[\phi_T J_0\mathbf{F}_0^{-1}(\mathbf{v}_T - \mathbf{v}_0)]\,, \tag{4.77}$$

where $\Gamma_T$ has to be given constitutively.

Coming to the momentum equation, following Section 7.3.1, the extracellular liquid is treated as an inviscid fluid in light of the usual assumptions used to get Darcy's law

$$\begin{cases} \mathbf{0} = -\phi_0\nabla P + \nabla\cdot\hat{\mathbf{T}}_0 - \mathbf{M}_{T0}(\mathbf{v}_0 - \mathbf{v}_T)\,, \\[2mm] \mathbf{0} = -\phi_T\nabla P + \nabla\cdot\hat{\mathbf{T}}_T + \mathbf{b}_T + \mathbf{M}_{\ell T}(\mathbf{v}_\ell - \mathbf{v}_T) - \mathbf{M}_{T0}(\mathbf{v}_T - \mathbf{v}_0)\,, \quad (4.78) \\[2mm] \mathbf{0}_\ell = -\phi_\ell\nabla P - \mathbf{M}_{\ell T}(\mathbf{v}_\ell - \mathbf{v}_T)\,. \end{cases}$$

As in Section 7.3.1, introducing $\mathbf{K}$ and $\mathbf{K}_0$ and writing the momentum equation of the mixture, one can write

$$\begin{cases} \mathbf{0} = -\phi_0\nabla P + \nabla\cdot\hat{\mathbf{T}}_0 - \mathbf{K}_0^{-1}(\mathbf{v}_0 - \mathbf{v}_T)\,, \\[2mm] \mathbf{v}_\ell - \mathbf{v}_T = -\mathbf{K}\nabla P\,, \tag{4.79} \\[2mm] \mathbf{0} = -\nabla P + \nabla\cdot(\hat{\mathbf{T}}_0 + \hat{\mathbf{T}}_T) + \mathbf{b}_T\,, \end{cases}$$

which can be written in the Lagrangian framework defined by the ECM. Having in mind the need to include pseudo-plastic effects, we have to mention that although for the ECM one can take

$$\hat{\mathbf{T}}_0 = \mu_0\phi_0[-J_0^{-q_0}\mathbf{I} + \mathbf{B}_0]\,, \tag{4.80}$$

to describe the behavior of the tumor cells we can distinguish these cases:

1. When and where the cell population is subject to a moderate amount of stress, the body behaves elastically; there are no plastic deformations, which means $\mathbf{F}_p = \mathbf{I}$ and then $\mathbf{F}_T = g_T\mathbf{F}_N$.

2. When and where the stress overcomes a threshold yield stress $\sigma$ the body behaves as a compressible liquid.

Referring to Figures. 7.13 and 7.14 one can argue that the resistance of a single bond is nearly constant, so that the threshold level distinguishing the appearance of plastic deformations is proportional to the amount of cells present in the sample. On this basis a constitutive equation of the following type can be suggested.

$$
\hat{\mathbf{T}}_T = \begin{cases} \mu\phi_T[-(J_N)^{-q}\mathbf{I} + \mathbf{B_N}], & \text{if } \sqrt{|I\!I_T|} < \phi_T\sigma, \\ \phi_T\left[-\Sigma\mathbf{I} + 2\left(\eta + \dfrac{\sigma}{\sqrt{|I\!I_{2D}|}}\right)\mathbf{D}\right], & \text{if } \sqrt{|I\!I_T|} \geq \phi_T\sigma, \end{cases} \tag{4.81}
$$

where $I\!I_T$ is the second invariant of the stress tensor with the isotropic pressure omitted.

We notice that in simple shear problems, the constitutive relation (4.81) is rewritten

$$
\hat{T}_{12} = \begin{cases} \mu\phi_T\dfrac{\partial u_x}{\partial Y}, & \text{if } \hat{T}_{12} < \phi_T\sigma, \\ \phi_T(\sigma + \eta\dot{\gamma}) & \text{if } \hat{T}_{12} \geq \phi_T\sigma, \end{cases} \tag{4.82}
$$

where $u_x$ is the displacement along the $x$-direction. In order to describe possible shear-thinning effects, $\eta$ can depend on $I\!I_D$

$$
\eta = m|I\!I_D|^{(n-1)/2}, \tag{4.83}
$$

where the coefficient $n$ is related to the slope of the shear stress behavior versus the shear rate. In this way we obtain a constitutive equation similar to the Herschel–Buckley model.

In order to be more realistic, one should take into account that ECM can be produced and cleaved, so that also for this constituent one should at least consider isotropic growth $\mathbf{G}_0 = g_0\mathbf{I}$. Luckily, as already mentioned in Section 7.2.2, the ECM is made of several constituents with different mechanical and chemical properties. Some of them continuously remodel, whereas others, such as elastin, barely turn over [RAa]. This biological fact may be very useful to still define a proper and useful Lagrangian reference frame fixed on an ECM constituent, so that one can consider at the same time ECM remodeling, tumor growth, and deformation of all constituents.

## 7.5   Future Perspective

It is becoming clearer and clearer that in addition to chemical signaling, mechanics plays an important role in tumor development not only to describe the mechanical interaction of the tumor with the surrounding tissues, but also for the interplay between mechanical properties of a tissue and the tumor developing in it. This opens up several new research directions that deserve further investigation.

Certainly, mathematical modeling still needs a characterization of the mechanical behavior of growing tissues, in order to quantify viscoelastic and plastic effect, to evaluate the importance of nonlinear effects and to identify the proper constitutive equation.

In this respect, for the sake of completeness, it need be mentioned that in addition to the examples presented above several authors proposed linear elastic-type models with the inclusion of suitable growth contributions [ARb, ARc, ARd, JOb], without splitting the deformation gradient into growth and deformation. More specifically, Jones and coworkers [JOb] propose

$$\mathbf{T}_T - \frac{1}{3}(\mathrm{tr}\mathbf{T}_T)\mathbf{I} = \frac{2}{3}E\left(\mathbf{E}_T - \frac{1}{3}g_T\mathbf{I}\right), \tag{5.1}$$

where $\mathbf{E}_T$ is the infinitesimal strain tensor and growth is assumed to be isotropic, or

$$\frac{D}{Dt}[\mathbf{T}_T - \frac{1}{3}(\mathrm{tr}\mathbf{T}_T)\mathbf{I}] - \mathbf{W}_T\mathbf{T}_T + \mathbf{T}_T\mathbf{W}_T = \frac{2}{3}E\left(\mathbf{D}_T - \frac{1}{3}(\nabla\cdot\mathbf{v}_T)\mathbf{I}\right), \tag{5.2}$$

where $D/Dt$ is the material time derivative following the tumor and $\mathbf{W}_T$ is the spin tensor, and Araujo and McElwain [ARb, ARc, ARd] proposed

$$\frac{\mathcal{D}}{\mathcal{D}t}\left[\mathbf{T}_T - \frac{1}{3}(\mathrm{tr}\mathbf{T}_T)\mathbf{I}\right] = 2\mu\left(\frac{\mathcal{D}\mathbf{E}_T}{\mathcal{D}t} - (\nabla\cdot\mathbf{v}_T)\mathbf{G}_T\right) - \phi_T\frac{\mathcal{D}P}{\mathcal{D}t}(3\mathbf{G}_T - \mathbf{I}), \tag{5.3}$$

where $\mathcal{D}/\mathcal{D}t$ is a convective derivative based on the tumor, and $\mathbf{G}_T$ is a diagonal growth tensor.

The use of linear elasticity probably gives more freedom because it circumvents the difficulties implied by the definition of a proper Lagrangian framework. However, there is incompatibility between the use of convective derivatives and linear elasticity. In addition, one has still to define what is a small deformation in a growing tumor, that is, a mass that starting from a single cell grows to a size of at least few millimeters. Probably, this is overcome by the introduction of plasticity and by the use of evolving natural configuration. In fact, as a first approximation then the deformation with respect to the configuration achieved after the occurrence of plastic phenomena can be assumed to be small. Actually, one knows that for tensions larger that 0.1 kPA adhesion bonds break up. It would then be very interesting to correctly frame the linear elastic approach suggested by the authors above using the concept of evolving natural configuration.

Other issues that need to be investigated further concern how a tumor remodels the surrounding environment and how, vice versa, the environment influences tumor growth. In order to do that, the multiphase models presented in this chapter should be generalized to include more constituents belonging to both the immune system and the stroma. In particular, the presence of macrophages and their migration toward the hypoxic regions is very important for the related immune response, the formation of cronic inflammations, and their angiogenic side effect. On the other hand, the possibility of developing therapies based on the use of engineered macrophages should be supported by suitable mathematical models.

In the stroma it might be important to include other types of cells such as fibroblasts related to the production of extracellular matrix or to distinguish the different constituents of the extracellular matrix, because, as already mentioned in Section 7.4.4, some of them are subject to stronger turnover and remodeling than others. The balance between the different constituents and their percentage influences the overall mechanical characteristics of the tissue and the formation of the different environments in which tumor cells live. As already stated, the entire remodeling process is strongly affected by the stresses and strains to which the tissue is subject.

Going to the inner characteristics of the cells, the tumor itself may contain several functionally different clones that differentiate in their genetic status, for example, cells with normal and abnormal expression of the tumor suppressor gene, p53, and hormone sensitive and insensitive cells. In this respect, one of the breakthroughs in modeling tumor growth consists in including what happens inside the cells and therefore in developing multiscale models that take into account the cascades of events recalled in Figure 7.3 possibly joined with those involving growth factors.

The need of working in a multiscale framework is an almost unconscious standard procedure in biology and medicine. In fact, in order to understand and describe the behavior of any biological phenomenon, researchers in life sciences tend to go to the smallest scale possible, because they know, for instance, that the behavior of a cell and the interactions that it has with the surrounding environment depend on the chemistry inside it and, after all, on the content of genetic information, on the particular genetic expression, on the activation of proper protein cascades, and on their cross influence. Of course, in order to do that one needs to have estimates on the affinity constants. In the absence of such measurements, one could initially start with Boolean reasoning. However, things become complicated when there are loops and intersections between different cascades triggered by different events.

For instance, in the models presented in this chapter the role played by nutrients and growth-promoting and inhibitory factors is considered secondary but it is not. So, it is important to develop models that consider both mechanical and chemical cues and establish the relative importance at the protein cascade level.

Another interesting problem that has not been studied yet is the growth of tumors in a mechanically heterogeneous environment, which includes network structures such as blood vasculature, airways, and lymphatic system, the interaction with physical barriers such as bones and cartilage, and the pressure on the surrounding tissues.

However, whenever developing all the generalizations above, one has to keep in mind the difficulties in obtaining specific measurements from the biologists. For instance, quantifying the dependence of the production rates of extracellular matrix and matrix-degrading enzymes from the level of stress and/or strain is not easy and data are not available yet, although the effect was put in evidence many years ago and is applied in clinical practice. Mathematical modeling urgently needs to be based on biological measurements. On the other hand, we are sure that experimental research can be stimulated by the development of mathematical models that on the basis of known experimental evidence and data showing the importance of mechanical aspects go one step beyond what is known in biology and medicine.

**Acknowledgments**
Partially supported by the European Community, through the Marie Curie Research Training Network Project HPRN-CT-2004-503661: Modelling, Mathematical Methods and Computer Simulation of Tumor Growth and Therapy.

## 7.6   References

[ADa]   Adam, J.A., and Bellomo, N., Eds. **A Survey of Models on Tumor Immune Systems Dynamics**, Birkhäuser, Boston (1996).

[AMa]   Ambrosi, D., and Guana, F., Stress modulated growth, *Math. Mech. Solids.*, (2005) at http://mms.sagepub.com/cgi/content/abstract/1081286505059739v1.

[AMb]   Ambrosi, D., and Mollica, F., On the mechanics of a growing tumor, *Int. J. Engng. Sci.*, **40** (2002), 1297–1316.

[AMc]   Ambrosi, D., and Mollica, F., The role of stress in the growth of a multicell spheroid, *J. Math. Biol.*, **48** (2004), 477–499.

[AMd] Ambrosi, D., and Preziosi, L., On the closure of mass balance models for tumor growth, *Math. Mod. Meth. Appl. Sci.* **12** (2002), 737–754.

[ARa] Araujo, R.P., and McElwain, D.L.S., A history of the study of solid tumor growth: The contribution of mathematical modelling, *Bull. Math. Biol.*, **66** (2004), 1039–1091.

[ARb] Araujo, R.P., and McElwain, D.L.S., A mixture theory for the genesis of residual stresses in growing tissues, I: A general formulation, *SIAM J. Appl. Math.*, **65** (2005), 1261–1284.

[ARc] Araujo, R.P., and McElwain, D.L.S., A mixture theory for the genesis of residual stresses in growing tissues, II: Solutions to the biphasic equations for a multicell spheroid, *SIAM J. Appl. Math.*, **65** (2005), 1285–1299.

[ARd] Araujo, R.P., and McElwain, D.L.S., A linear-elastic model of anisotropic tumour growth, *Eur. J. Appl. Math.*, **15** (2004), 365–384.

[BAa] Baumgartner, W., Hinterdorfer, P., Ness, W., Raab, A., Vestweber, D., Schindler, H., and Drenckhahn, D., Cadherin interaction probed by atomic force microscopy, *Proc. Nat. Acad. Sci. USA*, **97** (2000), 4005–4010.

[BEa] Becker, K.F., Atkinson, M.J., Reich, U., Becker, I., Nekarda, H., Siewert, J.R., and Hofler, H., E-cadherin gene mutation provide clues to diffuse gastric carcinoma, *Cancer Res.*, **54** (1994), 3845–3852.

[BEb] Bellomo, N., and De Angelis, E., Eds. Special issue on modeling and simulation of tumor development, treatment, and control, *Math. Comp. Modeling*, **37** (2003).

[BLa] Blatz, P.J., and Ko, W.L., Application of finite elasticity theory to the deformation of rubbery materials, *Trans. Soc. Rheology*, **6** (1962), 223–251.

[BRa] Breward, C.J.W., Byrne, H.M., and Lewis, C.E., The role of cell-cell interactions in a two-phase model for avascular tumor growth, *J. Math. Biol.*, **45** (2002), 125–152.

[BRb] Breward, C.J.W., Byrne, H.M., and Lewis, C.E., A multiphase model describing vascular tumor growth, *Bull. Math. Biol.*, **65** (2003), 609–640.

[BRc] Brewster, C.E., Howarth, P.H., Djukanovic, R., Wilson, J., Holgate, S.T., and Roche, W.R., Myofibroblasts and subepithelial fibrosis in bronchial asthma, *Am. J. Respir. Cell Mol. Biol.*, **3** (1990), 507–511.

[BRd] Brown, L.F., Guidi, A.J., Schnitt, S.J., van de Water, L., Iruela-Arispe, M.L., Yeo, T.-K., Tognazzi, K., and Dvorak, H.F., Vascular stroma formation in carcinoma in situ, invasive carcinoma and metastatic carcinoma of the breast, *Clin. Cancer Res.*, **5** (1999), 1041–1056.

[BUa] Bussolino, F., Arese, M., Audero, E., Giraudo, E., Marchiò, S., Mitola, S., Primo, L., and Serini, G., Biological aspects of tumour angiogenesis, in: **Cancer Modeling and Simulation**, Preziosi, L., Ed., Boca Raton, FL: Chapman & Hall/CRC Press, 1–22 (2003).

[BYa] Byrne, H.M., King, J.R., McElwain, D.L.S., and Preziosi, L., A two-phase model of solid tumor growth, *Appl. Math. Letters*, **16** (2003), 567–573.

[BYb] Byrne, H.M., and Preziosi, L., Modeling solid tumor growth using the theory of mixtures, *Math. Med. Biol.*, **20** (2004), 341–366.

[CAa] Canetta, E., Leyrat, A., Verdier, C., and Duperray, A., Measuring cell viscoelastic properties using a force-spectrometer: Influence of the protein-cytoplasm interactions, *Biorheology*, **42** (2005), 321–333.

[CAb] Carmeliet, P., and Jain, R.K., Angiogenesis in cancer and other diseases, *Nature*, **407** (2000), 249-257.

[CAc] Castilla, M.A., Arroyo, M.V.A., Aceituno, E., Aragoncillo, P., Gonzalez-Pacheco, F.R., Texeiro, E., Bragado, R., and Caramelo, C., Disruption of cadherin-related junctions triggers autocrine expression of vascular endothelial growth factor in bovine aortic endothelial cells. Effect on cell proliferation and death resistance, *Circ. Res.*, **85** (1999), 1132–1138.

[CAd] Cavallaro, U., Schaffhauser, B., and Christofori, G., Cadherin and the tumor progression: Is it all in a switch?, *Cancer Lett.*, **176** (2002), 123–128.

[CAe] Caveda, L., Martin-Padura, I., Navarro, P., Breviario, F., Corada, M., Gulino, D., Lampugnani, M.G., and Dejana, E., Inhibition of cultured cell growth by vascular endothelial cadherin (cadherin-5/VE-cadherin), *J. Clin. Invest.*, **98** (1996), 886–893.

[CHa] Chambers, A.F., and Matrisian, L.M., Changing views of the role of matrix metalloproteinases in metastasis, *J. Natl. Cancer Inst.*, **89** (1997), 1260–1270.

[CHb] Chaplain, M.A.J., Ed. Special issue *Math. Mod. Methods Appl. Sci.*, **9** (1999).

[CHc] Chaplain, M.A.J., Ed. Special issue on Mathematical Modeling and Simulations of Aspects of Cancer Growth, *J. Theor. Med.*, **4** (2002).

[CHd] Chaplain, M.A.J., **Mathematical Modelling of Tumour Growth**, Springer, New York (2006).

[CHe] Chaplain, M., Graziano, L., and Preziosi, L., Mathematical modelling of the loss of tissue compression responsiveness and its role in solid tumour development, *Math. Med. Biol.*, **23** (2006) 197–229.

[CHf]   Chiquet, M., Matthisson, M., Koch, M., Tannheimer, M., and Chiquet-Ehrismann, R., Regulation of extracellular matrix synthesis by mechanical stress, *Biochem. Cell Biol.*, **74** (1996), 737–744.

[CHg]   Christofori, G., and Semb, H., The role of cell-adhesion molecule E-cadherin as a tumors-suppressor gene, *Trends Biochem. Sci.*, **24** (1999), 73–76.

[COa]   Coats, S., Flanagan, W.M., Nourse, J., and Roberts, J., Requirement of p27kip1 for restriction point control of the fibroblast cell cycle, *Science*, **272** (1996), 877–880.

[CRa]   Craft, P.S., and Harris, A.L., Clinical prognostic-significance of tumor angiogenesis, *Annals of Oncology*, **5** (1994), 305–311.

[DEa]   Dejana, E., Lampugnani, M.G., Giorgi, M., Gaboli, M., and Marchisio, P.C., Fibrinogen induces endothelial cell adhesion and spreading via the release of endogenous matrix proteins and the recruitment of more than one integrin receptor, *Blood*, **75** (1990), 1509–1517.

[DEb]   Deleu, L., Fuks, F., Spitkovsky, D., Hörlein, R., Faisst, S., and Rommelaere, J., Opposite transcriptional effects of cyclic AMP-responsive elements in confluent or p27kip-overexpressing cells versus serum-starved or growing cells, *Molec. Cell. Biol.*, **18** (1998), 409–419.

[DIa]   Dietrich, C., Wallenfrang, K., Oesch, F., and Wieser, R., Differences in the mechanisms of growth control in contact-inhibited and serum-deprived human fibroblasts, *Oncogene*, **15** (1997), 2743–2747.

[DOa]   Dorie, M.J., Kallman, R.F., Rapacchietta, D.F., Van Antwerp, D., and Huang, Y.R., Migration and internalisation of cells and polystyrene microspheres in tumour cell spheroids, *Exp. Cell. Res.*, **141** (1982), 201–209.

[DOb]   Dorie, M.J., Kallman, R.F., and Coyne, M.A., Effect of Cytochalasin B Nocodazole on migration and internalisation of cells and microspheres in tumour cells, *Exp. Cell Res.*, **166** (1986), 370–378.

[ELa]   Elliot, C.M., The Stefan problem with a non-monotone constitutive relation, *IMA J. Appl. Math.*, **35** (1985), 257–264.

[EYa]   Eyre, D.R., Biochemistry of the invertebral disk, *Int. Rev. Connect. Tissue Res.*, **8** (1979), 227–291.

[FOa]   Folkman, J., Tumor angiogenesis, *Adv. Cancer Res.*, **19** (1974), 331–358.

[FOb]   Folkman, J., and Hochberg, M., Self-regulation of growth in three dimensions, *J. Exp. Med.*, **138** (1973), 745–753.

[FRa]   Franks, S.J., Byrne, H.M., King, J.R., Underwood, J.C.E., and Lewis, C.E., Modelling the early growth of ductal carcinoma in situ of the breast, *J. Math. Biol.*, **47** (2003), 424–452.

[FRb]   Franks, S.J., Byrne, H.M., Mudhar, H.S., Underwood, J.C.E., and Lewis, C.E., Mathematical modelling of comedo ductal carcinoma in situ of the breast, *Math. Med. Biol.*, **20** (2003), 277–308.

[FRc]   Franks, S.J., and King, J.R., Interactions between a uniformly proliferating tumor and its surrounding. Uniform material properties, *Math. Med. Biol.*, **20** (2003), 47–89.

[FRd]   Freyer, J.P., and Sutherland, R.M., Regulation of growth saturation and development of necrosis in EMT6/Ro multicellular spheroids by the glucose and oxygen supply, *Cancer Res.*, **46** (1986), 3504–3512.

[GOa]   Gottardi, C.J., Wong, E., and Gumbiner, B.M., E-cadherin suppresses cellular transformation by inhibiting $\beta$-catenin signalling in an adhesion-independent manner, *J. Cell. Biol.*, **153** (2001), 1049–1060.

[HAa]   Harja, K.M., and Fearon, E.R., Cadherin and catenin alterations in human cancer, *Genes Chromosomes Cancer*, **34** (2002), 255–268.

[HEa]   Helmlinger, G., Netti, P.A., Lichtenbeld, H.C., Melder, R.J., and Jain, R.K., Solid stress inhibits the growth of multicellular tumour spheroids, *Nature Biotech.*, **15** (1997), 778–783.

[HEb]   Helmlinger, G., Netti, P.A., Lichtenbeld, H.C., Melder, R.J., and Jain, R.K., Solid stress inhibits the growth of multicellular tumor spheroids, *Nature Biotech.*, **15** (1997), 778–783.

[HIa]   Hillen, T., Hyperbolic models for chemosensitive movement, *Math. Mod. Meth. Appl. Sci.*, **12** (2002), 1007–1034.

[JAa]   Jackson, T.L., and Byrne, H.M., A mathematical model to study the effects of drug resistance and vasculature on the response of solid tumours to chemotherapy, *Math. Biosci.*, **164** (2000), 17–38.

[JOa]   Johnson, P.R.A., Role of human airway smooth muscle in altered extracellular matrix production in asthma, *Clin. Exp. Pharm. Physiol.*, **28** (2001), 233–236.

[JOb]   Jones, A.F., Byrne, H.M., Gibson, J.S., and Dold, J.W., A mathematical model of the stress induced during solid tumour growth, *J. Math. Biol.*, **40** (2000), 473–499.

[KAa]   Kato, A., Takahashi, H., Takahashi, Y., and Matsushime, H., Inactivation of the cyclin D-dependent kinase in the rat fibroblast cell line, 3Y1, induced by contact inhibition, *J. Biol. Chem.*, **272** (1997), 8065–8070.

[KIa] Kim, S.-G., Akaike, T., Sasagawa, T., Atomi, Y., and Kurosawa, H., Gene expression of type I and type III collagen by mechanical stretch in anterior cruciate ligament cells, *Cell Struct. Funct.*, **27** (2002), 139–144.

[KJa] Kjaer, M., Role of extracellular matrix in adaptation of tendons and skeletal muscle to mechanical loading, *Physiol. Rev.*, **84** (2004), 649–698.

[KLa] Klominek, J., Robert, K.H., and Sundqvist, K.-G., Chemotaxis and haptotaxis of human malignant mesothelioma cells: Effects of fibronectin, laminin, type IV collagen, and an autocrine motility factor-like substance, *Cancer Res.*, **53** (1993), 4376–4382.

[KOa] Kowalczyk, R., Preventing blow-up in a chemotaxis model, *J. Math. Anal. Appl.*, **305** (2005), 566–588.

[LAa] Lawrence, J.A., and Steeg, P.S., Mechanisms of tumor invasion and metastasis, *World J. Urol.*, **14** (1996), 124–130.

[LEa] Levenberg, S., Yarden, A., Kam, Z., and Geiger, B., p27 is involved in N-cadherin-mediated contact inhibition of cell growth and S-phase entry, *Oncogene*, **18** (1999), 869–876.

[LEb] Levick, J.R., Flow through interstitium and other fibrous matrices, *Q. J. Cogn. Med. Sci.*, **72** (1987), 409–438.

[LIa] Liotta, L.A., and Kohn, E.C., The microenvironment of the tumor-host interface, *Nature*, **411** (2001), 375–379.

[MAa] MacKenna, D., Summerour, S.R., and Villarreal, F.J., Role of mechanical factors inmodulatin cardiac fibroblast function and extracellular matrix synthesis, *Cardiovasc. Res.*, **46** (2000), 257–263.

[MAb] Mantzaris, N., Webb, S., and Othmer, H.G., Mathematical modelling of tumour-induced angiogenesis, *J. Math. Biol.* **49** (2004), 111–187 (2004).

[MAc] Mao, J.J., and Nah, H.-D., Growth and development: Hereditary and mechanical modulations, *Amer. J. Orthod. Dentofac. Orthop.*, **125** (2004), 676–689.

[MAd] Matrisian, L.M., The matrix–degrading metalloproteinases, *Bioessays*, **14** (1992), 455–463.

[MAe] Maurice, D.M., The cornea and the sclera, in **The Eye**, Davson, H., Ed., Academic Press, (1984) 1–158.

[MOa] Mow, V.C., Holmes, M.H., and Lai, W.M., Fluid transport and mechanical problems of articular cartilage: A review, *J. Biomech.*, **17** (1984), 377–394.

[MOb] Mow, V.C. and Lai, W.M., Mechanics of animal joints, *Ann. Rev. Fluid Mech.*, **11** (1979), 247–288.

[NEa] Nelson, C.M., and Chen, C.S., VE-cadherin simultaneously stimulates ad inhibits cell proliferation by altering cytoskeletal structure and tension, *J. Cell Science*, **116** (2003), 3571–3581.

[ODa] Oda, T., Kanai, Y., Okama, T., Yoshiura, K., Shimoyama, Y., Birchmeier, W., Sugimura, T., and Hirohashi, S., E-cadherin gene mutation in human gastric carcinoma cell lines, *Proc. Natl. Acad. Sci. USA*, **91** (1994), 1858–1862.

[ORa] Orford, K., Orford, C.C., Byers, S.W., Exogenous expression of β-catenin regulates contact inhibition, anchorage-independent growth, anoikis, and radiation-induced cell cycle arrest, *J. Cell Biol.* **146** (1999), 855–867.

[PAa] Painter, K., and Hillen, T., Volume filling and quorum-sensing in models for chemosensitive movement, *Can. Appl. Math. Quart.*, **10** (2002), 501–543.

[PAb] Parson, S.L., Watson, S.A., Brown, P.D., Collins, H.M., and Steele, R.J.C., Matrix metalloproteinases, *Brit. J. Surg.*, **84** (1997), 160–166.

[PAc] Paszek, M.J., Zahir, N., Johnson, K.R., Lakins, J.N., Rozenberg, G.I., Gefen, A., Reinhart-King, C.A., Margulies, S.S., Dembo, M., Boettiger, D., Hammer, D.A., and Weaver, V.M., Tensional homeostasis and the malignant phenotype, *Cancer Cell* **8** (2005), 241–254.

[POa] Polyak, K., Kato, J., Solomon, M.J., Sherr, C.J., Massague, J., Roberts, J.M., and Koff, A., p27Kip1, a cyclin-Cdk inhibitor, links transforming growth factor-β and contact inhibition to cell cycle arrest, *Genes & Develop.*, **8** (1994), 9–22.

[PRa] Preziosi, L., Ed., **Cancer Modelling and Simulation**, CRC-Press/ Chapman Hall, Boca Raton, Fl (2003).

[PRb] Preziosi, L., and Farina, A., On Darcy's law for growing porous media, *Int. J. Nonlinear Mech.*, **37** (2001), 485–491.

[PUa] Pujuguet, P., Hammann, A., Moutet, M., Samuel, J.L., Martin, F., and Martin, M., Expression of fibronectin EDA+ and EDB+ isoforms by human and experimental colorectal cancer, *Am. J. Patho.*, **148** (1996), 579–592.

[RAa] Rao, I.J., Humphrey, J.D., and Rajagopal, K.R., Biological growth and remodeling: A uniaxial example with possible application to tendons and ligaments, *CMES*, **4** (2003), 439–455.

[RIa] Risinger, J.I., Berchuck, A., Kohler, M.F., and Boyd, J., Mutation of E-cadherin gene in human gynecological cancers, *Nature Genetics*, **7** (1994), 98–102.

[SEa]  Secomb, T.W., and El-Kareh, A.W., A theoretical model for the elastic properties of very soft tissues, *Biorheology*, **38** (2001), 305–317.

[STa]  St. Croix, B., Sheehan, C., Rak, J.W., Florenes, V.A., Slingerland, J.M., and Kerbel, R.S., E-cadherin-dependent growth suppression is mediated by the cyclin-dependent kinase inhibitor p27 (Kip1), *J. Cell Biol.*, **142** (1998), 557–571.

[STb]  Stetler-Stevenson, W.G., Hewitt, R., and Corcoran, M., Matrix metallo–proteinases and tumor invasion: From correlation to causality to the clinic, *Cancer Biol.*, **7** (1996), 147–154.

[STc]  Stockinger, A., Eger, A., Wolf, J., Beug, H., and Foisner, R., E-cadherin regulates cell growth by modulating proliferation-dependent $\beta$-catenin transcriptional activity, *J. Cell. Biol.*, **152** (2001), 1185–1196.

[SUa]  Sutherland, R.M., Cell and environment interactions in tumor microregions: The multicell spheroid model, *Science*, **240** (1988), 177–184.

[TAa]  Takeuchi, T., Misaki, A., Liang, S.-B., Tachibana, A., Hayashi, N., Sonobe, H., and Ohtsuki Y., Expression of T-cadherin (CDH13, H-cadherin) in human brain and its characteristics as a negative growth regulator of epidermal growth factor in neuroblastoma cells, *J. Neurochem.*, **74** (2000), 1489–1497.

[TAb]  Takeuchi, J., Sobue, M., Sato, E., Shamoto, M., and Miura, K., Variation in glycosaminoglycan components of breast tumors, *Cancer Res.*, **36** (1976), 2133–2139.

[TSa]  Tseng, S.C.G., Smuckler, D., and Stern, R., Comparison of collagen types in adult and fetal bovine corneas, *J. Biol. Chem.*, **257** (1982), 2627–2633.

[TZa]  Tzukatani, Y., Suzuki, K., and Takahashi, K., Loss of density-dependent growth inhibition and dissociation of $\alpha$-catenin from E-cadherin, *J. Cell. Physiol.*, **173** (1997), 54–63.

[TZb]  Tzukita, S., Itoh, M., Nagafuchi, A., Yonemura, S., and Tsukita, S., Submembrane junctional plaque proteins include potential tumor suppressor molecules, *J. Cell Biol.*, **123** (1993), 1049–1053.

[UGa]  Uglow, E.B., Angelini, G.D., and George, S.J., Cadherin expression is altered during intimal thickening in humal sapphenous vein, *J. Submicrosc. Cytol. Pathol.* **32** (2000), C113–C119.

[UGb]  Uglow, E.B., Slater, S., Sala-Newby, G.B., Aguilera-Garcia, C.M., Angelini, G.D., Newby, A.C., and George, S.J., Dismantling of cadherin-mediated cell-cell contacts modulates smooth muscle cell proliferation, *Circ. Res.*, **92** (2003), 1314–1321.

[WAa] Warchol, M.E., Cell proliferation and N-cadherin interactions regulate cell proliferation in the sensory epithelia of the inner ear, *J. Neurosci.*, **22** (2002), 2607–2616.

[WIa] Witelski, T.P., Shocks in nonlinear diffusion, *Appl. Math. Lett.*, **8** (1995), 27–32.

[WOa] Woo, S.L.-Y., Biomechanics of tendon and ligaments, in **Frontiers in Biomechanics**, G. W. Schmid-Schonbein, S. L.-Y. Woo and Zweifach, B.W., Eds., Springer-Verlag, New York, 180–195 (1986).

[YAa] Yang, C.-M., Chien, C.-S., Yao, C.-C., Hsiao, L.-D., Huang, Y.-C., and Wu, C.B., Mechanical strain induces collagenases-3 (MMP-13) expression in MC3T3-E1 osteoblastic cells, *J. Biol. Chemistry*, **279** (2004), 22158–22165.

[ZHa] Zhang, Y., Nojima, S., Nakayama, H., Yulan, J., and Enza, H., Characteristics of normal stromal components and their correlation with cancer occurrence in human prostate, *Oncol. Rep.*, **10** (2003), 207–211.

# 8

---

# *Inhomogeneities*
# *in Biological Membranes*

---

R. Rosso and E. G. Virga

*Dipartimento di Matematica*
*Università di Pavia*
*I-27100 Pavia, Italy*

ABSTRACT. This contribution is concerned with the mathematical modeling of biological membranes. In particular, it explores the role played by some structures that make them inhomogeneous.

---

## 8.1   Introduction

Modeling the complex behavior of biological membranes has been one of the major tasks of biophysics and biomathematics during the past three decades. "All biological membranes, including the plasma membrane and the internal membranes of eucaryotic cells, have a common overall structure: they are assemblies of lipid and protein molecules held together by noncovalent interactions" (see p.255 of [ALa]). Thus, the presence of inhomogeneities in a biological membrane cannot be easily overlooked, especially because they are responsible for the diverse functions performed by biological membranes, which range from transport of specific molecules across the membrane, to the reception of chemical signals from the extracellular environment. It is little surprise that such a complex structure could be only modeled by degrees, and that even crude approximations needed to be introduced at the beginning of the 1970s, when intensive research in the field began. Aims were ambitious, as easily inferred from the very title of Canham's [CAa] pioneering paper: "The Minimum Energy of Bending as a Possible Explanation of the Biconcave Shape of the Human Red Blood Cell." The ability to describe, at least qualitatively, the complex geometry of red blood cells, prompted many authors to propose mathematical models that could give a more accurate description of the complex morphologies of biological membranes. As a result, to account for the membrane's elastic properties, refined models were proposed that incorporated the structural inhomogeneities of the two leaflets forming the bilayer structure, the basic architecture of biological membranes. Moreover, because different species of lipids exist within a single membrane, the question arises as to whether a lateral inhomogeneous composition can affect the membrane morphology. Proteins are another basic ingredient of biological membranes that need a separate treatment. Because proteins are responsible for the effective functioning of biological membranes, their influence on the membrane's shape has been studied in depth. Particular emphasis has been put on the role of the membrane's elasticity in mediating forces among proteins, eventually leading to the formation of protein aggregates.

This cursory outlook already witnesses the richness of biological membrane modeling. In the following, we select a few topics, all arranged around the leading *leitmotiv* of inhomogeneity. After a general discussion on the elastic properties of biological membranes, contained in Section 8.2, Section 8.3 is focused on the coupling between the inhomogeneous composition of membranes and their elastic properties. Sections 8.4 to 8.6 are devoted to protein–protein membrane-mediated interactions. Section 8.4 is concerned with transmembrane proteins and with some mathematical problems they pose; in particular it is discussed how elasticity can either promote or hamper protein aggregation. In Section 8.5 thermal fluctuations are incorporated in the treatment of membrane-mediated interactions between proteins. Section 8.6 examines the peculiarities of peripheral proteins. Finally, Section 8.7 contains some conclusive remarks. Because the literature on the subject is enormous, many interesting topics such as adhesion and dynamics are missing in our review. The interested reader is urged to consult Seifert's review [SEb] for an account of the topics omitted here.

## 8.2   Bare Membranes

The basic constituents of biological membranes are lipid molecules, often called the *amphiphiles*, which are composed of two distinct parts: a polar hydrophilic headgroup, and one or two hydrophobic chains. This structure originates a conflict between the amphiphiles and the aqueous environment within which they live, and it is responsible for the rich variety of arrangements shown by their assemblies. Whereas amphiphilic molecules experience a hydrophobic attraction $F_\gamma$ that reduces the exposure of the hydrocarbon chains to water, the headgroups are subject to hydrophilic, ionic, or steric interactions giving rise to a repulsive force $F_h$. A further repulsive force $F_c$ among the hydrophobic chains is induced by their thermal motion, which overcomes attractive van der Waals forces. This competition drives the effective headgroup area $a$ towards an equilibrium value $a_0$ that minimizes the total interfacial energy.

Two more parameters play a basic role in determining the most favored aggregate: the volume $v$ spanned by the hydrocarbon chains, and the *effective chain length* $\ell_c$, which sets an upper bound on the extension of the chains. It turns out (see Chapters 16 and 17 of [ISa]) that spherical micelles require $v/a_0\ell_c < \frac{1}{3}$, nonspherical micelles are found when $v/a_0\ell_c \in (\frac{1}{3}, \frac{1}{2})$, bilayers when $v/a_0\ell_c \in (\frac{1}{2}, 1)$, whereas inverted structures need $v/a_0\ell_c > 1$. In a bilayer, two layers of amphiphiles are joined so that the hydrophobic parts are shielded by the polar headgroups. Because along the rims of the bilayer an unfavorable contact between hydrocarbon chains and water exists, bilayers can find it advantageous to bend themselves forming either hollow cylinders called the *tubules*, or closed structures, called the *vesicles*. Following Lee [LEa], the onset of curved monolayers can be explained by a balance of forces. When $F_\gamma$, $F_c$, and $F_h$ are in equilibrium, planar monolayers are formed, whereas structures curved in opposite directions are favored if either $F_c$ or $F_h$ prevails over the other forces. As to bilayers, frustration can occur whenever both monolayers would like to bend themselves in the same direction: when the elastic energy stored in the bilayer is too high, nonbilayer structures are preferred.

In the fluid phase $L_\alpha$, the chains of the amphiphiles are orthogonal to the ideal surfaces spanned by the headgroups and, hence, this makes a lipid membrane similar to a smectic-A liquid crystal. The width of a bilayer ($\simeq 10^{-9}$m) is much smaller than the lateral dimensions of either a vesicle or a tubule ($\simeq 10^{-6}$m), therefore it is natural to model the vesicle's membrane as a compact two-dimensional surface $\mathcal{S}$ embedded in a three-dimensional Euclidean space. Because lipid membranes are inextensible, the area of a vesicle or a tubule should be prescribed. The bilayer forming the membrane, although permeable to water, is impermeable to the dissolved ions, which makes it plausible to treat the volume enclosed by a vesicle as constant when the osmotic pressure between the inner and the outer fluid is large enough. No such constraint exists for tubules, inasmuch as water can freely flow through their open ends.

As recognized by Helfrich [HEc], lipid bilayers resist bending much more than shearing and tilting. For this reason, the early models for the elastic properties of lipid membranes were built upon a free-energy density depending on the curvature invariants of the membrane, that is, the *total curvature* $H$, defined as twice the mean curvature of $\mathcal{S}$ and the Gaussian curvature $K$. The simplest elastic free-energy functional is the Canham–Helfrich Hamiltonian, which is quadratic in the membrane's principal curvatures:

$$\mathcal{F}_{el} = \int_{\mathcal{S}} \left[ \frac{\kappa}{2} H^2 + \kappa_G K \right] \mathrm{d}A \,, \tag{2.1}$$

where $A$ is the surface-area measure, and the constants $\kappa$ and $\kappa_G$ are the bending rigidity and the Gaussian bending rigidity, which must obey the inequalities

$$\kappa > 0 \quad \text{and} \quad \kappa_G + 2\kappa > 0 \,,$$

to ensure that the free-energy density in (2.1) is positive-definite.

Now, the Gauss–Bonnet theorem states that $\int_{\mathcal{S}} K \mathrm{d}A$ is a null-Lagrangian, depending only on the differential-geometric properties of the boundary $\partial\mathcal{S}$ of $\mathcal{S}$. If $\mathcal{S}$ has no boundary then

$$\int_{\mathcal{S}} K \mathrm{d}A = 4\pi(1 - g) \,,$$

where the integer $g$ is the topological genus of $\mathcal{S}$ counting the number of holes in $\mathcal{S}$. Hence, provided that the topology of the vesicle's membrane is fixed, the Gaussian curvature $K$ gives a constant contribution to $\mathcal{F}_{el}$ and can be dropped.

The free energy (2.1), originally introduced by Canham [CAa], makes the tacit assumption that the two monolayers forming the membrane $\mathcal{S}$ are homogeneous, and so the undistorted configuration of an infinite bilayer would be flat. This assumption is too severe a simplification, because chemical inhomogeneities often exist between the monolayers. This was recognized by Helfrich who replaced the term $\int_{\mathcal{S}} H^2 \mathrm{d}A$ in (2.1) by

$$\mathcal{F}_{SC}[\mathcal{S}] := \frac{\kappa}{2} \int_{\mathcal{S}} (H - H_0)^2 \mathrm{d}A \,, \tag{2.2}$$

where $H_0$ is referred to as the membrane's spontaneous curvature. Although in principle $H_0$ can vary from point to point, it is often assumed to be constant. In the latter case, if no further constraint exists, the undistorted configuration of $\mathcal{S}$ would be a sphere of radius $2/H_0$.

Another variant of (2.1) is the *bilayer-couple* model, first proposed by Sheetz and Singer [SHa]. This model was introduced to describe the shape changes induced by drugs containing amphiphilic molecules that adhere to the membrane either on the outer or on the inner monolayer. The model stipulates that the two monolayers in the membrane, being asymmetrical, respond differently to external perturbations, although they remain coupled to each other. The name *bilayer-couple* was chosen to suggest an analogy with the behavior of a bimetallic couple when the temperature is changed. As noted by Waugh [WAa], introducing spontaneous curvature models of chemical asymmetries between the monolayers constituting the membrane, however, the bilayer-couple model focuses on monolayers differing in the number of molecules, so that a net area difference between them exists, because permeation of amphiphiles across the monolayers is unlikely to occur. Because the monolayers are parallel surfaces, fixing the area difference between them amounts, to first order, to prescribing the value of $\int_S H \mathrm{d}A$. Thus, a further constraint is introduced, which makes the functional $\mathcal{F}_{SC}$ the same as $\mathcal{F}_{el}$, because a constant $H_0$ in (2.2) eventually results in an additive constant to the elastic free energy.

Phase diagrams showing a rich variety of equilibrium shapes were obtained for vesicles in the last two decades, as reviewed by Seifert in [SEb]. Neither the spontaneous curvature model nor the bilayer-couple model, however, could faithfully account for the details of the *budding* transition, a morphological transition in which a variation in some external parameter, such as temperature, drives a spherical vesicle through a series of more and more elongated shapes culminating in a *vesiculated* shape, in which the original vesicle is connected to a smaller vesicle by a narrow neck [MIa].

To follow the various stages of the budding transition, the *area-difference* model was proposed which is somehow in between the spontaneous curvature and the bilayer-couple model. In this model, previously employed by Svetina and Žekš [SVa], a nonlocal contribution is inserted in the elastic free energy, which now becomes

$$F_\Delta = \frac{\kappa}{2} \int_S [(H - H_0)^2 + \alpha\beta(\Delta A - \Delta A_0)^2]\mathrm{d}A\,, \qquad (2.3)$$

where $\beta$ is a numerical prefactor depending on the bilayer's width and the membrane's area, whereas the positive coupling coefficient $\alpha$ measures the strength of the nonlocal bending rigidity, compared to the local bending rigidity $\kappa$. Finally, $\Delta A$ is the actual area-difference between the monolayers that can be different from its relaxed value $\Delta A_0$. The area-difference model reproduces the spontaneous curvature model in the limit where $\alpha \to 0$ and the bilayer-couple model when $\alpha \to +\infty$. The success of the area-difference model was not confined to budding: another of its achievements is the coherent description of shape transformations induced in membranes by microtubule growth [HEa].

Bending elasticity alone cannot account for some striking morphological transitions shown experimentally by Hotani [HOb], where a gradual reduction of the volume enclosed by a spherical vesicle, induced by appropriate variations in the osmotic pressure, makes the equilibrium configuration change from an axially symmetric shape to a multilobed shape. The attempts made to explain these morphological transitions by resorting solely to bending elasticity faced annoying inconsistencies [PAa], because buckling of a sphere into a prolate spheroid was predicted, in contrast with the oblate spheroids observed by Hotani. Moreover, previous analysis by Sekimura and Hotani [SEd] based on the minimization of $\mathcal{F}_{el}$ in (2.1) within a family of surfaces obtained by revolving some modified Cassini curves gave only a qualitative agreement with the observed deformation pathway. To recover consistency with observations, Pamplona and Calladine [PAa] then assumed that a lipid bilayer is also able to sustain a small amount of in-plane shear. However, the peanut shapes obtained in this way had only a vague resemblance to Hotani's experimental findings. A closer description was obtained in [PAb], where it was assumed that the elongated lobes found by Hotani had a tubular structure, more akin to a cylinder tube than to a sphere. A detailed analysis of several factors inducing extreme morphological changes in membranes was presented in [HOc]. As far as we know, no attempt has so far been made to justify the sequence of transformations obtained by Hotani by using the area-difference elasticity model. This approach should be further pursued, as shear elasticity is not expected to play a major role in lipid membranes, which unlike red blood cells lack any cytoskeleton.

A further elastic deformation is associated with tilting the amphiphiles with respect to the membrane's unit normal vector $\boldsymbol{\nu}$. Tilt deformations occur in the low-temperature crystalline phase $L_\beta$, and can also be induced by proteins or pores in the fluid phase $L_\alpha$. In the $L_\beta$ phase, amphiphiles are uniformly tilted by an angle that depends on temperature, making the bilayer close to a smectic-C liquid crystal, rather than to a smectic-A liquid crystal. As a consequence, a nontrivial vector field $\boldsymbol{m}$ is obtained by projecting the unit vector associated with the amphiphiles' direction onto the tangent plane to the mean membrane surface $\mathcal{S}$. If the vesicle has spherical topology, a theorem of Poincaré guarantees that the field $\boldsymbol{m}$ develops singularities somewhere on $\mathcal{S}$. These singular points, or defects, behave as electric charges of equal sign and tend to repel each other. This in turn leads to a reorganization of the in-plane order which is coupled with the membrane's curvature and hence causes shape modifications of the membrane, as discussed in [MAa, LUa]. Tilt of chiral amphiphiles was studied to explain the presence of helical structures exhibited by tubules and ribbons (see e.g. [SEe]). The proposed model could account for undulations in the profile of the tubules and yielded a coherent scenario for the kinetic evolution of flat membranes into tubules.

The role of tilt in the $L_\alpha$ phase was recently analyzed in depth by Hamm and Kozlov [HAa, HAb], and by May et al. in [MAf]. In [HAa] the role of tilt deformations was explored in connection with inverted hexagonal and cubic phases, whereas in [HAb] a general continuum model accounting for both tilt and bend deformations was proposed. The vector

$$t := \frac{n}{n \cdot \nu} - \nu,$$

everywhere orthogonal to $\nu$, is taken to measure the tilt of the end-to-end vector $n$ of a hydrocarbon chain with respect to the unit normal $\nu$ of the membrane. The magnitude $t$ of $t$ is nothing but the trigonometric tangent of the angle between $n$ and $\nu$. Accounting for bending, uniform tilt, and tilt gradient, the elastic free energy per unit area can be given the form

$$f = \frac{\overline{\kappa}}{2}(\tilde{H} - \tilde{H}_0)^2 + \overline{\kappa}_G \tilde{K} + \frac{\kappa_\theta}{2} t^2,$$

where the tilt vector $t$ and its gradient enter the renormalized total and Gaussian curvatures of the membrane, $\tilde{H}$ and $\tilde{K}$, and the new elastic term $\kappa_\theta t^2$, where $\kappa_\theta$ is called the *tilt modulus*. Finally, $\overline{\kappa}$ and $\overline{\kappa}_G$ are renormalized bending rigidities, and $\tilde{H}_0$ is the renormalized spontaneous curvature of the membrane.

A somehow complementary model to describe the tilt modulus was given in [MAf], where it was shown by methods of statistical mechanics that the tilt modulus consists indeed of two separate contributions, one of elastic origin due to stretching of hydrocarbon chains induced by tilt deformation, and the other of entropic origin reflecting the tilt-induced suppression of fluctuations in chain orientations. A model of tilt deformation that accounts for the bilayer structure was proposed by Seifert et al. [SEc], who explored the effects of a tilt difference between the two monolayers on the equilibrium shape of the membrane: a strong coupling between the tilt difference and the membrane's curvature can destabilize a flat bilayer or a spherical vesicle in favor of cylindrical or rippled structures. Finally, Fournier [FOa] proposed a model accounting for both tilt and dilation differences between the two monolayers forming a lipid membrane, albeit for shapes that are close to a referential planar configuration. The model was applied to study the effects of tilt difference and dilations on the interaction between proteins embedded in the membrane.

## 8.3  Inhomogeneous Membranes

"Understanding the lateral organization of proteins in a membrane, and the relevance of this organization for membrane functions is one of the hardest problems in membrane biology" (cf. p. 207 of [JEa]). To make a step towards the solution of this problem, the hypothesis of homogeneous membranes must be abandoned. In this section, before commencing a detailed analysis of inclusions embedded in a membrane, we consider the interplay between inhomogeneous composition and membrane geometry. We focus on membranes containing two distinct species of amphiphiles, say, $A$ and $B$.

The relationships between lateral organization in membranes and their equilibrium shape were studied long ago by Markin [MAc] who considered a cylindrical membrane formed by two interacting components. He assumed that the spontaneous curvature $H_0$ of the membrane could be written in terms of the concentration $c$ of, say, species $A$ as

$$H_0 = \xi_A c + \xi_B (1 - c),$$

where $\xi_A$ and $\xi_B$ are called the intrinsic spontaneous curvatures of species $A$ and $B$. Similarly, he assumed that the bending rigidity $\kappa$ of the composite membrane is related to the bending rigidities $\kappa_A$ and $\kappa_B$ of the individual components through

$$\frac{1}{\kappa} = \frac{c}{\kappa_A} + \frac{1-c}{\kappa_B}.$$

The free-energy density $F$ is then given by

$$F = \frac{\kappa}{2}(H - H_0)^2 + bc + kTc\ln c + kT(c_m - c)\ln(c_m - c) + wc^2,$$

where $k$ is the Boltzmann constant, $b$ and $w$ are constitutive parameters, $T$ is the absolute temperature, and $c_m$ is an upper bound on the admissible concentration of species $A$. The terms

$$bc + kTc\ln c + kT(c_m - c)\ln(c_m - c)$$

are customary for binary mixtures, whereas the last term $wc^2$ is associated with interactions between the two species. Markin showed that when $\kappa_A = \kappa_B$ and $\xi_A \neq \xi_B = 0$, the membrane's deformation increases with the spontaneous curvature $\xi_A$ and that phase $A$ tends to aggregate along the equator of the tubule. Increasing the temperature $T$ at a fixed value of $\xi_A$ affects the profile of $c$ which tends towards a constant value. No dramatic change occurs in the tubule's shape, also because the thermal differences between the monolayers are neglected in this model. Moreover, Markin showed that the interaction strength $w$ between the inclusion species is not influential on the membrane's shape, but tends to promote phase separation when $\gamma := 2wc_m/kT$ is below a critical negative value. Finally, for $\kappa_A \neq \kappa_B$ and $\xi_A = \xi_B$, Markin confirmed the intuitive expectation that membrane bending occurs in the softer region, where the bending rigidity is smaller. He also discussed the forces tending to displace particles along the membrane: when the particle is softer than the membrane, it moves towards the regions where curvature is higher, but this tendency can be contrasted if the particle induces a spontaneous curvature different from that of the bare membrane. In Markin's model, the coupling between inhomogeneities and curvature occurs via the parameters $H_0$ and $\kappa$. A direct coupling between the membrane's local curvature and the relative composition of the two species $A$ and $B$ was explored by Kawakatsu et al. [KAa], [TAb]. In [KAa] attention is focused on the strong segregation limit, where the two species $A$ and $B$ are highly immiscible, and so the membrane is partitioned into distinct one-component domains, separated by sharp domain walls. The opposite situation [TAb] occurs in the weak segregation limit, where immiscibility is only mildly hampered. For mathematical simplicity, in both papers the authors considered only tubules and axisymmetric vesicles. In addition to the bending elasticity (2.2), the free energy contained a contribution $F_\Phi$ associated with the inhomogeneity of the membrane and a coupling term $F_c$. By introducing the order parameter $\Phi := \Phi_A - \Phi_B$ which measures the relative concentration per unit area of species $A$ with respect to species $B$, $F_\Phi$ is given by the following Ginzburg–Landau expansion

$$F_\Phi := \int_S \left[ \frac{b}{2}(\nabla\Phi)^2 - \mu\Phi + \frac{A_2}{2}\Phi^2 + \frac{A_3}{6}\Phi^3 + \frac{A_4}{24}\Phi^4 \right] \mathrm{d}A,$$

where $\mu$ denotes the difference of the chemical potentials associated with the two species $A$ and $B$, and $b$ and the $A_i$s are phenomenological coefficients. The temperature $T$ enters $F_\Phi$ through the coefficient $b$ so that the homogeneous phase $\Phi \equiv 0$ minimizes $F_\Phi$ when $T > T_c^0$, where $T_c^0$ is called the *bare* transition temperature. Finally, the coupling term $F_c$ is given by

$$F_c := \Lambda \int_S \Phi H \mathrm{d}A,$$

where $\Lambda \geq 0$ is called the *coupling constant*. Thus, a nontrivial profile of the relative concentration $\Phi$ along the membrane induces an effective spontaneous curvature in the membrane. It should be noted, however, that the bending rigidity has one and the same value along the membrane.

For two-dimensional vesicles, when a membrane is effectively represented by a closed curve, and in the strong separation limit, it is shown that, when $\Lambda > 0$ is prescribed, the preferred number of domain walls depends on the pressure drop $\Delta P$ across the membrane, taken as positive when the pressure outside the membrane is larger than the pressure inside it. To prove this, the vesicle shape is first obtained when the position and the length of the domain walls are fixed. Then, the free energy is computed as a function of the number of walls $2n$, treating $\Delta P$ as a parameter. When $\Delta P > 0$, $n = 2$ is the most stable mode, whereas when $\Delta P < 0$ larger values of $n$ are more favorable. Moreover, when both the energetic cost of a domain wall and $\Delta P < 0$ are assigned, the value of $n$ in equilibrium increases with $\Lambda$. To explain this counterintuitive behavior, it should be noted that, when $\Delta P < 0$, the tubule's cross-section is likely to enclose large areas. This, in turn, makes high values of $n$ more favorable because, on increasing $n$, the equilibrium shape resembles more and more a circle, which encloses the largest area when the perimeter is prescribed by the inextensibility constraint. Similar results are obtained for axisymmetric vesicles.

In the weak segregation limit, the temperature is taken close to the bare transition temperature $T_c^0$, and the shifted order parameter $\Psi :=$ $\Phi + A_3/A_4$ is introduced to get rid of the cubic term in $F_\Phi$. Moreover, $\Psi$ is assumed to differ only slightly from a constant, average value $\langle \Psi \rangle$. By perturbing a circular shape of radius $r_0$, characterized by a value of $\Psi$ close to $\langle \Psi \rangle$, it is first proved that, when the rescaled pressure difference $\Delta p := r_0^3 \Delta P/\kappa$ exceeds 3, the circle is destabilized even in the absence of the coupling term $F_c$, as already known for homogeneous membranes. On the other hand, the presence of a coupling term is needed for the onset of instability when $\Delta p < 3$. In fact, by adopting a single-mode approximation for both the vesicle's contour and the shifted order parameter $\Psi$, and by setting for simplicity $\langle \Psi \rangle = 0$, it is proved that instability of a circular profile with $\Psi \equiv 0$ requires $\Delta p < 3$, at $T = T_c^0$, when the coupling term is present. When $\Psi$ does not vanish identically and $\Delta p < 0$, the destabilizing mode becomes more and more corrugated as the coupling constant $\Lambda$ increases, showing again the interplay between curvature and membrane composition at work to determine the equilibrium configuration. When the pressure drop has a fixed negative value, but the temperature is different from $T_c^0$, the transition temperature $T_c$ increases with the coupling coefficient $\Lambda$.

The complementary problem of phase segregation induced by shape transformations was studied within the area-difference-elasticity model by Seifert in [SEa]. Here, it is assumed that the local deviation of the composition from its mean value is different in the two monolayers, thus inducing an effective spontaneous curvature that depends on the position $r$ on the membrane

$$H_0(\Phi(\boldsymbol{r})) = \lambda \Phi(\boldsymbol{r}) + \overline{H}_0 \,,$$

where $\lambda$ is a coupling constant, and $\overline{H}_0$ is an overall, constant spontaneous curvature, in the spirit of Markin's approach [MAc]. By inserting $H_0(\Phi(\boldsymbol{r}))$ into $F_\Delta$ defined as in (2.3), the bending energy acquires a coupling term with the membrane composition $\Phi$, which also enters the Ginzburg–Landau term

$$F_\Phi = \frac{\kappa}{2} \varepsilon \int_{\mathcal{S}} [\Phi^2 + (\xi \nabla \Phi)^2] \mathrm{d}A \,,$$

where $\xi$ is a correlation length related to spatial fluctuations of the composition, and $\varepsilon$ is some molecular energy, divided by $\kappa$. By adding the constraint that no exchange of molecules occurs between the monolayers, which amounts to requiring $\int \Phi dA = 0$, by minimizing the functional $F_\Delta + F_\Phi + \mu \int \Phi dA$ with respect to $\Phi$ it is possible to map the free energy $F_\Delta + F_\Phi$ into a genuine area-difference elasticity model characterized by an effective bending rigidity $\hat{\kappa} = \kappa/(1 + \lambda^2 \varepsilon^{-1}) < \kappa$ and by a larger coupling coefficient $\hat{\alpha}$. In this way, the equilibrium problem is reduced to produce a phase diagram for one-component vesicles, with properly rescaled parameters. In particular, by assuming that the two monolayers have one and the same thermal expansion coefficient, a spherical vesicle characterized by a homogeneous composition profile $\Phi(\mathbf{r}) \equiv 0$ transforms itself into a vesiculated shape, as the enclosed volume decreases, and a curvature-induced phase segregation is found. The limiting shape at the end of the process consists of two homogeneous spherical vesicles connected by a narrow neck within which the composition suddenly changes.

A simple model to study shape transformations induced by intramembrane domains was proposed by Lipowsky [LIb], who considered a planar membrane in which phase $B$ forms an island of prescribed area $S$ in a sea of $A$ amphiphiles. The $A$–$B$ interface has an energetic cost, proportional to the length of the interface itself. Now, were the $B$-domain constrained to remain flat, it would be a circle, by the isoperimetric theorem. Lipowsky suggested that, to further reduce the interface energy, the $B$-domain could escape out of the membrane's plane so as to shorten the interface, at the price of paying some bending energy. If $\sigma$ is the line tension associated with the $A$–$B$ interface, and $L$ is a characteristic length related to the area $S$ of the $B$-domain by $\pi L^2 = S$, the energy becomes

$$E = \left[ \frac{L^2}{4}(H - H_0)^2 + \frac{L}{\xi}\sqrt{1 - (LH)^2} \right],$$

and it must be minimized within the class of configurations where the $B$-domain is a spherical cap. The dimensionless ratio $L/\xi$, where $\xi := \kappa/\sigma$ is the invagination length, determines whether budding occurs. When $L/\xi$ is small enough the free energy $E$ has three local minima. Two minima correspond to *complete* buds in which the $B$-domain is a sphere adhering at either side of the unperturbed membrane: these minima have the same energy if $H_0 = 0$ whereas, when $H_0 \neq 0$, the preferred complete bud is of course that reducing the frustration associated with the spontaneous curvature. The third minimum of $E$ corresponds to an *incomplete* bud which is a genuine spherical cap with finite radius, if $H_0 \neq 0$. On increasing $L/\xi$, the size of the $B$-domain reaches a critical value $L^*$ at which the incomplete budding has the same energy as the most favorable complete bud. Here, the incomplete bud could be thermally activated into a complete one. However, a simple estimate reveals that this is a rare event, and so a first-order transition to complete budding is unlikely to occur. Further increase of $L/\xi$ leads to the point where the incomplete budding is no longer a relative minimum of $E$, and complete budding finally occurs.

A more refined model accounting for closed membranes with spherical topology was proposed in [JUa] in the frame of the area-difference elasticity model. Here, two novel features emerged: when the membrane encloses a fixed volume, budding is hindered; and because there is no reason to assume that the elastic moduli have one and the same value in the $A$ and in the $B$ domains, it turns out that Gaussian curvature plays a nontrivial role, provided that the Gaussian rigidity modules of the domains differ from each other, because in that case the Gaussian rigidity difference enters through the boundary conditions along the $A$–$B$ interface.

An overview of possible factors leading to domain formations is contained in [LIa], where, in particular, the role of adhesion to a substrate in promoting intramembrane domains is discussed. Couplings between phase separation and membrane deformations with the same free energy as in [KAa] have been studied, among others, by Taniguchi [TAa], Jiang et al. [JIa], and Sunil Kumar et al. [LAa, SUa]. By resorting to a dissipative dynamics, in [TAa] Taniguchi studied the evolution of a two-component spherical vesicle that could form either a budded or an invaginated configuration, according to the average value $\langle \Phi \rangle$ of the order parameter at equilibrium from which the system is quenched. In [JIa], membranes with simple geometries such as cylinders, spheres, and tori are studied to gain more analytic insight. The equilibrium configurations have striped domains and the coupling between lateral inhomogeneity and bending elasticity causes characteristic deformations.

The elegant analysis of [JIa] was only applied to highly symmetric membranes, and this makes it impossible to follow the out-of-plane escape of intramembrane domains and the subsequent budding transition, as discussed in [LIb]. First remedies to this deficiency were found in [SUa] and [LAa], where the dynamics of domain growth and possible formation of buds was studied by use of Monte Carlo simulations. The budding transition occurs in several steps. First, intramembrane domains emerge that then grow within the membrane; second, single small buds spring out; and third, they coalesce into larger buds. The approach of [LAa] differs from that of [SUa] in that area-to-volume ratio, line tension, as well as hydrodynamic effects are accounted for. In particular, vesiculation is hampered by the volume constraint.

The deformation of a membrane follows a different avenue if the inhomogeneity is due to embedded inclusions, such as transmembrane proteins (see Section 8.3). Correlations between lateral distribution of embedded inclusions and membrane equilibrium shape were studied in [KRd], where a coupling between the geometry of the inclusions and the local curvature of the membrane is explored. In addition to the elastic free energy $F_\Delta$ in

(2.3) pertaining to the area-difference model, the term

$$F_\Sigma = -kTN \log \Sigma \tag{3.1}$$

is introduced to account for the inclusions. Here, $k$ is the Boltzmann constant, $T$ is the absolute temperature, $N$ is the constant number of inclusions present in the membrane, and the partition function $\Sigma$ is given by

$$\Sigma = \frac{1}{A} \int_S \exp[-E_i(\sigma_1, \sigma_2)]/kT \, \mathrm{d}A \,,$$

where $A$ is the membrane's area, and $E_i(\sigma_1, \sigma_2)$ is the interaction energy between membrane and inclusions, which depends upon the principal curvatures $\sigma_1$ and $\sigma_2$ of the membrane. When the inclusions are isotropic, $E_i$ is given the form

$$E_i = \kappa_i[(H - H_s)^2 - \frac{4}{3}K] \,,$$

where $\kappa_i$ measures the interaction strength, $K$ and $H$ are the Gaussian and the total curvatures of the membrane at the inclusion site $p$, and $H_s$ is the value of $H$ that the inclusion would prefer at $p$. The Boltzmann distribution appearing in (3.1) is obtained by minimizing the Gibbs free energy associated with an assembly of $N$ inclusions, at a fixed membrane's shape. Thence, the free energy $F = F_\Delta + F_\Sigma$ is minimized in a class of membrane shapes obtained by suitable rotations of modified Cassini ovals, which are general enough to embrace both biconcave and discotic shapes. As a result, the inclusions tend to accumulate in regions with higher curvature, and the equilibrium shape of the membrane lacks the axial symmetry typical of the homogeneous case.

When the inclusions are anisotropic, $E_i$ also depends upon their orientations. This case was worked out in [KRc], where the simpler functional (2.1) was chosen to model the elastic free energy. It turns out that the total free energy plotted against the area difference $\int_S H \mathrm{d}A$ between the monolayers attains a minimum when the membrane's equilibrium shape has a small bud. The fact that this minimum would be absent if the membrane were homogeneous indicates that anisotropic inclusions tend to stabilize budded configurations.

Results indicating the special role played by inclusions in driving a membrane towards exotic shapes were obtained in [BIc] where the membrane is viewed as a closed planar curve $\mathcal{C}$ and an inclusion is modeled as a rigid trapezium, that is, a truncated isosceles triangle. In the spirit of [PAc], the inclusion induces a preferred value $\vartheta_0$ of the *contact angle* $\vartheta$, the angle between the membrane's and the inclusion's unit tangent vectors. Thus the term

$$\mathcal{F}_a = -w\cos(\vartheta - \vartheta_0) \,,$$

where $w$ is a positive constant, is added to the elastic free energy in the spontaneous curvature model (2.2), which in two space dimensions reads as

$$\frac{\kappa}{2} \int_{\mathcal{C}} (\sigma - \sigma_0)^2 \mathrm{d}s \,,$$

where $\sigma$ is the curvature of $\mathcal{C}$, $\sigma_0$ is the spontaneous curvature and $s$ is the arc-length along $\mathcal{C}$. By considering a single inclusion, it was proved in [BIc] that the spontaneous curvature can drive the membrane to self-adhesion, inducing either protein segregation or protein absorption, depending on the value of $\sigma_0$. The presence of the inclusion together with the equilibrium value of the contact angle at the inclusion's boundary are essential to make the spontaneous curvature intervene in a two-dimensional setting, where it would otherwise contribute only a constant to the free energy. That embedded inclusions can promote changes in the vesicle's geometry was later confirmed in a three-dimensional model [BId], where the deformation of spherical and quasi-spherical vesicles with a single inclusion were explored. The results suggest that inclusions can promote transitions towards pear-shaped vesicles which are usually precursors of a budding transition. This paper also dwells on the role played by the vesicle impermeability in promoting transitions towards pear-shaped or stomatocytelike vesicles.

Another fascinating shape change is that accompanying the echinocytosis of red blood cells, the morphological transition leading to crenated equilibrium shapes where a number of *spicules*, that is, bumps of cylindric shape, emerge from the body of the vesicle, when some control parameters—such as the pH of the environment fluid—are changed. Some major differences should be noted between budding and echinocytosis. First, the neck connecting the spicules to the body of the vesicle is not too narrow, but spicules have a nearly constant cross-section, apart from the spherical cap surmounting them. Second, echinocytosis characterizes red blood cells which, unlike vesicles, are endowed with the *cytoskeleton*, an elastic network covering the membrane of the cell which confers to the cell a *shear* elasticity, in addition to the bending elasticity. The competition between bending and shear elasticities has been recognized as playing a crucial role in the onset of echinocytosis. In fact, the cytoskeleton inhibits the formation of spicules because the segregation of transmembrane proteins would lead to a redistribution of the cytoskeleton which has a high cost in shear elasticity. Hence, to model echinocytosis, [IGa] and [WAa] added the shear elastic energy $F_\mu$ of the cytoskeleton to the bending elasticity (2.3). For $F_\mu$, the approximate form

$$F_\mu = \mu \int_{\mathcal{S}} [\lambda^2 + \lambda^{-2} - 2] \mathrm{d}A$$

was considered, in which $\mu$ is the membrane shear modulus, and $\lambda$ is the surface extension ratio. Both approaches rely on a simplified treatment of the equilibrium shape of the membrane. Waugh [WAa] considered spicules that could originate from a circular patch, whereas in [IGa] the spicules were mounted upon a spherical membrane. In both models, the minimum of the total elastic free energy was sought in a narrow class of shapes.

Waugh proved that the area difference is much more effective than spontaneous curvature in promoting the formation of spicules and, more importantly, that modifications in the shear modulus $\mu$ can induce bumps that asymmetries alone in the membrane could not promote. In [IGa], the geometric parameters characterizing the spicules were determined by enforcing the constraints on the area of and on the volume enclosed by the membrane, and by fixing a value for the area difference between the monolayers. The number of spicules as well as their geometry was determined by minimizing the bending component of the free energy. Finally, by minimization of the total free energy as a function of the area difference, a stable echinocyte was obtained. Both these models treat the elastic network forming the cytoskeleton as incompressible; effects of compressibility were recently embodied in [MUa] by adding a stretching term

$$\frac{K_a}{2} \int_{\mathcal{S}} (\lambda_1 \lambda_2 - 1)^2 \mathrm{d}A \,,$$

where $\lambda_1$ and $\lambda_2$ are the principal extension ratios, related to $\lambda$ by $\lambda = \lambda_1/\lambda_2$, and $K_a$ is a positive elastic modulus. It is proved in [MUa] that the equilibrium shape of axisymmetric spicules depends upon the interplay between two length scales. The former, $\Lambda_{\mathrm{eff}} := 1/H_0^{\mathrm{eff}}$, is the inverse of the effective spontaneous curvature $H_0^{\mathrm{eff}} := H_0 - c(\Delta A - \Delta A_0)$ accounting for both sources of asymmetry between the membrane's monolayers: here, $H_0$, $\Delta A$, and $\Delta A_0$ are as in (2.3), and $c$ is a positive constant. The latter length scale $\Lambda_{el} := \sqrt{\kappa/\mu}$ compares the local bending rigidity with the shear modulus. When $\Lambda_{\mathrm{eff}}$ is positive and larger than $\Lambda_{el}$, equilibrium spicules have a rather smooth profile, whereas when either of the two lengths is of the same order of magnitude or $\Lambda_{el} > \Lambda_{\mathrm{eff}}$ the equilibrium spicules exhibit a sharper profile, close to those found experimentally. By looking for the equilibrium shapes when the number $n_s$ of spicules is varied, a satisfactory agreement with the experimental evidence exists for $n_s$ ranging from 30 to 60.

## 8.4  Transmembrane Proteins

As mentioned in the introduction, proteins embedded in membranes are responsible for the membrane's functionality, as they can, for instance, mediate information transfer, transport molecules, and perform metabolic activities (see e.g. [BEa], [JEa]). A distinction is usually made between *peripheral* and *transmembrane*, or *embedded*, proteins. Transmembrane proteins penetrate, at least partially, the thickness of the membrane, whereas peripheral proteins are less invasive. Besides this, a major difference between peripheral and transmembrane proteins resides in their binding to the membrane. Peripheral proteins can be anchored to the membrane via electrostatic forces, via weak forces associated with some nonspecific binding, or via a lipid extended conformation, where one of the acyl chains can flip outside the membrane to release internal stress. Transmembrane proteins are tightly bound to the membrane and, because of their hydrophobic properties, need a detergent to be solubilized out of the membrane (§2 of [BEa]).

In any case, transmembrane proteins are never completely buried in the membrane and indeed their lateral mobility is essential to guarantee membrane functions. Transmembrane proteins have a hydrophobic belt whose height could differ from the bilayer thickness. When a protein spans a multicomponent membrane, it tends to surround itself with the lipids yielding the best matching [GIa], and this can, in turn, lead to phase segregation among lipids when the protein concentration is large enough. On the contrary, instead of insisting on a forced matching with lipids, similar proteins could form aggregates that provide an easy way to reduce hydrophobic matching. Alternatively, if the hydrophobic matching is too large to be accommodated by simply stretching the lipid chains, the proteins can deform themselves [LEa]. The lateral pressure profile induced by the anisotropic structure across the membrane is strictly related to hydrophobic matching. Cantor [CAb, CAc] suggested that, when the protein is not a cylinder, the lateral pressure can cause conformational changes of the protein, affecting its functionality. This in turn opens the way to pharmacological applications, because drugs could be designed either to act directly upon the hydrophobic thickness of a membrane, or to modify the pressure profile on selected proteins. In a similar vein, Dan and Safran [DAb] showed that the conformation of transmembrane proteins is controlled by the lipid's characteristics through a membrane-induced line tension exerted on the proteins' lateral surface. The line tension is always a compression if the spontaneous curvature is zero, whereas it can be either a compression or a traction provided that the spontaneous curvature is large enough, and the protein resembles an hourglass. Although conformational changes are important in proteins, it is plausible to model these latter as rigid bodies, inasmuch as they are less deformable than a lipid bilayer [HAc].

The hydrophobic mismatch between transmembrane proteins and the membrane's thickness was early recognized as an important parameter to understand interactions between proteins in a lipid membrane. In fact, according to the *hydrophobic matching principle*, the lipid molecules in the vicinity of the proteins undergo either stretching or shrinking, to adjust their length so that the protein does not experience an unfavorable exposure of its hydrophobic belt to the aqueous medium. When two proteins are close enough, the membrane patches that they deform overlap so as to promote an indirect, or membrane-mediated interaction between proteins. Precisely, according to Dan et al. [DAa], these interactions are *short-ranged*, as they are due to local perturbations of the membrane's structure; in this respect they differ from the *long-range* membrane-mediated interactions induced by the suppression of long-wavelength bilayer fluctuations, which are treated in Section 8.5.

Before modeling membrane-induced protein–protein interactions, it is mandatory to understand how a single protein affects the membrane's profile and the energetics associated with this deformation. To simplify the study, it is common to model the membrane as a perturbation of a flat profile, so that the equilibrium equations are linear, and the signed distance $u(x, y)$ of the membrane from the referential $(x, y)$-plane is taken to describe the deformation. Because variations of the bilayer thickness are relevant at the length-scale involved in the problem, one should describe both surfaces that bound the bilayer. In this setting, a basic distinction is made between *symmetric* and *antisymmetric deformations*, according to whether the bilayer's midplane is left undeformed [MAd]. Clearly, symmetric and asymmetric membrane deformations are associated with symmetric and asymmetric proteins, so that cylindrical and conical proteins would be examples of the former and latter category, respectively. Symmetric modes [HUa] are the *compression-expansion* mode, where the bilayer changes its thickness, the *splay distortion* mode, where the amphiphiles are allowed to reorient themselves close to the protein, while remaining orthogonal to the membrane's surface, and the *surface-tension* mode associated with local variations of the monolayers' areas with respect to the flat reference configuration. Clearly, for symmetric modes, the shape of just one bounding surface suffices to describe the membrane's profile.

For a nearly flat membrane perturbed by a single cylindrical protein of radius $r_0$ and subject to symmetrical modes, it is natural to assume radial symmetry by setting $u = u(r)$, where $r$ is the distance from the protein's axis. Because the linearized equilibrium equation for the membrane associated with (2.1) is fourth order, four boundary conditions are required to solve the equilibrium problem. Two of them are quite natural:

$$u(r_\infty) = 0, \quad \text{and} \quad \left.\frac{du}{dr}\right|_{r_\infty} = 0,$$

where $r_\infty$ is a large value of $r$. Along the boundary of the cylindrical protein at $r = r_0$, the hydrophobic matching principle is accounted for by taking

$$u(r_0) = d_0 - \frac{\ell}{2}, \tag{4.1}$$

where $\ell$ is the length of the protein's hydrophobic belt, and $d_0$ is the monolayer's thickness when the protein is absent. Prescribing this condition amounts to saying that the exposure of the protein's hydrophobic belt to the aqueous ambient is energetically too costly, as compared with the typical elastic energy associated with bilayer's deformations.

Harroun et al. [HAc] checked the adequacy of this assumption in the case of the gramicidin channel, one of the most studied transmembrane proteins. By using experimental data they concluded that the assumption is realistic provided that the extension of the bilayer does not exceed a critical value $\delta \approx 5.3\text{Å}$. If the hydrophobic mismatch requires larger deformations, then "some slippage or incomplete matching is expected to occur" (p. 3178 of [HAc]). In fact, this approach had been pursued by Ring [RIa], who aimed at testing the liquid crystal model to explain correlations between lifetime of the gramicidin channel and the surface tension of certain membranes. In [RIa] he argued that, when the bilayer thickness $2d$ is larger than the channel length $\ell$, the lipid molecules closest to the channel are displaced along its axis. If $b$ is the nearest neighbor distance, it is assumed that deformation of the membrane starts at $r = r_0 + b\cos\vartheta$, where $r_0$ is the radius of the channel and $\vartheta$ is the contact angle. By setting $u_0 := d - (\ell/2)$ the boundary condition to be imposed is

$$u(r_0) = u_0 - b\sin\vartheta.$$

The contact angle $\vartheta$ involves the slope $u'$ of the membrane at $r = r_0$, upon which the fourth boundary condition is to be imposed. The conflicting proposals made for this condition can be essentially divided into two families. In [HUa], by comparisons with experimental results, Huang imposed a vanishing contact angle, that is,

$$\left.\frac{du}{dr}\right|_{r_0} = 0.$$

A different avenue was followed by Helfrich and Jakobsson [HEb], and later by Ring [RIa], who set

$$\left.\frac{du}{dr}\right|_{r_0} = s \tag{4.2}$$

and kept $s$ as a free parameter, to be determined by requiring that the free energy is minimized. The trial-and-error method of Helfrich and Jakobsson was revisited by Nielsen et al. [NIa] who noted that (4.2) is equivalent to penalizing the tilt of the amphiphiles away from the membrane's normal and that (4.2) is tantamount to imposing the natural boundary condition at $r = r_0$,

$$\frac{\partial}{\partial r} \Delta u \Big|_{r_0} = 0 \,,$$

where $\Delta$ is the Laplace operator. In [NIa], (4.2) was compared with the following clamped contact slope

$$\frac{du}{dr} \Big|_{r_0} = s \,,$$

where now $s$ has a prescribed value. Depending on the value of $s$, the equilibrium profile $u(r)$ of the membrane can fail to be monotonic. A more radical attitude towards (4.2) was taken by Harroun et al. [HAc] who noted that, if the slope of the membrane at the inclusion boundary is not prescribed, then boundary terms should be added to the energy as well. The theoretical analysis of [HUa], [HEb], [NIa], and [HAc] led to the equilibrium profiles of the membrane perturbed by a single protein, and to the evaluation of the energetic cost of each deformation mode. Although starting from different boundary conditions, the surface-tension mode invariably yields a negligible contribution. Nielsen et al. [NIa] also concluded that the splay-distortion mode prevails close to the protein, whereas the compression-expansion mode prevails away from the protein.

Interesting problems arise when membrane-mediated interactions between two transmembrane proteins are studied. Dan et al. [DAa] considered two parallel problems, according to whether the strong hydrophobic matching prescribes $u(r_0)$ (*stretched* boundary condition) or the slope of the membrane's profile $u'(r_0)$ (*sloped* boundary condition) at the protein's boundary. In both cases, the remaining boundary condition at $r = r_0$ is a natural one, stemming from of the minimization process. The elastic free energy is minimized within an hexagonal Wigner–Seitz cell surrounding a reference protein $\mathcal{P}$, by assuming the deformation $u_r(r)$ of the bilayer thickness relative to its unperturbed value $u_\infty$

$$u_r(r) := \frac{u(r) - u_\infty}{u_\infty}$$

as the independent variable. Here, $r$ is the distance from the axis of $\mathcal{P}$. By accounting for both stretching and bending elasticity, a natural correlation length

$$\rho := \sqrt[4]{\frac{4u_\infty^2 K}{B}}$$

is introduced, where $B$ and $K$ are the stretching and the bending moduli of each monolayer. For stretched boundary conditions, the perturbation in the membrane's shape decays on a length scale of $10\rho$, following a nonmonotonic profile. To compute the protein–protein interaction free energy, the equilibrium profile of the membrane is inserted into the free energy, which is then plotted against $L/\rho$, where $L$ is the protein spacing. When $L/\rho > 10$ the interaction free energy attains a plateau, because the perturbed regions do not overlap any longer. The absolute minimum of the interaction free energy occurs at $L = 0$, that is, when the two proteins form an aggregate. A secondary minimum at $L/\rho \simeq 3$ also exists that is separated from the minimum at $L = 0$ by an energy barrier which increases on decreasing $\rho$, that is, when bending the membrane is made easier than stretching it. In this latter case, protein aggregation is highly hampered and an equilibrium configuration with a finite spacing between proteins is preferred. Remarkably, when sloped boundary conditions are imposed, the interaction energy diverges at small separations and, hence, aggregation is forbidden, and proteins form a regular array where the typical interprotein spacing is proportional to $\rho$. Aranda-Espinoza et al. [ARa] showed that adding a spontaneous curvature $H_0$ can also drive the system towards equilibria with finite interprotein spacing when stretched boundary conditions are assumed.

Kralchevsky et al. [KRb] described a bilayer as an incompressible elastic medium, sandwiched between two Gibbs surfaces modeling the head group regions. Only cylindrical proteins and symmetric deformations were considered. By extending previous work [KRa] on capillary forces between colloidal particles in a thin liquid film, these authors used different constitutive laws, according to the deformation involved. The following hybrid constitutive law for the stress tensor $\mathbf{T}$ was postulated,

$$\mathbf{T} = 2\lambda\mathbf{P}\mathbf{E}\mathbf{P} - p(\mathbf{I} - \mathbf{P}),$$

where $\lambda$ is the shear elastic modulus, $p$ is the pressure, $\mathbf{E}$ is the symmetric part of the deformation gradient, and $\mathbf{P} := \boldsymbol{e}_z \otimes \boldsymbol{e}_z$ is the orthogonal projector along the midplane's normal. By also accounting for surface tension, Kralchevsky et al. showed that the attractive interaction is stronger when the bilayer thickness is smaller than the hydrophobic belt of the proteins. A further step was moved by Weikl et al. [WEb] who studied conical proteins in a nearly flat membrane by stressing the role of lateral tension as well as of the relative orientation of proteins, but neglecting spontaneous curvature. When no lateral tension is present, the interaction free energy $F_0$ is related to the distance $r$ between the inclusions through

$$F_0 = 4\pi\kappa(\alpha_1^2 + \alpha_2^2)\frac{a^4}{r^4} + O\left(\frac{1}{r^5}\right),$$

where $a$ is the radius of the inclusion's cross-section in the membrane's mid-plane, and $\alpha_1$ and $\alpha_2$ are the contact angles between the membrane and the two inclusions. Inasmuch as $F_0$ does not depend on the sign of $\alpha_1$ and $\alpha_2$, the relative orientation of the proteins is irrelevant, and the interaction is always repulsive. On the other hand, adding the lateral tension

$$\int_S \frac{\gamma}{2} (\nabla u)^2 \mathrm{d}A$$

to the free energy has a deep impact on the interaction free energy, which now reads as

$$F_\gamma = 2\pi\kappa\alpha_1\alpha_2(\xi a)^2 K_0(\xi r) + \pi\kappa(\alpha_1^2 + \alpha_2^2)(\xi a)^4 K_2^2(\xi r)\,,$$

where $K_0$ and $K_2$ are cylindrical Bessel functions, and $\xi := \sqrt{\gamma/\kappa}$. Here, when the proteins are equally oriented, and so $\alpha_1\alpha_2 > 0$, the interaction is always repulsive, whereas when the proteins have opposite orientations, the interaction energy fails to be monotonic in $r$, and attains its absolute minimum at $r = r^*(\gamma) > 0$.

May and Ben-Shaul [MAe] also considered the effects that protein size and shape have on both lipid-mediated protein–protein interactions and the phase transition between the fluid phase ($L_\alpha$) and the inverted hexagonal phase ($H_{II}$). Their approach is at the molecular level, because the elastic parameters such as the bending rigidity and the spontaneous curvature are expressed in terms of the molecular interactions. Moreover, the free energy is not minimized with respect to the membrane's shape, but with respect to the internal degrees of freedom of the lipid molecules, the chain length, and the tilt angle. In particular, this latter parameter is introduced to account for nonlamellar morphologies, and for proteins with a skewed boundary. So, for instance, when stretched boundary conditions are enforced on vaselike or barrellike inclusions, a finite spacing between proteins is predicted, instead of aggregation.

To close this section, we mention that all the approaches reviewed so far systematically neglect the closed topology of membranes, and its potential effects on interactions between transmembrane proteins. Elastic interactions between inclusions in a spherical vesicle were studied in [DOc]. In the spirit of the Monge gauge, the vesicle is treated as a small normal perturbation to a sphere, and the inclusions are modeled as equal cones with opening angle $\psi$. The long-range interaction is always repulsive, but its dependence on the angular separation $\Theta$ between the proteins is not universal. In fact, when the separation is small, the interaction force behaves as $\Theta^{-4}$, mimicking the $r^{-4}$ decay law found for nearly flat membranes. However, when $\Theta$ exceeds a critical value $\Theta_c \simeq \psi^{0.54}$, the interaction gets stronger than in the planar case, with a $\Theta^{-1/3}$ decay law.

The role of closed geometry was also stressed in the two-dimensional models employed in [BIa] and [BIb]. Although with this choice, the contribution of spontaneous curvature to the interaction cannot be accounted for completely, analytic solutions for the unperturbed shape of the membrane exist, at least when the osmotic pressure is low enough. Proteins are modeled as trapezia and different regimes occur, depending on whether both proteins have their short bases of length $a$ inside the membrane (*inner–inner* configurations), outside the membrane (*outer–outer* configurations), or the former inside and the latter outside (*inner–outer* configurations). The membrane-mediated force $F_{\mathrm{med}}$ is computed via a quasi-static approach, by differentiating the effective free-energy functional $\mathcal{F}_{\mathrm{eff}}$ with respect to the distance $L_1$ between the inclusions along the membrane,

$$F_{\mathrm{med}} = -\frac{\mathrm{d}\mathcal{F}_{\mathrm{eff}}}{\mathrm{d}L_1},$$

where $\mathcal{F}_{\mathrm{eff}}$ accounts for the inextensibility of the membrane, whose contour has a fixed length $L$. Moreover, for simplicity, the contact angles have one and the same fixed value $\vartheta_0$. The simplest scenario occurs for inner–inner configurations, where the interaction is always repulsive, tending to maximize the distance between the proteins. In the inner–outer configurations, the interaction is always attractive, provided that the inclusions are small enough ($a < 0.0744\ L$), it is repulsive when the inclusions are rather large ($a > 0.2\ L$) and, in between, there exists a value $L_{\mathrm{opt}} \in (0, L/2)$ of $L_1$ such that the interaction is attractive when $L_1 > L_{\mathrm{opt}}$ and repulsive otherwise. A somehow complementary situation arises for the outer–outer configurations because, for any value of $\vartheta_0$ and any value of $a$, a critical value $L_{\mathrm{cr}}(a, \vartheta_0)$ of $L_1$ exists such that the interaction is attractive when $L_1 < L_{\mathrm{cr}}$, and repulsive otherwise. We note in passing that the distinction between outer–outer and inner–inner configurations can be made only when the membrane has a closed geometry, and so escapes the analysis of [WEb] which was confined to nearly flat membranes.

The model presented in [BIa] was extended in [BIb] where the inclusions have different lengths $a_1$ and $a_2$ and can induce different contact angles $\vartheta_1$ and $\vartheta_2$. The membrane-mediated interactions exhibit a large variety of behaviors, depending on the values of $a_1$, $a_2$, $\vartheta_1$, and $\vartheta_2$. First, there is a regime where the interaction is always attractive, regardless of the distance between the inclusions; there is also a regime where the inclusions repel each other and try to maximize their separation along the membrane; finally, there are three mixed regimes where the interaction free energy $\mathcal{F}$ can have a minimum, a maximum, or both at some value of $L_1/L \in (0, 1/2)$. If $\mathcal{F}$ has a minimum, an optimal equilibrium configuration exists with finite spacing between the proteins. If $\mathcal{F}$ has a maximum, the equilibrium configuration at $L_1/L \in (0, 1/2)$ is unstable, and either the proteins aggregate, or they are pushed towards antipodal points. Finally, if $\mathcal{F}$ has both a minimum and a maximum, an unstable equilibrium exists at a short separation between the inclusions, and a stable one at a larger separation. Hence, if the inclusions are at short distances, they will aggregate, otherwise they will stay at a finite distance.

## 8.5   The Role of Thermal Fluctuations

The role of thermal fluctuations in the interactions between fluid membranes was recognized by Helfrich [HEd] who noted that the suppression of fluctuating modes occurring when two undulating, nearly parallel membranes are close together causes an entropy decrease and, hence, induces a repulsive steric interaction comparable with the van der Waals attractive forces. In our context, Schröder [SCa] recognized that transmembrane proteins cause the suppression of fluctuations of an appropriate order parameter. Clearly, the major difficulty in studying fluctuations is the explicit computation of the appropriate partition function. On the other hand, it is worth noting that, being these interactions are long-range, the membrane thickness is always neglected.

An important contribution was given by Goulian et al. [GOb], who explored three different regimes. In the first, where the thermal energy is negligible with respect to the typical bending energy of the membrane, the interaction energy is proportional to $r^{-4}$, $r$ being the distance between the proteins. The interaction is attractive or repulsive, depending on the relative strength of the bending rigidities: in particular, a nontrivial contribution from the Gaussian bending rigidity is obtained. In the remaining cases, the fluctuations are important and a further distinction is made between the perturbative regime, where the bending and the Gaussian rigidities of the inclusions differ by small amounts $\delta\kappa$ and $\delta\kappa_G$ from that of the membrane, and a strong coupling regime, where the proteins are modeled as rigid bodies. By assuming that the contact angle of the membrane at every inclusion is dictated by its shape, it is shown in both cases that the interaction energy decays as $r^{-4}$. However, whereas in the perturbative regime the force is attractive if $\delta\kappa\delta\kappa_G < 0$, and repulsive otherwise, in the strong coupling limit, the interaction is always attractive.

Orientational interactions were explored in [GOa], where rodlike proteins were considered. The attractive interaction energy, besides decaying like the fourth power of the center-to-center distance, also depends on the rods' relative orientations and attains its minimum when the rods are either parallel or perpendicular to one another. The analysis of [GOa] applies when the inclusion centers are at a distance $r$ much larger than the rod's length $L$. The case $r < L$ was analyzed in [HOa] by resorting to Monte Carlo simulations where the continuum description of the membrane is replaced by a discrete one. When the inclusions are parallel, the short-range interaction is attractive and it is proportional to the membrane's bending rigidity.

Park and Lubensky [PAc] examined, among other things, ellipsoidal proteins that break the up–down symmetry. They introduced in the free-energy functional a suitable coupling term with the membrane's elasticity and showed that the interaction free energy $F$ is attractive and given by

$$F = -\frac{(AdQ_2)^2}{16\pi\kappa r^2}\cos 2(\theta_1 + \theta_2),$$

for proteins with an elliptic cross-section along the tangent plane of the membrane and with axes at angles $\theta_1$ and $\theta_2$ relative to the center-to-center vector. $F$ also depends on the cross-sectional area $A$ of the proteins, a coupling coefficient $d$, and the scalar-order parameter $Q_2$ associated with the second-rank tensor $\mathbf{Q}$ that models the protein's anisotropy. It is worth noting that the interaction now decays like $r^{-2}$, instead of the more common $r^{-4}$ decay law.

A different view was introduced by Dommersnes and Fournier [DOa] who modeled transmembrane proteins as pointwise constraints that locally fix the membrane's curvature tensor. If no further requirement is imposed, the interaction splits into two contributions: a Casimir interaction $F_C$ due to fluctuations, which exhibits the ubiquitous $r^{-4}$ decay, and a mean field force $F_{MF}$, which has its roots in the membrane's average elastic deformations. The leading terms of $F_{MF}$ are

$$F_{MF} \simeq 4\pi\kappa\left(c_1^2 + c_2^2 - 4c_1c_2\frac{a^2}{r^4}\right)\frac{a^6}{r^4},$$

where $a$ is a microscopic cut-off introduced to regularize the Green function of the biharmonic operator at the inclusion site, and $c_1$, $c_2$ are the curvatures enforced by the proteins.

Proteins are often subject to external torques either due to mechanical agents, as is the case when a protein is bound to the cytoskeleton, or impressed by an applied field. A sufficiently strong torque can hold the orientation of the proteins fixed. This further constraint modifies the scenario, because now the attractive Casimir interaction has a much slower decay:

$$F_C = \frac{k_BT}{2}\log\left[\left(3 + 2\log\frac{2r}{a}\right)\left(1 + 2\log\frac{2r}{a}\right)\right].$$

Moreover, the elastic part of the interaction also undergoes a deep change as it now decays as $r^{-2}$, when the inclusions have one and the same orientation, whereas the decay is further slowed down by a logarithmic term, when the orientations are different. Inasmuch as the Casimir and mean-field interactions have different functional forms, it is not surprising that stable equilibria exist where the proteins form patterns with a finite spacing.

A controversial issue concerns the dependence of the interaction energy on the bending rigidity $\kappa$. As seen above, behaviors as diverse as $F \propto \kappa$ and $F \propto \kappa^{-1}$ are both encountered. Such a diversity was noted by Marchenko and Misbah [MAb], who analyzed the elastic interactions between two pointlike inclusions, by stressing the effects that the inclusion symmetries exert upon the free-energy decay. These authors claim that their analysis departs from tradition in certain major ways. First, are boundary conditions. Most authors introduce the effects of inclusions only through the boundary conditions, so that the equilibrium equations are homogeneous, however, here the inclusions enter the equilibrium equation directly as pointwise sources of deformation. Second, are inclusion symmetries. At variance with [PAc], the requirement that the order tensor $\mathbf{Q}$ be traceless is dropped for inclusions with a $C_2$ symmetry. Consequently, the orientational interaction energy, which still decays as $r^{-2}$, acquires a new term that was absent in [PAc]. Moreover, orientational interactions decaying as $r^{-n}$ are found for inclusions with $C_n$ symmetry. By exploiting a formal analogy with electrostatics, Marchenko and Misbah expand the interaction free energy in power series of the proteins' separation, retaining only the first nontrivial term, which depends on the inclusions symmetry. The corresponding direct or linear interaction vanishes if the inclusions are isotropic. Again by analogy with electrostatics, where an ion interacts with a neutral atom with a symmetric charge distribution only through the induced polarization, here the presence of a second inclusion induces an effective anisotropy on the first inclusion, which can be formally obtained by expanding the order tensor $\mathbf{Q}$ in powers of the curvature tensor. The resulting interaction energy behaves as $r^{-4}$ and is proportional to $\kappa^{-2}$, in contrast with the $\kappa^{-1}$ dependence obtained by Fournier's group.

In a subsequent paper [BAa], Bartolo and Fournier reconciled the approaches of [DOa] and [MAb] by showing that the discrepancies in the results were apparent, as they simply refer to different limit cases. If $\sigma_1^i$ and $\sigma_2^i$ are the principal curvatures of the membrane at the $i$th inclusion site, the interaction free energy consists of two terms, one for each inclusion, that are written as

$$U_i = \frac{1 + 2\epsilon}{2} \Gamma \left[ (\sigma_1^i - c)^2 + (\sigma_2^i - c)^2 \right] - \frac{\epsilon}{2} \Gamma (\sigma_1^i - \sigma_2^i)^2 ,$$

where $\epsilon$ is related to the inclusion anisotropy and is subject to the limitation $\epsilon > -1/2$, to avoid local instabilities. The constant $c$ represents the value of the curvature induced by the inclusion, and $\Gamma > 0$ is the rigidity of the potential. The interaction between inclusions is found through the minimization of the free energy $F = F_m + U_1 + U_2$, where $F_m$ is the elastic free energy of the membrane, which also accounts for higher-order elastic moduli. As a first step, $F$ is minimized with respect to the membrane shape, by keeping the curvature tensor at the inclusion sites frozen. Then, the resulting free energy is further minimized with respect to the components of the curvature tensor of the membrane at the inclusion sites. The resulting interaction $F_{\text{int}}(r)$ as a function of the distance $r$ between the inclusions is given the form

$$F_{\text{int}}(r) = \frac{\kappa \Gamma^3 c^2 (1 + 2\epsilon)^2}{2\pi^2(\kappa + 2\Gamma b^{-2})[\kappa + 4\Gamma b^{-2}(1 + 2\epsilon)]^2} \frac{1}{r^4} + O\left(\frac{1}{r^6}\right),$$

where $b$ is a nanometric length that stems from the ratio between the bending elasticity $\kappa$ and higher-order elastic moduli. In the limit of hard inclusions, corresponding to $\Gamma/\kappa \gg b^2$, the interaction free energy reduces to

$$F_{\text{int}}^{\text{hard}}(r) \simeq \frac{\kappa b^2 c^2}{64\pi^2}\left(\frac{b}{r}\right)^4$$

and so proportionality with $\kappa$ is recovered in this limit, in agreement with [PAc] and [DOa], among others. In the limit case of soft inclusions, where $\Gamma/\kappa \ll b^2$, the interaction free energy becomes

$$F_{\text{int}}^{\text{soft}}(r) \simeq \frac{\Gamma^3 c^2}{2\pi^2 \kappa^2 r^4}(1 + 2\epsilon),$$

exhibiting the same $\kappa^{-2}$ dependence found in [MAb].

The repulsive behavior often exhibited by long-range interactions of elastic origin seems, however, in contrast with the natural tendency of proteins to form aggregates. Although thermal fluctuations induce attraction between proteins, the presence of protein clusters in membranes seems to be independent of fluctuations. This led Kim et al. [KIa] to reexamine curvature-mediated interactions in a nearly flat membrane, by considering the interactions among more than two proteins. By use of complex analytic methods, these authors proved that curvature-mediated interactions are nonpairwise additive and that the terms responsible for lack of additivity are always attractive. Dynamical simulations showed that sufficiently large aggregates of proteins give rise to stable equilibrium patterns, the smallest of which is composed of five proteins arranged at the vertices of a regular pentagon. If proteins have elliptic, rather than circular cross-sections, then stable aggregates are formed even for three proteins [KIb]. In [DOb], where proteins are modeled as pointlike constraints on the membrane's curvature, multibody effects were found that lead to the formation of regular arrays and induce an egg-carton structure in the membrane, as sometimes observed in experiments.

## 8.6  Peripheral Proteins

The complex structure of biological membranes makes it difficult to obtain reliable experimental data concerning the membrane-mediated protein–protein interactions. To overcome this intrinsic difficulty, Koltover et al. [KOa] mimicked proteins by using latex beads and showed that, if two beads are close together, an attractive force exists that is neither of colloidal origin, nor is due to thermal fluctuations, but is mediated by the membrane's elasticity. These results prompted several authors to study membrane deformations induced by colloidal binding. For instance, Deserno [DEa] studied deformations of axisymmetric membranes caused by a single colloidal particle by adding to the spontaneous curvature model (2.2) a lateral tension $\sigma$ and an adhesion free energy $-wa$, where $w$ is a positive constant called the *adhesion potential*, and $a$ is the area of the wrapped portion of the particle. Although the lateral tension contrasts the wrapping of the particle, adhesion would favor complete wrapping. When $w$ increases, for fixed $\sigma$, a free particle becomes first partially wrapped, and then fully wrapped. Precisely, after the continuous transition from free to partially wrapped particle, a further increase of $w$ makes the fully wrapped configuration locally stable. However, a large energy barrier of $20kT$ exists that separates the partially from the totally wrapped regimes, so that a particle is totally wrapped when a spinodal line is crossed, at large values of $w$. On decreasing $w$, a specular sequence is encountered because the particle unbinds itself from the membrane when a further spinodal line is crossed. As a result, a hysteresis loop is predicted that depresses the partially wrapped regime, which can even be completely skipped if the energy barriers are high enough.

A computation of membrane-mediated forces between colloidal particles based upon surface geometry was recently pursued by Guven et al. [MUb, CAd]. They expressed the surface stress tensor of a membrane surface $\mathcal{S}$ by resorting to the first variation of a generalized Helfrich Hamiltonian whose density is an unspecified function of the geometric invariants of $\mathcal{S}$. If $\boldsymbol{\nu}$ is the outer unit normal vector to $\mathcal{S}$, the first variation is given the form

$$\delta H = \int_{\mathcal{S}} \mathcal{E}(\mathcal{H})\boldsymbol{\nu} \cdot \boldsymbol{u} \mathrm{d}A + \int_{\mathcal{S}} \mathrm{div}_{\mathrm{s}}\mathbf{B}\mathrm{d}A \,,$$

where $\boldsymbol{u}$ is the surface displacement, $\mathcal{E}(\mathcal{H})$ is a scalar function, and the vector field $\mathbf{B}$ depends on both the tangential component of $\boldsymbol{u}$ and its gradients; finally, $\mathrm{div}_{\mathrm{s}}$ denotes the surface-divergence operator. If $\mathbf{B}$ is written as

$$\mathbf{B} = -\mathbf{T}\boldsymbol{u} + \text{terms containing gradients of } \boldsymbol{u} \,,$$

where $\mathbf{T}$ is a tensor field, by requiring that $\delta H$ vanishes identically when $\boldsymbol{u}$ is a rigid translation, and exploiting the equilibrium equation $\mathcal{E}(\mathcal{H}) = 0$ that holds for any admissible perturbation, the conservation law

$$\mathrm{div}_{\mathrm{s}}\mathbf{T} = \mathbf{0}$$

is obtained, a form in which $\mathbf{T}$ can be identified as the surface stress tensor. The knowledge of $\mathbf{T}$ makes it possible to compute forces acting on particles bound to lie on $\mathcal{S}$. $\mathbf{T}$ is not necessarily symmetric and, as a consequence, when invariance against rotations is imposed [CAd], a couple stress tensor emerges to ensure the balance of torques. Although computations remain difficult in general, in [MUb] the sign of the force between two parallel adsorbed cylinders was discussed in detail and it was shown that the membrane-mediated interaction can be either repulsive or attractive, according to whether the cylinders lie on the same side, or on opposite sides of a nearly flat membrane. The same result was obtained by Weikl [WEa] with a traditional approach, in which the equilibrium profile of the membrane is first obtained as a function of the adhesive length of the cylinders, that is, the length of the membrane's segment wrapped around the cylinders. This makes the elastic free energy an ordinary function of the adhesive segment, which is then minimized to obtain the complete equilibrium configuration from which the behavior of the interaction follows.

A different approach to model the interaction between a peripheral protein and a membrane was proposed by Rosso et al. [ROa] in a two-dimensional setting, and then generalized in [ROb] to three space dimensions. A peripheral protein—or, equivalently, an adhesive bead—is modeled by a suitable central symmetric potential and the dynamics of the complete system is described by a dissipation principle. In the general case, the evolution equations cannot be dealt with analytically, but when the ambient fluid surrounding both the membrane and the bead is so viscous that deformations of the membrane are negligible, it is possible to describe how the distribution of curvature along the membrane influences the motion of the bead. Precisely, the normal component of the force acting on the bead depends on which side the bead approaches the membrane, that is, on whether it "sees" either its concave or its convex side. When a bead adheres to the membrane, it is acted upon by a force with a tangential component that depends on the curvature's gradient. In two dimensions, the bead migrates towards regions with large curvature, in agreement with the results of Odell and Oster [ODa] on the curvature segregation of integral proteins in the membranes of the Golgi apparatus. In the complete, three-dimensional setting employed in [ROb], the role of curvature gradients as a leading factor for the bead's motion was confirmed, but the interplay between the force felt by the bead and membrane's geometry is much more involved. Because no special requirement was placed in [ROa] upon the geometry of the membrane, it would be interesting to explore the membrane-mediated interactions between two beads, to see how the results obtained by Weikl [WEa] for nearly flat membranes could be generalized.

## 8.7 Closing Question and Prospects

After this journey into the world of models for biological membranes, a natural question arises: how close are the theoretical predictions to experiments performed by biologists? Needless to say, the overwhelming complexity of biological systems is far from having been captured by the necessarily simple models employed in the past three decades since membrane modeling has become a major topic in biophysics. However, the results are encouraging, and the basic mechanisms of biological membranes have probably been correctly unravelled. Much work still remains to be done to propose more realistic models for both inhomogeneous membranes and protein–protein interactions. It is not difficult to predict that knowledge and methods from both statistical and continuum mechanics, combined with techniques from the dynamics of phase transition will be needed to achieve this goal. Such a genuine interdisciplinary character confers to this branch of mathematical modeling a particular, perhaps unique fascination that we hope will continue to attract researchers for a long time.

## 8.8 References

[ALa] Alberts, B., Bray, D., Lewis, J., Raff, M., Roberts, K., and Watson, J.D., **Molecular Biology of the Cell**, Garland (1983).

[ARa] Aranda-Espinoza, H., Berman, A., Dan, N., Pincus, P., and Safran, S., Interaction between inclusions embedded in membranes, *Biophys. J.*, **71** (1996), 648–656.

[BAa] Bartolo, P.G. and Fournier, J.-B., Elastic interaction between "hard" or "soft" pointwise inclusions on biological membranes, *Eur. Phys. J. E*, **11** (2003), 141–146.

[BEa] Benga, G. and Holmes, R.P., Interactions between components in biological membranes and their implications for membrane function, *Prog. Biophys. Mol. Bio.*, **43** (1984), 195–257.

[BIa] Biscari, P., Bisi, F., and Rosso, R., Curvature effects on membrane-mediated interactions of inclusions, *J. Math. Biol.*, **45** (2002), 37–56.

[BIb]   Biscari, P. and Bisi, F., Membrane-mediated interactions of rod-like inclusions, *Eur. Phys. J. E*, **7** (2002), 381–386.

[BIc]   Biscari, P. and Rosso, R., Inclusions embedded in lipid membranes, *J. Phys. A: Math. Gen.*, **34** (2001), 439–459.

[BId]   Biscari, P., Canevese, S.M., and Napoli, G., Impermeability effects in three-dimensional vesicles, *J. Phys. A: Math. Gen.*, **37** (2004), 6859–6874.

[CAa]   Canham, P.B., The minimum energy of bending as a possible explanation of the biconcave shape of the human red blood cell, *J. Theor. Biol.*, **26** (1970), 61–81.

[CAb]   Cantor, R.S., Lateral pressures in cell membranes: A mechanism for modulation of protein function, *J. Phys. Chem. B*, **101** (1997), 1723–1725.

[CAc]   Cantor, R.S., The influence of membrane lateral pressures on simple geometric models of protein conformational equilibria, *Chem. Phys. Lip.*, **101** (1999), 45–56.

[CAd]   Capovilla, R., and Guven, J., Stresses in lipid membranes, *J. Phys. A: Math. Gen.*, **35** (2002), 6233–6247.

[DAa]   Dan, N., Berman, A., Pincus, P., and Safran, S.A., Membrane-induced interactions between inclusions, *J. Phys. II France*, **4** (1994), 1713–1725.

[DAb]   Dan, N. and Safran, S.A., Effect of lipid characteristics on the structure of transmembrane proteins, *Biophys. J.*, **75** (1998), 1410–1414.

[DEa]   Deserno, M., Elastic deformation of a fluid membrane upon colloid binding, *Phys. Rev. E*, **69** (2004), 031903.

[DOa]   Dommersnes, P.G. and Fournier, J.-B., Casimir and mean-field interactions between membrane inclusions subject to external torques, *Europhys. Lett.*, **46** (1999), 256–261.

[DOb]   Dommersnes, P.G. and Fournier, J.-B., The many-body problem for anisotropic membrane inclusions and the self-assembly of "saddle" defects into an "egg-carton," *Biophys. J.*, **83** (2002), 2898–2905.

[DOc]   Dommersnes, P.G., Fournier, J.-B., and Galatola, P., Long-range elastic forces between membrane inclusions in spherical vesicles, *Europhys. Lett.*, **42** (1998), 233–238.

[FOa]   Fournier, J.-B., Microscopic membrane elasticity and interactions among membrane inclusions: interplay between the shape, dilation, tilt and tilt-difference modes, *Eur. Phys. J. B*, **11** (1999), 261–272.

[GIa] Gil, T., Ipsen, J.H., Mouritsen, O.G. , Sabra, M.C., Sperotto, M.M., and Zuckermann, M.J., Theoretical analysis of protein organization in lipid membranes., *BBA-Rev. Biomembranes*, **1376** (1998), 245–266.

[GOa] Golestanian, R., Goulian, and Kardar, M., Fluctuation-induced interactions between rods on membranes and interfaces, *Europhys. Lett.*, **33** (1996), 241–245.

[GOb] Goulian, M., Bruinsma, R., and Pincus, P., Long-range forces in heterogeneous fluid membranes, *Europhys. Lett.*, **22** (1993), 145–150.

[HAa] Hamm, M., and Kozlov, M.M., Tilt model of inverted amphiphilic mesophases, *Eur. Phys. J. B*, **6** (1998), 519–528.

[HAb] Hamm, M. and Kozlov, M.M., Elastic energy of tilt and bending of fluid membranes, *Eur. Phys. J. E*, **3** (2000), 323–335.

[HAc] Harroun, T.A., Heller, W.T., Weiss, T.M., Yang, L., and Huang, H.W., Theoretical analysis of hydrophobic matching and membrane-mediated interactions in lipid bilayers containing gramicidin, *Biophys. J.*, **76** (1999), 3176–3185.

[HEa] Heinrich, V., Božič, Svetina, S., and Žekš, B., Vesicle deformation by an axial load: from elongated shapes to tethered vesicles, *Biophys. J.*, **76** (1999), 2056–2071.

[HEb] Helfrich, P. and Jakobsson, E., Calculation of deformation energies and conformations in lipid membranes containing gramicidin channels, *Biophys. J.*, **57** (1990), 1075–1084.

[HEc] Helfrich, W., Elastic properties of lipid bilayers: theory and possible experiments, *Z. Naturforsch. C*, **28** (1973), 693–703.

[HEd] Helfrich, W., Steric interaction of fluid membranes in multilayer systems, *Z. Naturforsch. A*, **33** (1978), 305–315.

[HOa] Holzlöhner, R., and Schoen, M., Attractive forces between anisotropic inclusions in the membrane of a vesicle, *Eur. Phys. J. B*, **12** (1999), 413–419.

[HOb] Hotani, H., Transformation pathways in liposomes, *J. Molec. Biol.*, **178** (1984), 113–120.

[HOc] Hotani, H., Nomura, F., and Suzuki, Y., Giant liposomes: from membrane dynamics to cell morphogenesis, *Curr. Opin. Coll. Interf. Sci.*, **4** (1999), 358–368.

[HUa] Huang, H.W., Deformation free energy of bilayer membrane and its effect on gramicidin channel lifetime, *Biophys. J.*, **50** (1986), 1061–1070.

[IGa] Iglič, A., Kralj-Iglič, V., and Hägerstrand, H., Amphiphile induced echinocyte-spheroechinocyte transformation of red blood cell shape, *Eur. Biophys. J.*, **27** (1998), 335–339.

[ISa] Israelachvili J., **Intermolecular and Surface Forces**, Academic (1992).

[JIa] Jiang, Y., Lookman, T., and Saxena, A., Phase separation and shape deformation of two-phase membranes, *Phys. Rev. E*, **61** (2000), R57–R60.

[JEa] Jensen, M.Ø. and Mouritsen, O.G., Lipids do influence protein function–the hydrophobic matching hypothesis revisited, *BBA-Biomembranes*, **1666** (2004), 205–226.

[JUa] Jülicher, F. and Lipowsky, R., Shape transformation of vesicles with intramembrane domains, *Phys. Rev. E*, **53** (1996), 2670–2683.

[KAa] Kawakatsu, T., Andelman, D., Kawasaki, K., and Taniguchi, T., Phase transitions and shapes of two component membranes and vesicles I: Strong segregation limit, *J. Phys. II France*, **3** (1993), 971–997.

[KIa] Kim, K.S., Neu, J., and Oster, G., Curvature-mediated interactions between membrane proteins, *Biophys. J.*, **75** (1998), 2274–2291.

[KIb] Kim, K.S., Neu, J., and Oster, G., Effect of protein shape on multi-body interactions between membrane inclusions, *Phys. Rev. E*, **61** (2000), 4281–4285.

[KOa] Koltover, I., Rädler, J., and Safinya, C.R., Membrane mediated attraction and ordered aggregation of colloidal particles bound to giant phospholipid vesicles, *Phys. Rev. Lett.*, **82** (1999), 1991–1994.

[KRa] Kralchevsky, P.A. and Nagayama, K., Capillary forces between colloidal particles, *Langmuir*, **10** (1994), 23–36.

[KRb] Kralchevsky, P.A., Paunov, V.N., Denkov, N.D., and Nagayama, K., Stresses in lipid membranes and interactions between inclusions, *J. Chem. Soc. Faraday Trans.*, **91** (1995), 3415–3432.

[KRc] Kralj-Iglič, V., Heinrich, V., Svetina, S., and Žekš, B., Free energy of closed membrane with anisotropic inclusions, *Eur. Phys. J. B*, **10** (1999), 5–8.

[KRd] Kralj-Iglič, V., Svetina, S., and Žekš, B., Shapes of bilayer vesicles with membrane embedded molecules, *Eur. Biophys. J.*, **24** (1996), 311–321.

[LAa] Laradji, M. and Sunil Kumar, P.B., Dynamics of domain growth in self-assembled fluid vesicles, *Phys. Rev. Lett*, **93** (2004), 198105.

[LEa] Lee, A.G., How lipids affect the activities of integral membrane proteins, *BBA-Biomembranes*, **1666** (2004), 62–87.

[LIa] Lipowsky, R. and Dimova, R., Domains in membranes and vesicles, *J. Phys: Condens. Matt.*, **15** (2003), S31–S45.

[LIb] Lipowsky, R., Budding of membranes induced by intramembrane domains, *J. Phys. II France*, **2** (1992), 1825–1840.

[LUa] Lubensky, T.C. and Prost, J., Orientational order and vesicle shape, *J. Phys. II France*, **2** (1992), 371–382.

[MAa] MacKintosh, F.C. and Lubensky, T.C., Orientational order, topology, and vesicle shapes, *Phys. Rev. Lett.*, **67** (1991), 1169–1172.

[MAb] Marchenko, V.I. and Misbah, C., Elastic interaction of point defects on biological membranes, *Eur. Phys. J. E*, **8** (2002), 477–484.

[MAc] Markin, V.S., Lateral organization of membranes and cell shapes, *Biophys. J.*, **36** (1981), 1–19.

[MAd] May, S., Theories on structural perturbations of lipid bilayers, *Curr. Opin. Coll. Interf. Sci.*, **5** (2000), 244–249.

[MAe] May, S. and Ben-Shaul, A., Molecular theory of lipid-protein interaction and the $L_\alpha$-$H_{II}$ transition, *Biophys. J.*, **76** (1999), 751–767.

[MAf] May, S., Kozlovsky, Y., Ben-Shaul, A., and Kozlov, M.M., Tilt modulus of a lipid monolayer, *Eur. Phys. J. E*, **14** (2004), 299–308.

[MIa] Miao, L., Seifert, U., Wortis, M., and Döbereiner, H.-G., Budding transitions of fluid-bilayer vesicles: The effect of area-difference elasticity, *Phys. Rev. E*, **49** (1994), 5389–5407.

[MUa] Mukhopadhyay, R., Lim, G.H.W., and Wortis, M., Echinocyte shapes: Bending, stretching, and shear determine spicule shape and spacing, *Biophys. J.*, **82** (2002), 1756–1772.

[MUb] Müller, M.M., Deserno, M., and Guven, J., Geometry of surface-mediated interactions, *Europhys. Lett.*, **69** (2005), 482–488.

[NIa] Nielsen, C., Goulian, M., and Andersen, O.S., Energetics of inclusion-induced bilayer deformations, *Biophys. J.*, **74** (1998), 1966–1983.

[ODa] Odell, E. and Oster, G., Curvature segregation of proteins in the Golgi, in **Lectures on Mathematics in the Life Sciences**, Vol. **24**, Goldstein, B. and Wolfsy, C. Eds., (1994), pp. 23-36.

[PAa] Pamplona, D.C. and Calladine, C.R., The mechanics of axially symmetric liposomes, *J. Biomech. Eng. T-ASME*, **115** (1993), 149–159.

[PAb] Pamplona, D.C. and Calladine, C.R., Aspects of the mechanics of lobed liposomes, *J. Biomech. Eng. T-ASME*, **118** (1996), 482–488.

[PAc] Park, J.-M. and Lubensky, T.C., Interactions between membrane inclusions on fluctuating membranes, *J. Phys. I France*, **6** (1996), 1217–1235.

[RIa]  Ring, A., Gramicidin-channel induced lipid membrane deformation energy: influence of chain length and boundary conditions, *BBA-Biomembranes*, **1278** (1996), 147–159.

[ROa]  Rosso, R., Sonnet, A.M., and Virga, E.G., Dynamics of kinks in biological membranes, *Continuum Mech. Thermodyn.*, **14** (2002), 127–136.

[ROb]  Rosso, R., Curvature effects in vesicle-particle interactions, *Proc. R. Soc. London A*, **459** (2003), 829–852.

[SCa]  Schröder, H., Aggregation of proteins in membranes. An example of fluctuation-induced interactions in liquid crystals, *J. Chem. Phys.*, **67** (1977), 1617–1619.

[SEa]  Seifert, U., Curvature-induced phase segregation in two component vesicles, *Phys. Rev. Lett.*, **70** (1993), 1335–1338.

[SEb]  Seifert, U., Configurations of fluid membranes and vesicles, *Adv. Phys.*, **46** (1997), 13–137.

[SEc]  Seifert, U., Shillcock, J., and Nelson, P., Role of bilayer tilt difference in equilibrium membrane shapes, *Phys. Rev. Lett.*, **77** (1996), 5237–5240.

[SEd]  Sekimura, T. and Hotani, H., The morphogenesis of liposomes viewed from the aspect of bending energy, *J. Theor. Biol.*, **149** (1991), 325–337.

[SEe]  Selinger, J.V., MacKintosh, F.C., and Schnur, J.M., Theory of cylindrical tubules and helical ribbons of chiral lipid membranes, *Phys. Rev. E*, **53** (1996), 3804–3818.

[SHa]  Sheetz, M.P. and Singer, S.J., Biological membranes as bilayer couples. A molecular mechanism of drug-erythrocyte interactions, *Proc. Nat. Acad. Sci. USA*, **71** (1974), 4457–4461.

[SUa]  Sunil Kumar, P.B., Gompper, G., and Lipowsky, R., Budding dynamics of multicomponent membranes, *Phys. Rev. Lett*, **86** (2001), 3911-3914.

[SVa]  Svetina, S. and Žekš, B., The elastic deformability of closed multi-layered membranes is the same of that of a bilayer membrane, *Eur. Biophys. J.*, **21** (1992), 251–255.

[TAa]  Taniguchi, T., Shape deformation and phase separation dynamics of two-component vesicles, *Phys. Rev. Lett.*, **76** (1996), 4444–4447.

[TAb]  Taniguchi, T., Kawasaki, K., Andelman, D., and Kawakatsu, T., Phase transitions and shapes of two component membranes and vesicles II: Weak segregation limit, *J. Phys. II France*, **4** (1994), 1333–1362.

[WAa] Waugh R.E., Elastic energy of curvature-driven bump formation on red blood cell membrane, *Biophys. J.*, **70** (1996), 1027–1035.

[WEa] Weikl, T.R., Indirect interactions of membrane-adsorbed cylinders, *Eur. Phys. J. E*, **12** (2003), 265–273.

[WEb] Weikl, T.R., Kozlov, M.M., and Helfrich, W., Interaction of conical membrane inclusions: Effect of lateral tension, *Phys. Rev. E*, **57** (1998), 6988–6995.

Printed in the United States of America